the BIG and the SMALL

Volume 1

Vignettes in Physics
A Series by G Venkataraman

A Hot Story
At the Speed of Light
Bhabha and His Magnificent Obsessions
Bose and His Statistics
Chandrasekhar and His Limit
Quantum Revolution (3 vols.)
1. *The Breakthrough*
2. *QED: The Jewel of Physics*
3. *What is Reality?*
Raman and His Effect
Saha and His Formula
The Many Phases of Matter
Why Are Things the Way They Are?

Forthcoming
The Big and the Small Vol. II: From the Microcosm to the Macrocosm—The Fascinating Link between Particle Physics and Cosmology

the BIG and the SMALL

Volume 1

Journey into the Microcosm

The Story of

Elementary Particles

G Venkataraman

Universities Press

THE BIG AND THE SMALL

UNIVERSITIES PRESS (INDIA) PRIVATE LIMITED

Registered Office
3-6-747/1/A & 3-6-754/1, Himayatnagar, Hyderabad 500 029, Telangana, India
info@universitiespress.com; www.universitiespress.com

Distributed by
Orient Blackswan Private Limited

Registered Office
3-6-752 Himayatnagar, Hyderabad 500 029, Telangana, India

Other Offices
Bengaluru, Chennai, Guwahati, Hyderabad, Kolkata,
Mumbai, New Delhi, Noida, Patna, Visakhapatnam

First published in India by
Universities Press (India) Private Limited 2001
Reprinted 2014, 2017, 2023

ISBN 978-81-7371-227-2

501894

Typeset at
Orient BlackSwan

Printed in India at
Yash Printographics, Noida 201 301

Published by
Universities Press (India) Private Limited
3-6-747/1/A & 3-6-754/1, Himayatnagar, Hyderabad 500 029, Telangana, India

Contents

Preface

To the adult reader

This book and others in this series written by me are inspired by the memory of my son Suresh who left this world soon after completing school. Suresh and I often used to discuss physics. It was then that I introduced him to the celebrated Feynman Lectures.

Hans Bethe has described Feynman as the most original scientist of this century. To that, perhaps, may be added the statement that Feynman was also the most scintillating teachers of physics in this century.

The Feynman Lectures are great but they are at the textbook level and meant for serious reading. Moreover, they are a bit expensive, at least for the average Indian student. It seemed to me that there was scope for small books on diverse topics in physics which would stimulate interest, making at least some of our young students take up later a serious study of physics and reach for the Feynman as well as the Landau classics.

Small books inevitably remind me of Gamow's famous volumes. They were wonderful, and stimulated me to no small extent. Times have changed, physics has grown and we clearly need other books, though written in the same spirit.

In attempting these volumes, I have chosen a style of my own. I have come across many books on popular science where elaborate sentences often tend to obscure the scientific essence. I have therefore opted for simple English, and I don't make any apologies for it. If a simple style was good enough for the great Enrico Fermi, it is also good enough for me. I have also employed at times a chatty style. This is deliberate. Feynman uses this with consummate skill, and I have decided to follow in his footsteps (whether I have succeeded or not, is for readers to say). This book is meant to be read for fun and excitement. It is a book you can even lie down in bed and read. Without going to sleep, I hope!

Naturally I have some basic objectives, the most important of which is to stimulate the curiosity of the reader. Here and there the reader may fail to grasp some details, and in fact I have deliberately pitched things a bit high on occasions. But if the reader is able to experience at least in some small measure the *excitement* of science, then my purpose would have been achieved. Apart from excitement, I have also tried to convey that although we might draw boundaries and try to compartmentalise Nature into different subjects, she herself knows no such boundaries. So we can always start anywhere, take a random walk and catch a good glimpse of Nature's glory. Where she is concerned, all topics are 'fashionable'. There is today an unnecessary polarization of the young towards subjects that are supposed to be fashionable. To my mind this is unhealthy, and I have tried to counter it.

This series is essentially meant for the curious. With humility, I would like to regard it as some sort of a 'Junior Feynman Series', if one might call it that. With much love, and sadness, it is dedicated to the memory of Suresh who inspired it.

To the young reader

Have you ever wondered what we are made up of? Perhaps you would answer: "Molecules". Well, what are molecules made up of? Do I hear you say: "Atoms!"? Do you know what atoms are made up of? O.K., you do. How far down the line can you go?

Are you aware how far physicists have actually gone in this direction? That's what this book is all about. The surprising thing is that by probing deeper and deeper into the innermost recesses of the atom, physicists have been able to obtain not only a better understanding of the atom itself but of the vast Cosmos also, and indeed of the origin of the Cosmos itself!

Get set then for reading about the greatest intellectual adventure in human history—man's attempt to reconstruct Creation by a combination of the most daring flights of imagination and the most mind-boggling experiments ever!

Remarks

While the physics underlying this book is unquestionably exciting, some parts of the text may appear rather abstract, abstruse and technical. Unfortunately, these aspects cannot be avoided nor can they be completely brushed aside. You may, if you wish, skip such portions during a first reading. Missing these, however, should not seriously handicap you in obtaining a broad overview of the subject. And an overview is what is important in the first instance. But make sure you do come back to what you skipped over at the start, so that your understanding is eventually more complete.

1

Reconstructing Creation — An Introduction

From time immemorial, man has pondered over questions like, "What is the nature of the Universe? Where did it come from? Who am I? What is my role?", etc. Sages, seekers and philosophers have pursued such an enquiry, and what have emerged as answers, form the core of human knowledge. For example, the ancient texts of India, the *Vedas*, discuss both Creation as well as the nature of man.

Knowledge in ancient times had two strands — the spiritual and the secular. The two were closely intertwined, and, in fact, people often did not make a distinction between the two; this was true of all ancient civilizations, including ours. *Ayurveda* provides a good example of this close linkage. You are perhaps under the impression that *Ayurveda* is nothing but ancient Indian medicine; not quite. Rather, it is the science of well-being and deals with all aspects of health, both mental as well as physical. The stress is largely on being healthy, both physically and spiritually.

After the Renaissance, science developed at a rapid pace, becoming increasingly involved with empirical details. At this stage, people pursuing scientific enquiry slowly stopped providing a spiritual base to their investigations; and the separation between science and religion took place, which is a feature of modern science. However, if you do research in any subject and earn a doctorate, the degree awarded is still the degree of Doctor of Philosophy (Ph.D.). This is an echo back to older times, when all branches of science were subsumed under the umbrella of *philos* (love of) *sophos* (wisdom).

Till the end of the last century physics was primarily concerned with macroscopic phenomena, as is clearly reflected by the developments in mechanics, thermodynamics and optics. Meanwhile, the chemist John Dalton boldly invoked the ancient concept that the atom was the basic building block of matter, thus laying the foundation of modern chemistry. Following his lead, chemists took to the atomic theory but in marked contrast, physicists generally kept away from the atomic concept of matter, barring notable exceptions like Maxwell and Boltzmann. I have narrated elsewhere (in ***A Hot Story***) how Boltzmann faced a rough time for promoting the atomic concept of matter, and eventually committed suicide. Mercifully, saner counsel soon prevailed and physicists also accepted the atom as the basic building block of matter. And, like chemists of the time, they, too, believed that the atom was indestructible.

Beginning in 1896, a series of developments took place which not only overthrew the concept of the indestructible atom but in fact paved the way for a glimpse of Creation itself. When J.J. Thomson discovered the electron towards the end of the last century, little did he know that he was raising the curtain on the most thrilling drama that mankind has ever witnessed. As far as he was concerned, the atom could be split and the electron liberated. The atom was not indivisible as people had assumed till then. On the contrary, the atom had something inside it; it had an internal structure, which included the electron.

Now the electron is negatively charged, whereas the atom as a whole is electrically neutral. Therefore, there ought to be some positive charge also inside the atom which neutralises the electrons present.

How exactly was this positive charge distributed inside the atom? That question received a definitive answer from Rutherford in 1911 when he discovered the atomic nucleus. The game now became absorbing, with every answer provoking new questions. For example, one question was: "Does the nucleus have an internal structure of its own?" Chadwick provided a clear answer to that one in 1932 when he discovered the neutron, and soon it became evident that the atomic nucleus was made up of an appropriate number of protons and neutrons. Once again fresh questions arose as to whether the proton and the neutron had internal structures of their own. Indeed, one began to wonder if this business of internal structure went on endlessly, like a box within a box within a box ..., or whether it terminated somewhere (see Figs. 1.1 and 1.2).

The development of physics during the last fifty years or so is largely the story of breathtaking discoveries made in finding answers to tantalizing questions like the ones just indicated. The questions might look simple but finding the answers was far from simple! It involved literally thousands of scientists from all over the globe, not to mention billions of dollars as well! In a manner of speaking, the whole drama can be likened to a symphony orchestra with the Creator Himself wielding the baton and squeezing bits and pieces of sublime music out of the players. The concert is not over yet, and if you are lucky you too can get a piece of the action!

Meanwhile, the revelations that have come from these scientific discoveries have almost given us a glimpse of Creation itself. In short, the story of high-energy physics may aptly be called the Second Creation or man's attempt to reconstruct Creation. That drama is unfolded in the following chapters. Read on!

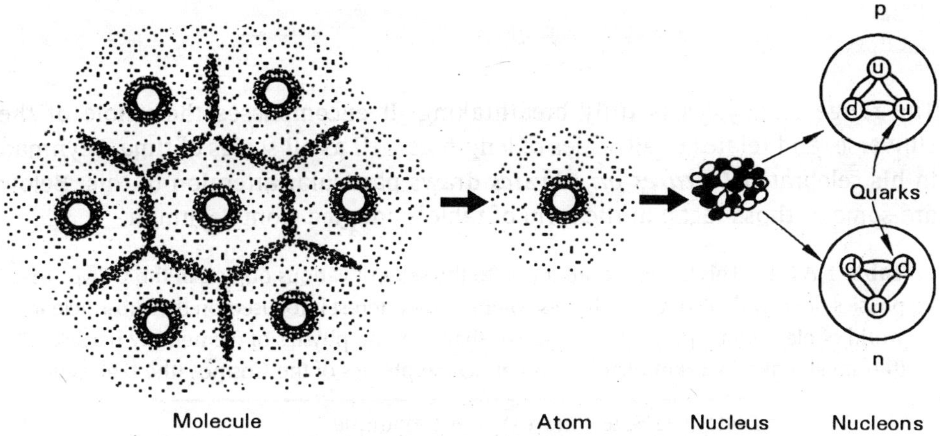

Fig.1.1 The inward-bound journey has been roughly like this: Structure of liquids/solids → structure of molecules → structure of atoms → structure of the nucleon → quarks and leptons. (Don't worry about the words quarks and leptons; they will soon be explained.) Each step opened up a world in itself, and many physicists stayed in each of these fields; thus one has solid state physicists, molecular physicists and so forth. But there are also the frontiersmen who keep pushing — one is reminded of the way the American West was opened up. Today, we have arrived at the stage of saying that the proton and the neutron are made up of quarks. However, one is not quite sure if quarks and leptons represent the end of the road or if the road has any end at all!

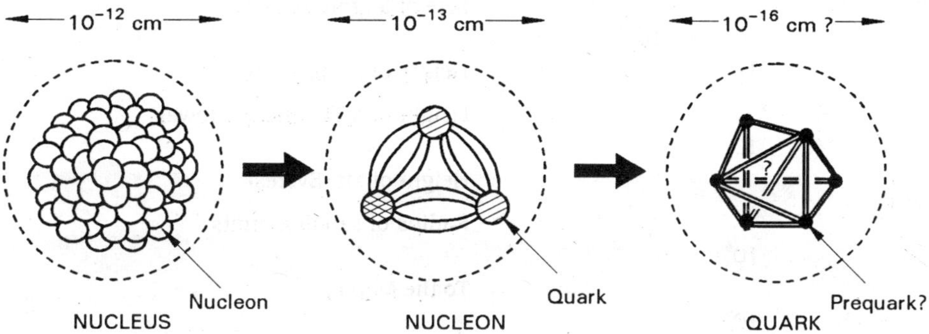

Fig.1.2 This is an amplification of the previous figure, and illustrates the tiny nooks and corners people are exploring. The common belief today is that nucleons are made up of quarks, and that all matter is made up of quarks and leptons in suitable combinations — that is all there is to it. This is a modernisation of the belief of the thirties that all matter is made up of electrons, protons and neutrons. But already there are some who are questioning the current proposition by suggesting that the quarks themselves are built up of prequarks or preons. If prequarks exist, then what about pre-prequarks and so on, i.e., $(\text{pre})^n$quarks? Sounds like opening Pandora's box, doesn't it? Well, at least it will keep a lot of people busy!

Appendix 1 to Chapter 1

The range of physics is truly breathtaking. It encompasses the whole of the Universe, and relates to all scales of length as well as all scales of time. Feynman in his celebrated ***Lectures on Physics*** draws particular attention to this. Below are some of those facts, in the form of tables, adapted from Feynman.

Table 1.A1.1 This table calls attention to the sizes of various objects in the Universe; physics deals with all of these. In this volume, our interest is focused on the sub-nuclear world of elementary particles. But, as we shall see, the physics of elementary particles (the small) links, in a remarkable manner, to the physics of the Cosmos (the big) itself!

Distances/Size (metres)	An example
	Size of a quark?
10^{-18}	
	Size of a nucleus
10^{-15}	
10^{-12}	
	Size of an atom
10^{-9}	
	Size of a virus
10^{-6}	
	Size of a grain of salt
10^{-3}	
1	Height of a small boy
	Height of a TV antenna tower
10^{3}	
	Height of Mt. Everest
	Radius of satellite orbits
10^{6}	
	To the Moon
10^{9}	
	To the Sun
10^{12}	
	Radius of Pluto's orbit
10^{15}	
	To the nearest star
10^{18}	
	To the nearest galaxy
10^{24}	
	Edge of the Universe
Even greater distances	?

Table 1.A1.2 This table summarises the time scales of various phenomena in the Universe. Once again, the range is staggering. Clocks do not exist for measuring very small time scales; but they can be accurately inferred by various measurements.

Time (seconds)	Example of a process/phenomena	Mean life of
10^{18}	Age of the Universe	
	Age of the Earth	^{238}U
10^{15}		
	Appearance of man	
10^{12}	Vedic period	
		^{226}Ra
10^{9}	Average life of a human	
		^{3}H
10^{6}		
	Duration of a day	
10^{3}		
	Time taken for light to come to Earth from the Sun	
		Neutron
1	Time for one heartbeat	
10^{-3}	Period of a sound wave	
10^{-6}	Period of a radio wave	Muon
		Charged pion
10^{-9}	Time taken for light to travel ~1 metre	
10^{-15}	Period of atomic vibration	
10^{-18}		
	Time taken for light to cross an atom	
10^{-21}		
	Period of nuclear vibration	
10^{-24}	Time for light to cross a nucleus	Strange particles

Appendix 2 to Chapter 1

In the formulae of elementary particle physics, the fundamental constants c and $\hbar$ occur frequently. Theoretical particle physicists are lazy or clever (!), depending on your point of view, and seldom include the c's and the $\hbar$'s in their formulae. They simply say, "We shall set $c = \hbar = 1$", and get on with their business. They leave you, if you are keen, to insert the missing constants yourself.

You might wonder, "If both c and $\hbar$ are set equal to unity, what on earth does it do to the dimensions of the various physical quantities?" The answer to this is linked to the fact that particle physicists not only set the numerical value of c and of $\hbar$ equal to 1 but also treat these quantities as dimensionless! Why all these bizarre actions? There is an advantage, which we shall presently see.

In usual physics, we have three basic units — mass, length and time. Here also one has three, namely c, $\hbar$ and one more. Of these, c and $\hbar$ have been declared to be dimensionless! This leaves one more unit, which is usually chosen to be energy, and is expressed in terms of MeV or even GeV — see ***Watt*** for a description of energy units. Once this is done, it turns out that the dimensions of all other physical quantities can be simply written as $[\text{GeV}]^n$, where n is either positive or negative, integer or a fraction. How does one figure all this out?

Consider $\hbar$. It has the dimension of [Energy] · [Time]. In particle physics, one makes it dimensionless. Therefore,

$$[\text{Energy}] \cdot [\text{Time}] = \text{No dimension}$$

whence,

$$[\text{Time}] = [\text{Energy}]^{-1}$$

Similarly, since $c = 1$ and has the dimension [Length] · $[\text{Time}]^{-1}$, we have

$$[\text{Length}] = [\text{Time}] = [\text{Energy}]^{-1}$$

Recall the values of $\hbar$ and $\hbar c$. They are

$$\hbar = 6.582 \times 10^{-25} \text{ GeV. sec}$$

$$\hbar c = 1.973 \times 10^{-16} \text{ GeV. metre}$$

However, particle physicists are setting both these values equal to unity, creating thereby a crazy system of units. But one can still go from the crazy units (CU) to the traditional sane units (SU). For example, if the length of an object in CU is $\eta \text{ GeV}^{-1}$, then the length in SU, i.e., length in metres is given by $\eta \cdot 1.973 \times 10^{-16}$.

Thus we have the following:

Quantity	Dimension in CU	Units
$\hbar$	None	1
c	None	1
Mass	[E]	GeV
Length	$[E]^{-1}$	GeV^{-1}
Time	$[E]^{-1}$	GeV^{-1}
e	None	—

Notice that in the CU system, e has no dimensions but in the normal (say CGS) system, e has a dimension [E.L]. Hence when working with normal or sane units, it is customary to divide e^2 by $[\hbar . c]^2$ to make it dimensionless. The value of this quantity is (1/137). Likewise, the weak charge and the colour charge (these quantities will be defined as we go along), are also made dimensionless. One more remark about units. There are many options for SU and in deciding which SU to use, the question of the dimension of e also arises. In the old days, people used the *Gaussian* system. In this, e is measured in *electrostatic units* (esu) and Coulomb's law is written as

$$F = \frac{q_1 q_2}{r^2}.$$

In the SI units which are more popular these days, e is measured in coulombs and Coulomb's law is written as

$$F = \frac{1}{4\pi\varepsilon_0} \cdot \frac{q_1 q_2}{r^2}.$$

In particle physics, when the pundits finally get down to putting in numbers, they cannot avoid the restoration of the $\hbar$s and the cs. At this stage, they also have to use an SU, with a proper convention for Coulomb's law. They usually prefer the *Heaviside–Lorentz* system in which Coulomb's law is written as

$$F = \frac{1}{4\pi} \cdot \frac{q_1 q_2}{r^2}.$$

The electric charges in the three systems are related as

$$q_{\mathrm{HL}} = \sqrt{4\pi}\; q_{\mathrm{G}} = \frac{1}{\sqrt{\varepsilon_0}}\; q_{\mathrm{SI}}.$$

2

Then and Now

This chapter gives a brief overview of and compares the situation in the field of elementary particles in the thirties on the one hand and at the present time on the other. At both times, people were in search of the basic building blocks of matter; only, the view, which then was simplistic, now has a firmer basis. The modern perspective is summarised in what is often called the Standard Model (SM). The tip of the iceberg, the SM, is exposed in this chapter.

From time to time, people have felt that the end of the road has been reached as far as physics is concerned. This happened, for example, towards the end of the nineteenth century although Lord Kelvin warned that it was not so (see ***A Hot Story***). Again, after the advent of quantum mechanics, both Dirac and Max Born felt in their respective ways, that it was all over (see ***QR-I***); but how wrong they were! Lately such pronouncements have become rare, although some noises were made recently when someone came up with a *Theory of Everything* (TOE). However, most people do not believe in such predictions. And experience (see Fig. 1.1) also suggests that the end is not quite in sight. Even then, at the moment one has reached a plateau of sorts and so it is worth taking stock of the situation.

Let us go back about sixty years and see what the situation was like then. This is summarised below in Table 2.1. Life was simple then but far from realistic; we knew of just a handful of what may be called *elementary particles*.

Table 2.1 Elementary particles as they were known around 1933.

Name	Symbol	Mass
Electron	e^-	m_e
Proton	p	$\sim 1836\, m_e$
Neutron	n	$\sim 1836\, m_e$
Neutrino	ν	0
Photon	γ	0

Remarks

1. As mentioned in the text, there are anti-particles corresponding to e, p, n and ν. Their existence was known even then but they are not shown here. Anti-particles are discussed in Appendix 2 to this chapter.
2. The photon is included above, in the light of our present experience. Way back in the thirties, few would have spoken of the photon in the same breath as the other elementary particles.
3. By the thirties, it was known that particles in Nature obey either Bose–Einstein or Fermi–Dirac statistics, i.e., they are either bosons or fermions (for more concerning these, see ***Bose***). The four particles listed above are all fermions. The photon on the other hand is a boson.
4. Presently, the masses of the various particles are given in terms of the electron mass m_e. Later, we shall give all masses in energy units.

Corresponding to the four particles then known, there were also four anti-particles (see Appendix 2 to this chapter), namely the positron, the anti-proton, the anti-neutron and the anti-neutrino. As explained in ***QR-I***, the idea of the anti-particle was an off-shoot of the Dirac equation. In brief, back in the early thirties there were only four elementary particles known (and of course their four anti-counterparts) to worry about. Now a house cannot be built with bricks alone; some cement to hold the bricks together is also needed. Thus, while struggling to determine what the basic building blocks of matter are, people have also worried about the forces that hold the various particles together. Today we recognise four fundamental forces and I have referred to them several times in the companion volumes; here they will appear on the centre stage.

What role do the fundamental forces play? At the level of the Cosmos, the dominating force is gravity. Next comes the electromagnetic force, the kingpin of everything here on Earth from biology, geology and chemistry right down to atomic physics. Till 1905 it was believed that the electromagnetic force was purely classical, but that year, Einstein drew attention to its quantum nature. Today, we visualise the good old electromagnetic field in terms of exchange of photons, the quanta of the e.m. field (see ***QR-II***). The other two forces that make up the family of four fundamental forces are the strong force, also referred to as the strong interaction (S.I.), and the weak force, also referred to as the weak interaction (W.I.).

In those days, one knew very little about these two forces except that:

- The strong force provided the glue which kept the neutrons and the protons bound together in the nucleus.
- The weak force was responsible for radioactivity.

- Both the strong and weak forces were of short range — the guess was that the range was less than $\sim 10^{-12}$ cm. By contrast, both gravity and the electromagnetic forces are of long range.

Like the photon which is the quantum of the electromagnetic field, were there any quanta associated with the strong and the weak forces? In the thirites, no one knew for sure. Yukawa speculated that for the strong force there was a 'messenger' particle, the meson (see ***Bhabha***). No such mediators were proposed then in the case of the weak force.

Let us go now to the early nineties. In the intervening sixty years or so, a tremendous amount of information was gathered, by distilling which a clearer and firmer picture could be constructed. The theoretical fabric that presents such a picture goes by the name of the *Standard Model* (SM). You will hear more about this SM as we go along, and here, let me give you a brief introduction.

The SM, as Nobel awardee Glashow once put it, is like a tapestry woven by many hands. A more down-to-earth description would be that it is a curious amalgam of two of the best theories of the day, 'bolted' together in a somewhat clumsy manner. Nevertheless, it does a fairly good job, and this it does by relying on quantum field theory (this is where a previous reading of ***QR-II*** would be helpful) and an esoteric idea called *gauge symmetry* (to which you will be properly introduced later). The SM sees the world as built up of two sets of particles, the quarks and the leptons, each described by its own theory so to say, the former by *quantum chromodynamics* (QCD) and the latter by the *electro-weak theory* (don't get scared! You will soon learn what these are all about). Oh yes, mediating between these particles or controlling the interactions between them are the electromagnetic, the weak, and the strong forces, together with their associated quanta or 'messenger' particles. That, in brief, is the SM package.

Notice that gravity does not figure in the SM. This does not mean that gravity has no place; it has and a very important one too. But bringing in gravity is no easy task, and in fact that is where the future lies. In other words, though the SM has answered a number of questions of the past and has given us some temporary respite, it raises many questions of its own; there is therefore a lot of work that still remains to be done. So don't get disappointed and think that the show is over. When you graduate from college, there will still be a lot of challenges left. But remember, the game is getting harder all the time which means that you've got to be smarter!

In a sense, this volume is essentially all about how the SM came into being. It is a fascinating story, and I hope I will succeed in capturing at least some of that excitement for you.

To set the ball rolling, a few tables have been presented in Appendix 1 to this chapter. The story of physics during the last sixty years or so is, in a large measure, one of going from Table 2.1 to the ones in the appendix, which essentially represent a digest of the SM. You may not be able to grasp the contents of these tables in the very first reading itself, but that should not bother you. You will have to keep coming back to these tables again and again as you progress through the book, and when you do that you will find that your understanding would have improved considerably.

Appendix 1 to Chapter 2

This appendix is a collection of various odd facts that would aid progress.

In the thirties, the world was 'simple' since only four elementary particles were known, as listed in Table 2.1. The belief, or at least the hope, at that time was that everything in the Universe could be explained with the help of the particles listed in that table. But this did not prove true. As the years went by, many discoveries were made and, it turned out that life wasn't as simple as Table 2.1 suggests. By the late fifties the population of the so-called elementary particles had risen to over fifty, and by the sixties the count was so high that one wag remarked that while in earlier days the discoverers of elementary particles were rewarded with the Nobel Prize, the time had come now to fine them $10,000/- for each such discovery! Apart from this, one wondered if the word elementary, as applied to these particles, had any meaning at all. Certainly not in the strict sense of the word, but it was still a convenient adjective to describe what we today refer to as baryons, mesons and leptons; and so the usage continued. However, people realised that the numerous elementary particles that were turning up everyday in high energy physics experiments were rather like atoms of different elements. So the search was on to see if all these hundreds of so-called elementary particles could be explained in terms of some basic entities, just as one could understand various molecules as being built up of different atoms and the atomic nuclei themselves arising as the result of different combinations of neutrons and protons. Not only were such basic building blocks eventually found, but it was also realised that most of the known facts could be neatly tied up in a package called the Standard Model. The important aspects of this model are captured in the various tables in this appendix.

Before we get started, a bit of terminology. In the beginning, particles were classified according to their mass — thus one talked of light particles (*leptons*), medium particles (*mesons*) and heavy particles (*baryons*). Gradually, the original meanings attached to the terms leptons, etc., lost their meaning but it was nevertheless found that leptons, mesons and baryons did form categories of their own. Later we shall see the reason for this. The term *hadron* is now used to comprehensively refer to both the mesons as well as the baryons.

The first of these tables, i.e., Table 2.A1.1, lists the leptons. As just indicated, historically, the term *lepton* was used to indicate a light particle (i.e., light compared to the proton), and the name did seem appropriate for e^- and ν. Later, other leptons were discovered which were rather 'bulky' but the name stuck. Leptons are fermions, and come in two varieties — charged and uncharged. All of these dutifully participate in weak interactions (W.I.); in addition, the charged leptons also participate in electromagnetic (e.m.) interactions. Later, we shall discuss this in greater detail. All leptons are assigned what is called a *lepton number*, which is conserved in reactions, the way an electric charge is. This too will come up for further discussion shortly.

Table 2.A1.1 The lepton family

Name	Symbol	Charge	Mass
Electron	e^-	−1	0.511 MeV
Electron neutrino	ν_e	0	~0
Muon	μ^-	−1	106.6 MeV
Muon neutrino	ν_μ	0	~0
Tau	τ^-	−1	1784 MeV
Tau neutrino	ν_τ	0	< 164 MeV

Remarks

1. All leptons are fermions.
2. The negative superscript means that the particle is negatively charged.
3. Corresponding to each particle listed above, there is an anti-particle; the latter, however, are not shown here.
4. Historically, first there was only the neutrino. Subsequently, guided by Dirac's theory, people said that there must also be an anti-neutrino. Still later, it was realised that the electron, the muon and the tau each have their own brand of 'special, personal' neutrinos!
5. In the past, mass used to be given in terms of the electron rest mass m_e. We follow here the current practice of giving it in energy units.
6. The neutrino mass is expected to be zero from theoretical considerations. What is shown above are the estimates from experiments. In two cases, experiments show that the mass is almost zero. In the third case, the estimate available is rather crude; all that can be said is that the mass must be less than 164 MeV.
7. Finally, it is believed (at least at present) that leptons have no internal structure; in other words, they are 'point' particles. Therefore, one may take them to be fundamental.

We turn next to Table 2.A1.2 which features the quarks. They help us to view all hadrons — baryons as well as mesons, as suitable combinations of quarks and anti-quarks (as will be explained in detail later).

Table 2.A1.2 The quark family

Name	Symbol	Charge (in electron units)	Mass
Up	u	2/3	310 MeV
Down	d	−1/3	310 MeV

Contd...

Table 2.A1.2 (*Contd.*) The quark family

Charm	c	2/3	1500 MeV
Strange	s	−1/3	505 MeV
Top	t	2/3	177 GeV
Bottom	b	−1/3	5 GeV

Remarks

1. All quarks are fermions.
2. The anti-quarks are not shown.
3. Observe that quarks have *fractional* charges. This is one of their unusual features. More about this later.
4. Earlier, the t and the b quarks used to be called Truth and Beauty respectively, but somehow, the more mundane labels Top and Bottom have replaced them.
5. As of now, quarks are regarded as fundamental (just as the leptons are). But who knows?! (Recall Fig. 1.2.)
6. Notice that there are as many families of quarks as there are of leptons. Is this an accident? You will hear about that later.
7. One usually speaks of three generations in the lepton and the quark families; each generation has two members.

Table 2.A1.3 shows the three generations of leptons and quarks.

Table 2.A1.3 Three generations of leptons and quarks

Generation	Leptons	Quarks
1	e^-	Up (u)
	ν_e	Down (d)
2	μ^-	Strange (s)
	ν_μ	Charm (c)
3	τ^-	Bottom (b)
	ν_τ	Top (t)

What about interactions between these various species of basic blocks? That information is provided in Table 2.A1.4.

Table 2.A1.4 The basic forces

	Range	Strength	Messenger	Spin	Charge	Mass
Gravity	Infinity	$\sim 10^{-38}$	Graviton	2	0	0
Electro-magnetism	Infinity	1/137	Photon	1	0	0

Contd...

Table 2.A1.4 (*Contd.*) The basic forces

			Vector boson			
Weak	$< -10^{-14}$ cm	10^{-5}	W^+	1	+1	80 MeV
			W^-	1	−1	80 MeV
			Z^0	1	0	91 MeV
Strong	$<10^{-13}$ cm	1	Gluons	1	0	0

Remarks

1. The messenger particles are sometimes collectively referred to as gauge bosons.
2. The vector bosons are sometimes also referred to as intermediate bosons.
3. Gluons come in many varieties; more about them later.
4. Gravity acts on everything. The electromagnetic force acts only on charged particles. The weak force acts on leptons, mesons and baryons, while the strong force acts only on mesons and baryons.
5. At present, we have only a classical theory of gravity (due to Einstein). If and when a quantum theory of gravity is developed, one expects that the quantum of that field would be the *graviton*. At present, the graviton is only conjectured but everybody is ready to bet on it!
6. The strength is indicated by e^2 and quantities analogous to that are expressed in dimensionless form.
7. Quantities like $e^2/\hbar c$ and their counterparts are called *dimensionless coupling constants*.
8. While in the case of e.m. interactions, the dimensionless coupling constant is given by $(e^2/\hbar c) \sim (1/137)$; for strong interactions it is $(f^2/\hbar c)$, where f obviously plays a role analogous to e; and for gravitational interactions, it is $[G_{\text{Newton}} \cdot (m_p^2/\hbar c)]$, where m_p is the mass of the proton and G_{Newton} is the gravitational constant.
9. Feynman diagrams can be drawn to indicate interactions to various orders (see ***QR-II***), the order being determined by the number of times the interaction vertices appear. And while computing the contribution from a given Feynman diagram, one notes that the coupling constant enters to the power of the order. For example, in second order e.m. interactions, the strength would be proportional to $(e^2/\hbar c)^2$.
10. Normally we think of electrical interactions in a static manner. For example, we write the interaction between two electrons as $\sim (e^2/r)$. This implicitly assumes that if one electron moves by some amount, the other electron knows about this movement *instantly*. What if the two electrons are separated by one billion kilometres? Even then the other electron would know about the change instantly. Wouldn't this get us into trouble with relativity? Ah, you have got the point. Maxwell, who worried about the way charges interact, etc., carefully analysed the earlier work of Faraday, Ampere and others, and came to the conclusion that the interaction between charged particles is mediated by electromagnetic fields and that field signals travel at the speed c of light. Later Einstein realised that this speed c is in fact a ceiling to the speed with which information can be transmitted. Einstein similarly showed that gravitational

interactions too are mediated by fields and that gravitational signals also are transmitted with a velocity *c*.

Many of the elementary particles discovered in the fifties appeared to behave in a 'strange' way compared to, for example, the more familiar neutron and the proton. What exactly was strange about their behaviour will be explained shortly. These particles therefore came to be called *strange particles*. It was also discovered that these particles had attributes other than the familiar mass, charge and angular momentum (spin). Since you will be coming across some of these quite frequently, it is better to get introduced to them right now. This list includes *electron number, muon number, tauon number, isotopic spin* (I), *strangeness* (S), *baryon number* (B) and *charm* (C); the list is by no means complete but this sample should do for the present. Some of the quantum numbers associated with leptons are indicated in Table 2.A1.5.

Table 2.A1.5 Various numbers associated with leptons

	Lepton number	Electron number	Muon number	Tauon number
Leptons				
e^-	1	1	0	0
ν_e	1	1	0	0
μ^-	1	0	1	0
ν_μ	1	0	1	0
τ^-	1	0	0	1
ν_τ	1	0	0	1
Anti-leptons				
e^+	−1	−1	0	0
$\bar{\nu}_e$	−1	−1	0	0
μ^+	−1	0	−1	0
$\bar{\nu}_\mu$	−1	0	−1	0
τ^+	−1	0	0	−1
$\bar{\nu}_\tau$	−1	0	0	−1

These various numbers play a role analogous to electric charge conservation in reactions. Let me give an example. Consider the reaction

$$\mu^- \longrightarrow e^- + \nu_e + \nu_\mu .$$

You can easily convince yourself that the sum of the muon numbers on both sides is the same, as also the sums of the electron numbers. It works the same way with baryon numbers, strangeness, etc.

A word now about strangeness. As a preliminary, we must note that most elementary particles decay — notable exceptions to these are the electron and the proton; about the proton it must be said that so far one believed that it was stable, meaning it would never spontaneously decay. However, recent theories suggest that the proton could possibly decay; more about this later, but for the moment we can take comfort in the fact that the lifetime for decay is $\sim 10^{30}$ years, which is more than a million trillion times the age of the Universe! So we are quite safe; we would be long gone before the protons in our body disappear through decay.

Let us get back to this decay business, which is usually the norm in the world of elementary particles. Barring the neutron whose decay time is of the order of a few minutes, almost all the other elementary particles decay within a fraction of a second. However, the decay times fall into three broad classes:

(a) Decay times between 10^{-25} and 10^{-20} second.

(b) Decay times between 10^{-13} and 10^{-18} second.

(c) Decay times between 10^{-5} and 10^{-9} second (see also Fig. 2.A1.1).

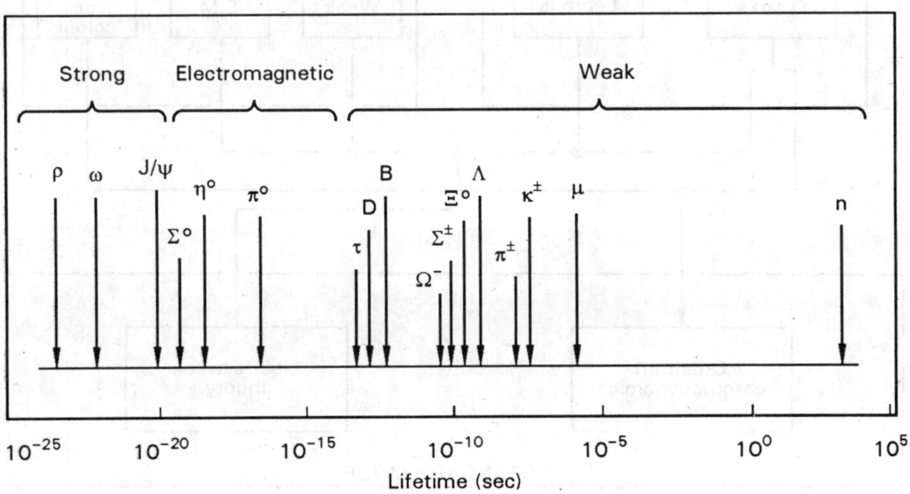

Fig. 2.A1.1 This shows the lifetimes of various particles decaying by strong, electromagnetic, and weak interactions. The various Greek symbols represent different elementary particles. Except the neutron and the muon, the rest were discovered after the fifties.

Decay processes in category (a) are characteristic of strong interactions; likewise, decays in categories (b) and (c) are associated respectively with electromagnetic and weak interactions.

Whenever an experimentalist studies an elementary particle reaction, two things are observed:

(a) the time for the formation of a particle, and

(b) the time for its decay.

Depending upon the observed decay time, it can be said that the decay proceeds via S.I., or e.m. or W.I.. With this background, we can now understand why people called certain particles observed in the fifties as 'strange particles'. Consider the reaction

$$\pi^- + p \rightarrow K^0 + \Lambda .$$

It turns out that this reaction proceeds via S.I., which means it takes place within about 10^{-24} sec. What about K^0 and Λ? It turns out that both of them decay with lifetimes around 10^{-10} sec, i.e., they decay via W.I.. This was puzzling — the formation of K^0 and Λ is via the S.I. but their subsequent decay is via W.I..

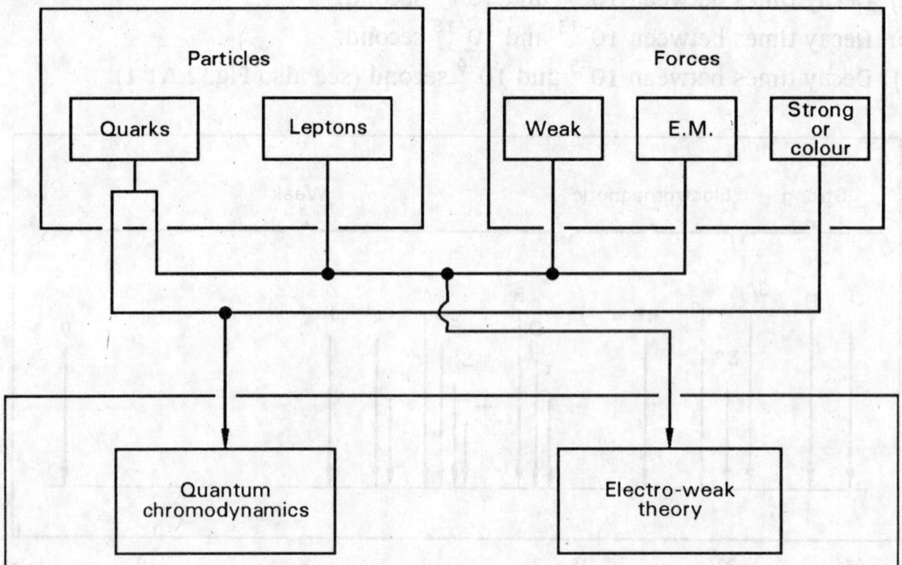

Fig. 2.A1.2 This figure shows how the SM is a composite of two different theories — the electro-weak theory that unifies the weak and the electromagnetic interactions, and QCD which is a theory for S.I. These two theories have still to be neatly integrated into one unified package. Several schemes have been suggested but none has proved itself yet. It should be mentioned that in modern terminology, the strong force is referred to as the colour force. This is the force that acts between quarks. Colour is a property like electric charge and just as particles with charge interact, particles with colour do likewise.

Thus, particles like K^0 were dubbed strange particles. Soon after this came the concept of the strangeness quantum number, to which a reference has already been made. Let us now see how this strangeness operates. The 'parent' particles π^- and p have $S = 0$; $S = +1$ for K^0 while for Λ it is -1. Thus strangeness is conserved in this reaction; in general, strangeness is conserved in S.I., whereas in W.I., it is not!

And finally, Fig. 2.A1.2 gives a comprehensive overview of the SM. The SM is not an integrated model like the Bohr model of the atom, for instance. It is the jargon used for referring collectively to the electro-weak theory on the one hand and quantum chromodynamics (QCD) on the other. As will be explained later, the search is still on for a unified fabric of these two distinct theories.

I guess this much should do for the present. Perhaps you are quite dizzy with all the name-dropping I have done! But don't worry, for as we go along, all this will become clear, even though it might not quite be so at present.

Appendix 2 to Chapter 2

Generally speaking, for every particle, there exists an anti-particle. Used as we are to the positron e^+ which is the anti-particle of the electron, we might think that an anti-particle is one which has the opposite charge. This is alright for charged particles but what about electrically neutral particles like the neutron, for example? Here we have to be more careful. It turns out that every particle can be assigned a number of descriptors/quantum numbers, like the lepton number, baryon number and so on. Let $\{N_i\}$ denote the complete set of such numbers. There must be at least one number N_α in this set such that for the anti-particle, it has the opposite sign, i.e., if for the particle it has the value N_α, then for the anti-particle it has the value $-N_\alpha$. For example, for the neutron, the baryon number is 1 whereas for the anti-neutron it is –1; similarly, for the neutrino the lepton number is 1 whereas for the anti-neutrino it is –1. It is in this manner that one distinguishes between the particle and its anti-particle when it is electrically neutral.

A number of particles like the photon belong to the class of messenger particles. It turns out that electrically neutral messenger particles are their own anti-particles. Included in this list are the photon, the Z^0 and the gluons.

A word now about the notation for the anti-particles. The standard notation is to use an overbar; for example, if p denotes the proton then $\bar{p}$ denotes the anti-proton. There are a few exceptions to this rule though; for example, the positron which is the anti-particle of the electron is denoted by e^+ because we denote the electron as e^-. Watch out for these few exceptions as we go along.

3

A Pot-Pourri of Some Basic Facts

This chapter is intended to expose you to some of the basics of conservation laws and their relationship to symmetry. You must already be familiar with the laws of conservation of linear momentum, of angular momentum and of energy, which are the legacies of classical physics. In the world of elementary particles, there are many other similar conservation laws like that for the lepton number, the baryon number and so forth. Historically, these laws emerged from careful experimental observations, just as the law of conservation of energy and the law of conservation of electric charge were deduced earlier. Now, however, we know that all such laws have a deep and fundamental basis. The main purpose of this chapter is to make you aware of this substratum.

3.1 Introduction

The first two chapters were in a sense concerned with the preliminaries; these being over, we can now get down to business. As already mentioned, the story of high energy physics is, in a substantial measure, the story of the emergence of the SM. In a sense, this story encompasses everything from the deepest recesses of the atom to the vast Universe itself. But before we set out on this exciting journey of exploration, it is useful to equip ourselves with a few facts.

We will start with the scales of some of the entities found in the Universe, illustrated in pictorial form in Figs. 3.1–3.3. I strongly urge you to spend some time absorbing and digesting these diverse and extraordinary facts; they supplement those already presented in Chapter 1. What is remarkable is that the nature of the 'big' is closely intertwined with that of the 'small'. Philosophers sensed this a long time ago and prophetically declared that the microcosm is in the macrocosm, and vice versa. However, it is only in recent times that the precise nature of this linkage has been revealed, and believe me, it has been a truly stunning revelation.

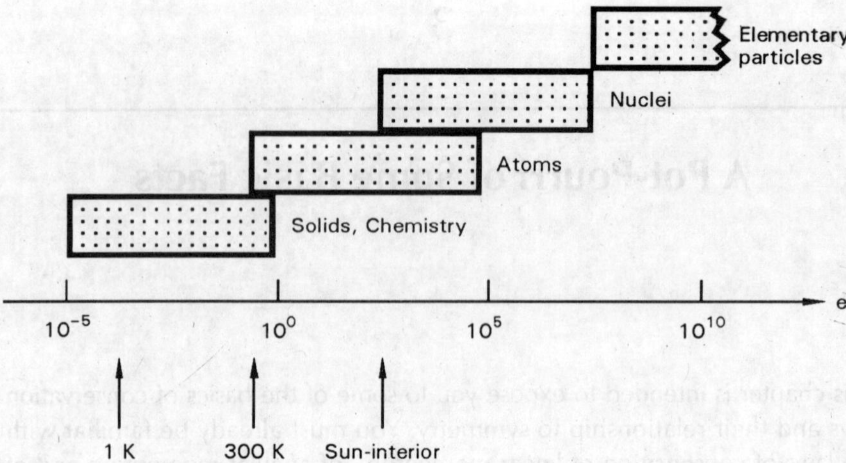

Fig. 3.1 Illustrated here are the excitation energies per particle in various cases. Also indicated in a few cases are the corresponding temperatures. Note that the energies indicated do *not* refer to the *total* energy of a physical system. If we consider that, then the energy of objects like the quasars would be truly mind-boggling.

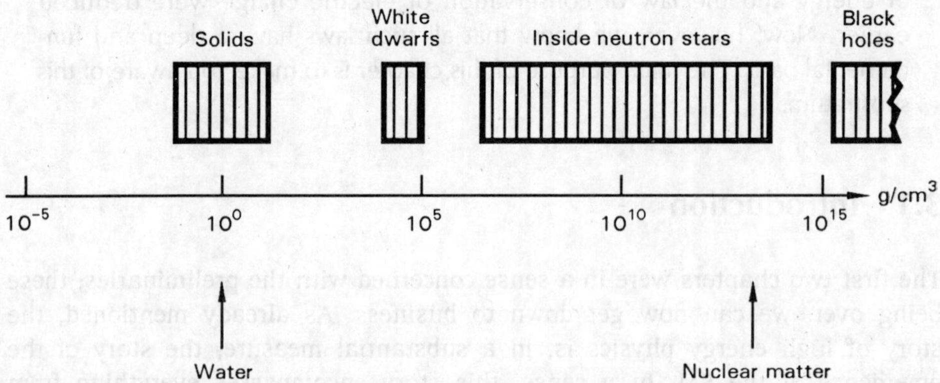

Fig. 3.2 The density of matter occurring in the Universe also varies over a wide range. We are scarcely aware of this because our daily experience is restricted to solids, liquids and gases. It takes a lot of imagination to figure out properties of matter under extreme conditions like in white dwarfs. That is precisely what Chandrasekhar did, winning undying fame (see ***Chandra***). Today we are aware of matter under even more extreme conditions such as those obtained in the interior of neutron stars and black holes.

3.2 Species in the particle zoo

From the 1930's onwards, hundreds of elementary particles came to be discovered, and for some time the situation was quite bewildering. Even so, physicists realised that two broad categories of particles could be identified, the leptons and the hadrons; the latter in turn were subdivided into mesons and baryons; later, the difference between these two subcategories would become clear. Besides all these, there were the 'messenger' particles or the gauge bosons; all these have already been briefly introduced to you. The first step in understanding these various species was to classify and organise them into families. For this purpose, various attributes were employed.

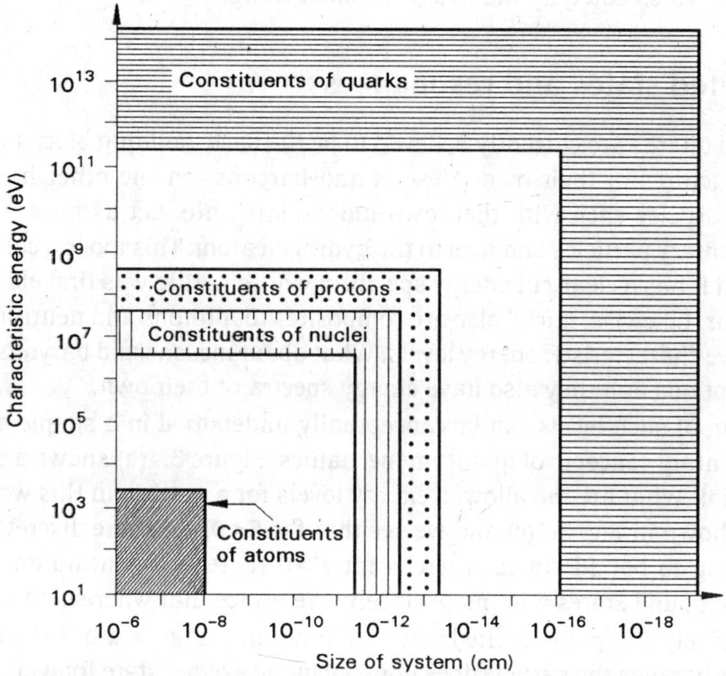

Fig. 3.3 You might have noticed in Fig. 3.1 that the smaller the system, the greater is the excitation energy. This reciprocal relationship is highlighted in this figure. As we shall see later, this is what brings about a link between the macrocosm and the microcosm.

3.3 Some standard attributes

Elementary particles have many attributes, all of which together help in their identification. Prominent among these — and these were also the first to be recognised — are the mass (m), the charge (Q), and the intrinsic angular momentum or *spin* (J). I must caution you that the term spin does not imply that the particle is actually spinning. It does nothing of that kind! In that case, why use the word spin? Ah! That is a historical accident, the details of which may be found in some of the companion volumes. Let us just note here that spin is an intrinsic property. It *appears* to be like angular momentum associated with ordinary rotational motion but there is no rotation involved. Rather, it is related to rotational symmetry in an 'inner' space; more on that later.

Spin is always a multiple of $\hbar$ or $\hbar/2$. Particles with half-integer spin (i.e., spin angular momentum equal to $\hbar/2$, $3\hbar/2$, $5\hbar/2$,...) are *fermions* while particles with integer spin (i.e., spin angular momentum equal to 0, $\hbar$, $2\hbar$, ...) are *bosons*. In this language, leptons are fermions while mesons are bosons. I presume I do not have to tell you to which category the gauge bosons belong!

3.4 Excited states and resonances

Leptons and quarks are currently believed to be the basic building blocks with no internal structures of their own. Mesons and baryons, on the other hand, are composite particles each with their own unique structure. Let us for a moment forget elementary particles and turn to the hydrogen atom. This too is a composite particle, and it has a clear-cut energy spectrum whose nature was first elucidated by Niels Bohr. Likewise, nuclei also are composites (of protons and neutrons) and they too have characteristic energy levels. What about mesons and baryons? They are no exceptions and they also have energy spectra of their own.

The origin of such levels can be conceptually understood in a simple manner using elementary concepts of quantum mechanics. Figure 3.4(a) shows a square-well potential. What are the allowed energy levels for a particle in this well? The answer is shown in Fig. 3.4(c) and we see that for $E<0$, there are discrete levels corresponding to bound states whereas for $E>0$ there is a continuum. Let us examine the bound states a bit more closely. We notice that whereas the ground state is infinitely sharp, the excited states all have an energy width $\Delta E = \Gamma$. This width arises because the particle does not stay in the excited state forever; it loses energy and makes a downward jump. As a result, excited states have finite lifetimes τ related to the energy width Γ by the uncertainty principle, i.e., $\tau\Gamma \sim \hbar$. Caution: different levels can have different widths.

Turning next to the continuum states, one would think there is nothing interesting in them — not true! See Fig. 3.5. Suppose, now, a beam of particles with energy $E > 0$ is fired at the potential. Classically, the particles would just 'overfly' the well; but in quantum mechanics things are different. Firstly there is a certain amount of reflection, and secondly the transmission shows peaks or resonances at certain energies (see Fig. 3.6). In view of Fig. 3.5, you should not be puzzled that resonances appear where a continuum of states exist. The situation depicted in Figs. 3.4 and 3.6 is no doubt idealised but the essential features apply to mesons and baryons. What I mean is this: suppose I take a proton (which is a baryon) and shoot at it a pion (which is a meson). The proton will present a potential well to the pion. Under certain conditions one should expect resonances, and indeed these are observed. We shall return later to particle resonances.

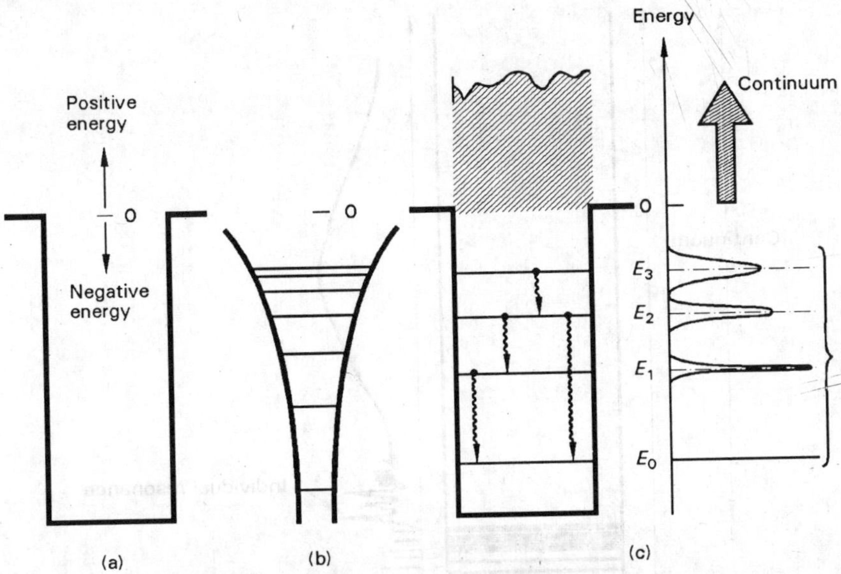

Fig. 3.4 Schematic diagram of the energy levels of a particle in a potential well. (a) shows an idealised square well whereas (b) depicts a more realistic one such as is presented by an ion to an electron. The allowed energy levels for the square well are sketched in (c). For $E < 0$, they are discrete. Of these the ground state is sharp while the others have a finite width. For $E > 0$, there is a continuum. A few possible transitions are also shown.

3.5 Symmetry and conservation laws

Conservation laws form the backbone of physics, and some of you might be aware of how much juice can be squeezed out of well-known conservation laws like those that relate to linear momentum or angular momentum or energy. The three conservation laws just mentioned are not the only ones known. There are several others, and a good many of them were discovered from the analysis of experimental data on elementary particles; these include the conservation of isospin, lepton number, etc. Incidentally, you might be concerned that I am doing a lot of name-dropping without much of an explanation. Don't worry! The explanations will come in due time, and at present I am merely setting the stage. So, patience please!

What exactly do we mean by a conservation law? Essentially that during a physical process there is a quantity whose value remains unchanged. When, for example, two billiard balls collide, the total linear momentum remains unchanged or, in other words, the sum of the linear momenta of the two balls *before* and *after* the collision is the same. The same sort of thing holds in particle physics also.

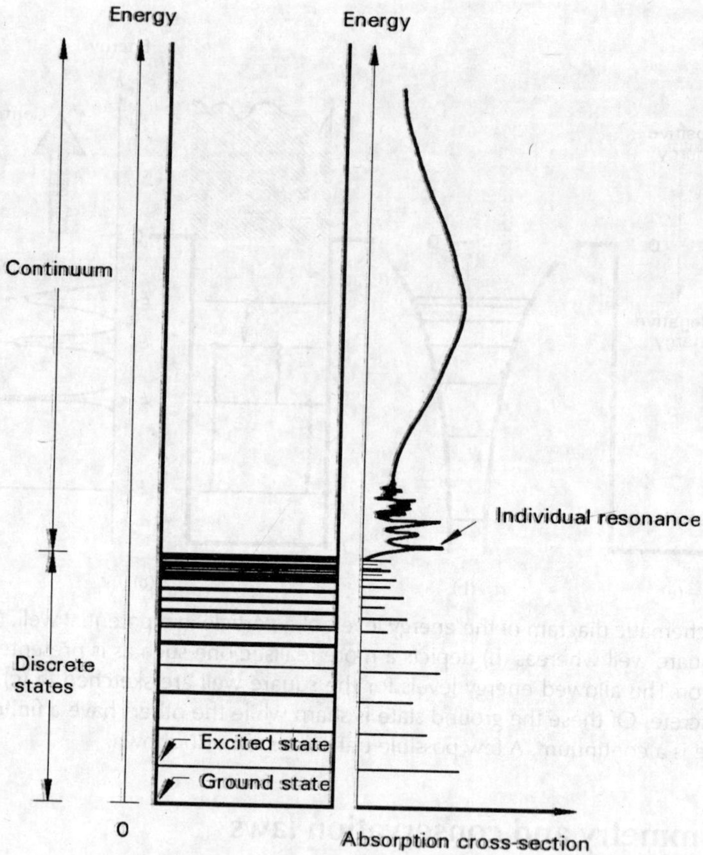

Fig. 3.5 This figure is a schematic diagram of the excited states of a nucleus, which is a composite system like the ones we are currently interested in. Suppose the nucleus is in the ground state and it is allowed to absorb electromagnetic energy. The cross-section for absorption would vary as shown on the right. The resonances associated with the discrete levels are easy to understand. It is seen, however, that there are a few resonances even in the continuum region. This is because the so-called continuum is really a conglomeration of closely spaced levels. Each level has a finite width. As the excitation energy increases, not only do the levels overlap but the widths also increase. On both these counts the resonances then get washed out, leading to a continuum.

Consider the reaction

$$\Sigma^+ \longrightarrow \pi^0 p,$$

also sometimes written as

$$\Sigma^+ \longrightarrow \pi^0 + p.$$

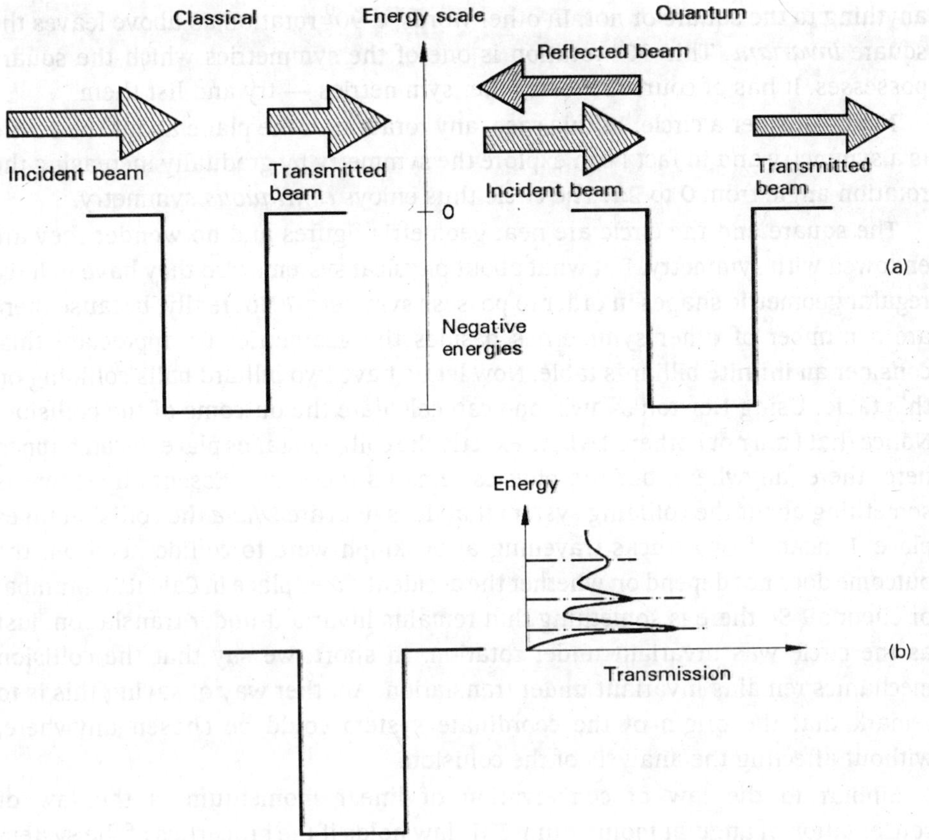

Fig. 3.6 Suppose a beam of particles with $E > 0$ is directed at the square well. What happens in the classical and quantum cases is shown in (a). Observe the possibility of reflection in the quantum case. The finer aspects of quantum transmission are highlighted in (b), especially the resonances.

What this says is that the elementary particle Σ^+ decays into a neutral pion (π^0) and a proton (p). It turns out, among other things, that the *baryon number* should be conserved in the reaction and this implies that the baryon number of Σ^+ must equal the sum of the baryon numbers of the π^0 and the p. The baryon number is thus one of the conserved quantities. The discovery of such conservation laws was the outcome of a large number of experiments performed by a large number of people.

Why on earth does Nature have conservation laws? This is an important question as it leads us to the profound connection between conservation laws and the symmetries inherent in the system. A few words first about symmetries.

Consider a square. Everyone would agree it is a nice, symmetrical figure. Suppose we turn the square by 90^0 (in the plane and about the centre); the square then goes into itself. In fact, if you were not looking and I had quickly turned the square before you looked at it again, you would not know whether I had done anything to the square or not. In other words, a 90^0 rotation as above leaves the square *invariant*. This 90^0 rotation is one of the symmetries which the square possesses. It has of course several other symmetries — try and list them.

Next, consider a circle. In this case, any rotation in the plane about the centre is a symmetry and in fact I can explore the symmetry by gradually increasing the rotation angle from 0 to 2π. The circle thus enjoys *continuous* symmetry.

The square and the circle are neat geometric figures and no wonder they are endowed with symmetry. But what about physical systems? Do they have to have regular geometric shapes in order to possess symmetry? Not really, because there are a number of other symmetries besides the geometric. To appreciate this, consider an infinite billiards table. Now let us have two billiard balls colliding on this table. Using Newton's laws, one can calculate the outcome of the collision. Notice that I am not bothered where exactly the collision takes place. It can happen here, there, anywhere, but the physics remains the same. Essentially, there is something about the colliding system that does not care *where* the collision takes place. I mean if two trucks travelling at 60 kmph were to collide head on, the outcome does not depend on whether the accident takes place in Calcutta, Mumbai or Chennai! So there is something that remains invariant under translation, just as the circle was invariant under rotation. In short, we say that the collision mechanics remains invariant under translation. Another way of saying this is to remark that the origin of the coordinate system could be chosen anywhere, without affecting the analysis of the collision.

Similar to the law of conservation of linear momentum is the law of conservation of angular momentum. This law holds if the properties of the system do not depend upon the orientation of the system; the concept is explained further in Fig. 3.7. To put it differently, isotropy of space leads to the law of conservation of angular momentum.

There is one other symmetry which also is better appreciated by considering the colliding truck analogy. You will agree that the outcome of the accident does not depend on when it takes place — it could be yesterday, today or tomorrow. In other words, invariance under *translation of time* is also possible. The famous conservation laws of classical physics, namely the law of conservation of linear momentum, the law of conservation of angular momentum and the conservation of energy arise respectively from the invariance under continuous spatial translation, continuous rotation and continuous time translation. In this context, don't miss Box 3.1.

3.6 A deeper background to conservation

Quantum mechanics (see also *QR-I*) provides a neat explanation of the connection between invariance/symmetry and conservation laws. Let H be the

Box 3.1

Our ideas concerning the role of symmetry in physics owe much to Amalie Emmy Noether. It was she who first showed that if there is a symmetry, a conservation law cannot be far behind.

Emmy was born in 1882, in Erlangen, Germany, where her father was a professor of mathematics. In those days, education for girls was frowned upon and the University of Erlangen did not allow girl students to matriculate. However, through the influence of her father, Emmy was permitted to attend classes. In 1904, the University abandoned its archaic rules and four years later Emmy earned the doctorate degree in mathematics, with honours. But now there was a new problem — she could not get a job anywhere, either as a teacher or as a research assistant. She applied to Gottingen, where David Hilbert (to whom we owe, among other things, the Hilbert space) was the presiding deity, but without luck. (By the way, Gottingen also boasted of Arnold Sommerfeld, several of whose students later won the Nobel Prize.) Emmy offered to work at Gottingen without salary but her offer was rejected. Hilbert was furious and exploded, "I don't see the sex of the candidate is an argument against her admission to *Privatdozent* (teaching assistant). After all, we are a University and not a bathing establishment." Finally in 1919, Emmy was given a teaching position and four years later she began to be paid for her work. To her credit, Emmy bore it all stoically.

In 1933 came fresh trouble when Hitler began hounding out Jews. Emmy was a Jew and she had to flee. Like many others she landed in America, where once again the job problem cropped up which was later partially solved. Finally came a medical problem which claimed her life (in 1935). Delivering her memorial address, the celebrated mathematical physicist Herman Weyl said, "She was not clay, pressed by the artistic hands of God into harmonious form, but rather a chunk of primary human rock into which He had blown His creative breath of life."

Hamiltonian of a physical system and Ψ an eigenfunction. Let an operation U acting on the system transform Ψ to Ψ', i.e.,

$$\Psi' = U\Psi.$$

We assume U to be an infinitesimal (unitary) operation which can be represented as $U = (1 + i\varepsilon F)$, where i is a parameter and $\varepsilon F << 1$. Finite operations are built up by repeating the infinitesimal one repeatedly and in a suitable manner (as will be explained shortly). The operator F is called the *generator*.

Suppose U leaves H invariant. This means

$$H = U^{\dagger} H U$$

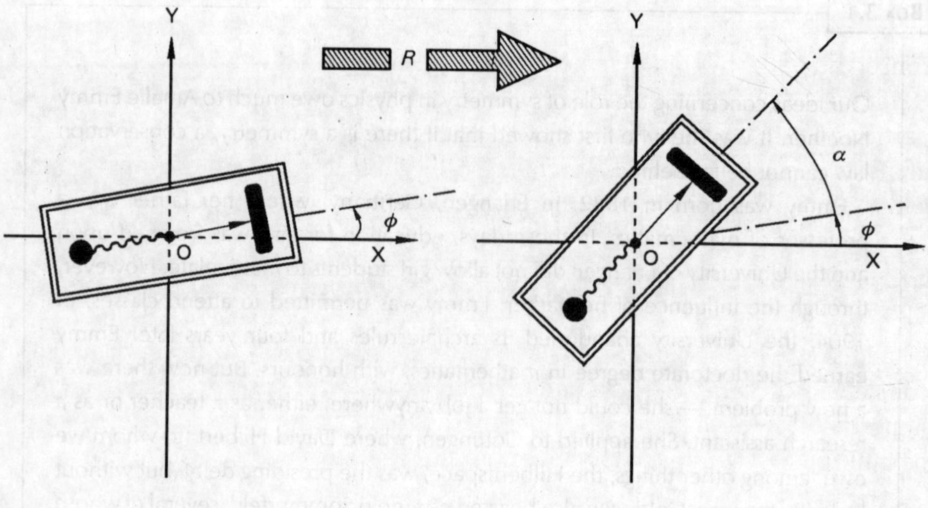

Fig. 3.7 Let us say that there is an experimental apparatus of some kind. OX and OY define a set of axes with respect to the lab and let ϕ denote the angle which describes the orientation of the apparatus. Suppose the apparatus is rotated by the angle α as shown. Rotational invariance means that the apparatus is not affected by this rotation or, in other words, the outcome of an experiment done with the apparatus remains the same as it was when the experiment was done in the earlier orientation. Rotational invariance leads to the conservation of angular momentum. Since the invariance is related to angular isotropy, one often says that the conservation of angular momentum is the consequence of the isotropy of space. In brief, conservation of angular momentum is linked to the fact that it does not matter which way you turn an object around in space.

$$\text{or} \quad HU - UH = 0, \quad \text{i.e.,} \quad [H, U] = 0, \qquad [3.1]$$

where [,] denotes the commutator. Remembering the form of U we then have

$$H(1 + i\varepsilon F) - (1 + i\varepsilon F)H = 0$$

or, since ε is arbitrary,

$$[H, F] = 0. \qquad [3.2]$$

Now we know that $<F>$, the expectation value of F, is conserved if F commutes with H. Thus, the conservation of F is linked with invariance of the Hamiltonian under the transformation U. An example might help. In this case we can write

$$U = 1 + \varepsilon \frac{d}{dx} = 1 + i\varepsilon \left(\frac{1}{i}\frac{d}{dx}\right) = 1 + i\varepsilon \left(-i\frac{d}{dx}\right) = 1 + i\varepsilon \left(\frac{p_x}{\hbar}\right),$$

remembering that the operator $p_x = -i\hbar(d/dx)$. We thus see that if H commutes with U, or equivalently the operator p_x, then linear momentum is conserved.

The message of this example is the following: if a quantity in Nature is observed to be conserved in a physical process, look for a symmetry underlying the conservation law. This symmetry would correspond to the invariance of the Hamiltonian under a particular transformation or set of transformations. In particle physics, people discovered many a conservation law from a painstaking analysis of experimental data and then got a glimpse of the underlying symmetry, which almost invariably was internal.

3.7 Discontinuous transformations/symmetries

In addition to continuous translational and rotational transformations earlier considered, *discontinuous* transformations are also possible, reflection being an important example — I mean, either there is a reflection or there is not; it cannot be built up by adding a whole lot of infinitesimal reflections.

Among the important discontinuous transformations we need to know about are:

(a) Parity,

(b) Charge conjugation, and

(c) Time reversal.

Discontinuous symmetries were not seriously considered in classical physics, but in the quantum world they play a key role.

3.8 Parity transformation

Applying a parity transformation to a physical system essentially implies creating a 'mirror image' of the system concerned. For a long time physicists implicitly believed that the behaviour of the direct world and the 'mirror' world were the same. And this invariance of behaviour under parity transformation was simply referred to as the *conservation of parity*. By the way, sometimes the parity transformation is described as an inversion operation rather than as a mirror reflection. However, as can be seen from Fig. 3.8, inversion and reflection are closely related. To understand better what parity conservation really means, consider the particle reaction

$$a + b \longrightarrow c + d.$$

This, say, is an allowed reaction in the direct world. Suppose we create a 'mirror' world. Will there be a corresponding reaction in that world? If there is, we say parity is conserved and if there is not, we say that parity is not conserved.

Can we make this a bit more precise? Yes we can. Let $|a\rangle$, $|b\rangle$, etc., denote the states of the various particles and P the parity transformation. Remember: $|\;\rangle$ is

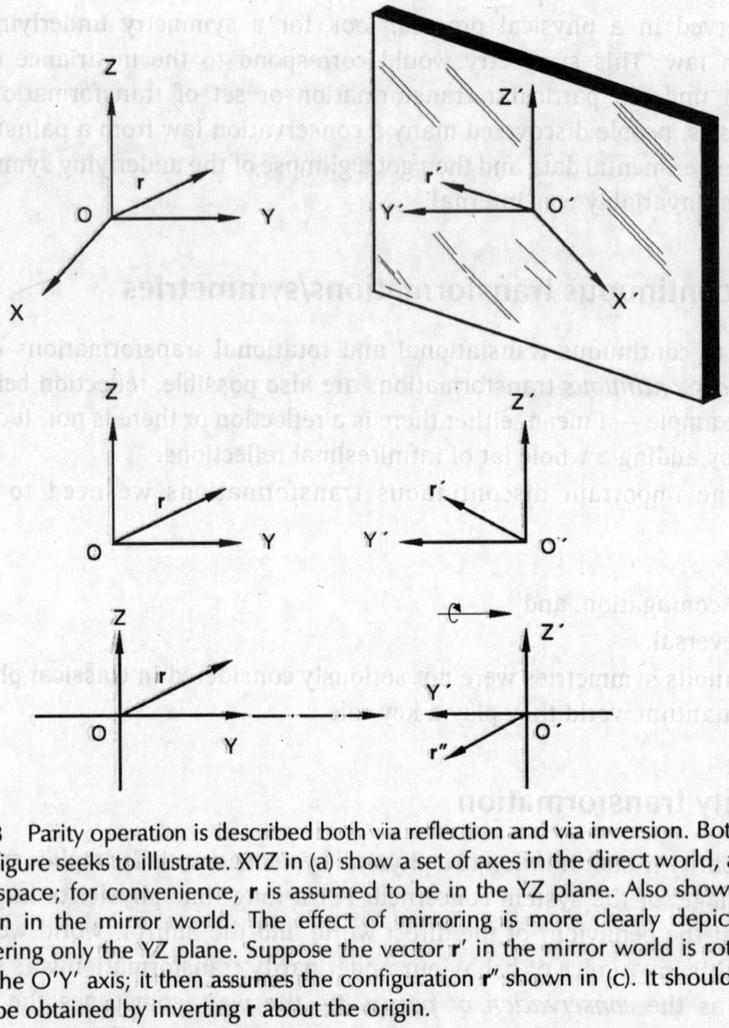

Fig. 3.8 Parity operation is described both via reflection and via inversion. Both are related, as this figure seeks to illustrate. XYZ in (a) show a set of axes in the direct world, and **r** a vector in this space; for convenience, **r** is assumed to be in the YZ plane. Also shown in (a) is the situation in the mirror world. The effect of mirroring is more clearly depicted in (b) by considering only the YZ plane. Suppose the vector **r**′ in the mirror world is rotated by 180° about the O′Y′ axis; it then assumes the configuration **r**″ shown in (c). It should be clear that **r**″ can be obtained by inverting **r** about the origin.

the quantum mechanical notation for the state of the particle. We can now write the initial and the final states as

$$|\text{initial}\rangle = |a\rangle.|b\rangle.|\text{relative motion of } a \,\&\, b\rangle$$

$$|\text{final}\rangle \;\; = |c\rangle.|d\rangle.|\text{relative motion of } c \,\&\, d\rangle.$$

Now apply the operator P on the initial as well as the final states, according to the rule

$$P|\text{initial}\rangle = P|a\rangle.\,P|b\rangle.\,P|\text{relative motion of } a \,\&\, b\rangle$$

$$P|\text{final}\rangle \;\; = P|c\rangle.\,P|d\rangle.\,P|\text{relative motion of } c \,\&\, d\rangle.$$

If parity is conserved, it simply means that

$$P|\text{initial}> = P|\text{final}>.$$

Perhaps all this might sound a bit terse and abstract but later when we discuss the Wu experiment, you will see how all these dry arguments come alive.

Incidentally, you may wonder what exactly $P|a>$ stands for. Without going into details, let me just mention that $P|a> = \pi_a|a>$, where π_a is either +1 or −1 (those of you who are familiar with *QR-I* will recognise π_a as the eigenvalue of the parity operator P). This π_a is called the *intrinsic parity* of the particle a. Every elementary particle has an intrinsic parity assigned to it. For the proton and the neutron it is +1. Sometimes, this is simply abbreviated as '+'. Likewise, intrinsic parity of −1 is abbreviated as '−'.

Notice that if the parity transformation is applied twice, we get back the original system, i.e., $P^2 = 1$. In other words, the mirror image of the image world is the original world itself.

3.9 Charge conjugation

If parity takes us to the mirror world, charge conjugation is even more exotic for it takes us to the 'anti-world'! Maybe I should take you through this slowly.

Let us start with the hydrogen atom. As you know, it is a system in which a proton is bound to an electron. Now consider the anti-particles of these two, i.e., the anti-proton and the positron. Can we have an anti-hydrogen atom in which a positron is bound to an anti-proton? Why not? In fact, very recently, such anti-hydrogen atoms have been produced in the laboratory. Starting from the anti-H atom, one could think of a whole variety of anti-matter including anti-planets, anti-stars and what not (including anti-people!).

Would physics in the anti-world be the same as in ours? If it is, we speak of *C invariance*. In a nutshell, charge conjugation implies transforming a particle into its anti-particle and C invariance means that any physical process and its charge-conjugated counterpart cannot be distinguished from each other. Note incidentally that as in the case of P, $C^2 = 1$. Also, like π in the case of P, one sometimes talks of *charge parity* — it is the eigenvalue of the operator C.

3.10 Time reversal

Time reversal (T) is rather like space inversion. Whereas in the latter we have $\mathbf{r} \to -\mathbf{r}$, with T we have $t \to -t$. More precisely,

$$\mathbf{r} \xrightarrow{T} \mathbf{r}, \quad t \xrightarrow{T} -t.$$

In turn, this implies

$$\mathbf{p} \xrightarrow{T} -\mathbf{p},$$

where **p** is the momentum of the particle. For this reason, time reversal is better described as *motion reversal*. More generally, it implies not only the reversal of translational motion but also of rotational motion.

What does all this really mean? I have discussed this matter in some detail in *A Hot Story*, but a brief recall is pertinent. Consider Fig. 3. 9(a) which shows three stages of a collision of two billiard balls. If we shoot a movie, we can capture the continuous change from the position in Fig. 3.9(a(i)) to that in Fig. 3.9 (a(iv)).

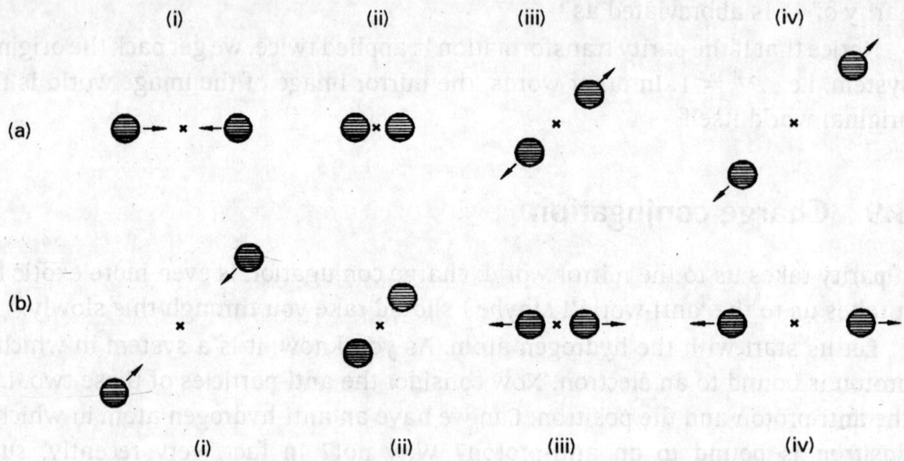

Fig. 3.9 (a) shows the evolution of a collision sequence of two billiard balls bumping into each other. (b) shows the time-reversed sequence. In the macro world, time flows strictly in one direction and time-reversed scenarios are not observed. In the micro world, time reversal symmetry is permitted (except in very rare cases).

Now consider the sequence in Fig. 3.9(b). If you think about it, you will realise that the starting point in the second scenario is obtained simply by reversing the motions in the final stages of scenario one. This is the sequence we would see if the film were run backwards. The laws of mechanics allow the sequence 3.9(b(i)) to 3.9(b(iv)) to freely evolve without any problem; in other words, *this second sequence is also one of the allowed solutions of the equations of motion of the system*. Now reflect carefully on this: we have two possible collision scenarios. Both are allowed. In fact, they are time-reversed copies of each other. Thus, the same equation is able to predict the solution applicable to the direct as well as the time-reversed world, and for this reason we speak of *T* invariance. To put it differently, the basic equations of classical physics, be they Newton's laws or Maxwell's equations, are unaffected by the replacement of *t* by –*t*. A given evolution scenario and its time-reversed counterpart are both permitted by the equations of motion and are thus *both* physically realisable. In quantum mechanics the situation is a bit more complicated, but here too one can speak of T invariance if $[H, T] = 0$.

You might wonder, "Does motion reversal always work? Suppose I introduce a drop of ink in a glass of water. The ink molecules diffuse and soon make the water appear uniformly coloured. If now the motions of all the ink molecules are reversed, will they all collect back into a drop? We never see the ink drop re-forming. So, what happens to T invariance now?" This is a valid doubt, and the answer is that in the *macroscopic* world, time *does* flow in only one way — the young get old but the old do not get young! There is an important reason for this, which I have discussed in ***A Hot Story***. With an ink drop and water, we are dealing with the macroscopic world where time always flows only in one direction but in the *microscopic* world (and that includes the world of elementary particles), we expect T invariance to hold.

3.11 The *CPT* theorem

I have so far been arguing that (in the world of elementary particles) one has every reason to expect separately, P invariance, C invariance and finally, T invariance as well. It is obvious then that one must also have the combined *CPT* invariance. Stated this way, it might seem that *CPT* invariance (the order of C, P and T does not matter — they could be applied in any order) is a trivial corollary. Not quite. Luders and later Pauli proved *without* specifically invoking individual invariances, that *CPT* invariance *always* holds under quite general conditions. In other words, the Luders–Pauli theorem effectively says that physics in our world and in the parity-reflected, time-reversed, anti-world is the same!

Suppose P is violated (as happens in weak interactions). Since *CPT* invariance holds, it implies that there must be a compensating violation of either T or C. That is what makes the Luders–Pauli theorem important — it provides clues about other possible violations.

3.12 Discrete symmetry in formal language

Analogous to Eqn. [3.2], we have, if P invariance holds,

$$[H, P] = 0.$$

If C invariance holds,

$$[H, C] = 0.$$

And if T invariance holds,

$$[H, T] = 0.$$

Here H could stand for H_{strong}, H_{weak} or $H_{\text{e.m.}}$ as the case may be. Before Lee and Yang made their famous discovery regarding P violation (more about that later), people believed that $[H_{\text{weak}}, P] = 0$. But experiment showed that parity is violated in weak interactions, which meant that $[H_{\text{weak}}, P] \neq 0$. On the other hand, the Luders–Pauli theorem calls for comprehensive CPT-invariance, implying $[H_{\text{weak}}, CPT] = 0$. In turn this implies that either $[H_{\text{weak}}, C] \neq 0$ or $[H_{\text{weak}}, T] \neq 0$, i.e., parity violation is accompanied by a matching C or T violation, in order to protect the Luders–Pauli theorem.

3.13 Discrete versus continuous symmetries

Let us pause for a moment and absorb the import of some of the important ideas just discussed. Basically, we came across two distinct kinds of symmetries — continuous symmetry and discrete symmetry. Rotation is an example of the former, and reflection of the latter. We say rotation symmetry is continuous not only because we can have any rotation, but also because any given rotation by any angle θ can be built by a cumulation of successive infinitesimal rotations — it is like turning a knob from 0° to the value θ continuously. Reflection is not like this; it cannot be built up gradually via a succession of infinitesimal reflections, for there is no such thing — one either has a reflection or one does not. Thus, reflection is a discrete symmetry. We have also noted that in both cases there are conservation laws and in both cases there are quantum numbers. To give an example, in the case of rotation we have the conservation of angular momentum and the relevant quantum number is the angular momentum quantum number. Likewise, in the case of reflection we have the conservation of parity and the relevant quantum number is the parity quantum number, referred to simply as parity. It is interesting to note that while continuous symmetries lead to *additive conservation laws*, discrete symmetries lead to *multiplicative conservation laws*. What I mean is simply this: suppose we have two particles undergoing a reaction, and we want to examine the conservation of angular momentum in the process. For obtaining the angular momentum of the initial state, we would simply add the angular momentum quantum numbers of the two particles reacting with each other. Likewise, for obtaining the angular momentum of the final state, we would simply add the angular momentum quantum numbers of all the reaction products. The conservation of angular momentum implies that the sum of the angular momentum quantum numbers before the reaction is the same as the sum of similar quantum numbers after the reaction.

Contrast this with the case of parity. If you recall Section 3.8, the parity of the initial state is simply given by the product of the intrinsic parities of the particles participating in the reaction (ignoring for the moment the effects of orbital motions); the parity of the final state also is determined in a similar fashion. Parity conservation implies in this case that the product *before* equals the product *after*.

Let us go a little more deeply into this business. We know that if F is an operator corresponding to a quantum observable, and if $[H, F] = 0$, it implies that

$$(d/dt) <F> = 0$$

i.e., the expectation value of F in a given quantum state is a constant, implying that F is conserved. This is fine as a theorem, but the question arises as to how conserved quantities like F can be found. This is where symmetry, invariances, etc., enter into the picture.

Suppose there is a unitary transformation U which leaves H invariant, i.e.,

$$U^{\dagger} HU = H$$

$$\text{or} \quad UH = HU$$

which is the same as $[H, U] = 0$. If now U can somehow be related to F, then we can say that when U leaves H invariant, F would be conserved. It turns out that for continuous transformations, U can be written in the form

$$U = \exp (i\varepsilon F)$$

where ε is a real parameter, and F is called the generator of U. If U is vanishingly small, then

$$U \sim 1 + i\varepsilon F \quad \text{and} \quad \varepsilon F << 1.$$

By repeated application, U becomes

$$U = \exp (i\varepsilon F)$$

as already noted. I hope you can figure that one out.

In quantum field theory — we shall see later that both the electro-weak theory and quantum chromodynamics mentioned earlier are examples of quantum field theories — it often happens that field variables transform under U as below:

$$\psi \longrightarrow \psi' = \exp (i\varepsilon N)\psi .$$

Here ψ is the field variable and N is a quantum number like the lepton number, baryon number, strangeness, etc. I shall have more to say on this matter later but for the present I hope you have got a glimpse of how these mysterious quantum numbers arise; they do so because the system (technically it is the Lagrangian, but that will come later) is invariant under a unitary transformation in the so-called inner space. It is this internal symmetry that leads to all these exotic quantum numbers. Before high energy experiments were performed, no one had any clue that Nature had hidden internal spaces and symmetries.

3.14 Concerning interactions

I have used the word 'interaction' several times, without precisely indicating what it means, except perhaps in a vague sort of way. It's now time to rectify that.

By interaction, we mean the force which operates between the two elements that are interacting. In classical physics, the concept of interaction is familiar to us from the examples of gravitation and electrostatics, where one discusses it in terms of either a potential or a force, as convenient. In particle physics, the word interaction is used in a more comprehensive sense, but it implies the same thing. An interaction between two particles could lead to binding, scattering or annihilation. When discussing interactions between elementary particles, we must remember that they are either bosons or fermions, and further that while bosons can be created or annihilated singly, fermions are always absorbed or emitted in pairs—these constraints follow from lepton and baryon conservation laws.

Interactions are conveniently represented via Feynman diagrams, and we shall draw quite a few of them as we go along. I won't go into the technicalities of Feynman diagrams here (some of it can be found in ***QR-II***) but intuitively one can see what they imply, from the examples given in Fig. 3.10. The ranges and the strengths of the four fundamental interactions vary widely, as Fig. 3.11 shows.

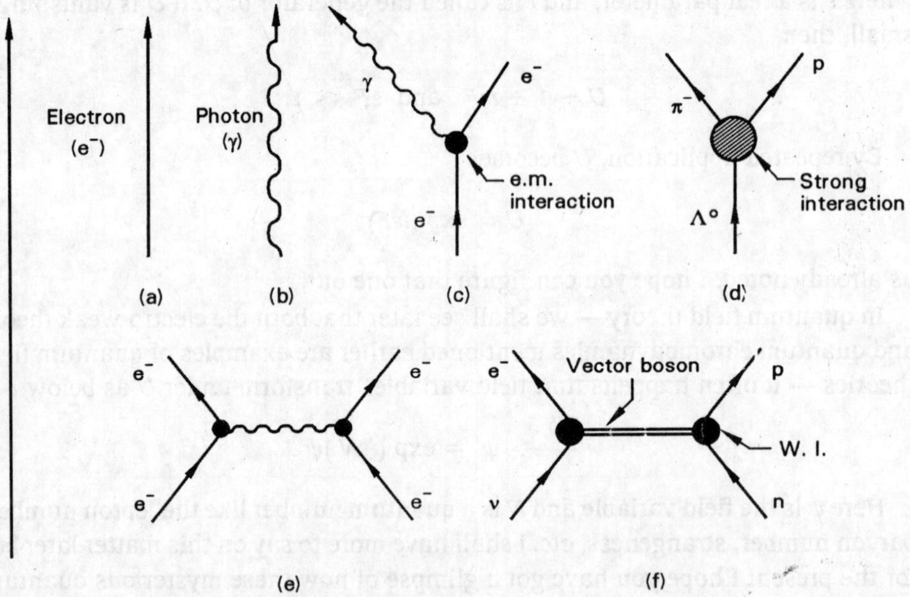

Fig. 3.10 The Feynman diagrams provide a convenient way of picturing complex interactions. Basically they are built up of elementary modules. Illustrated here are a few simple diagrams. Observe that (e) depicts a second order process involving two electromagnetic interactions like that in (c). This is how complexity is generated.

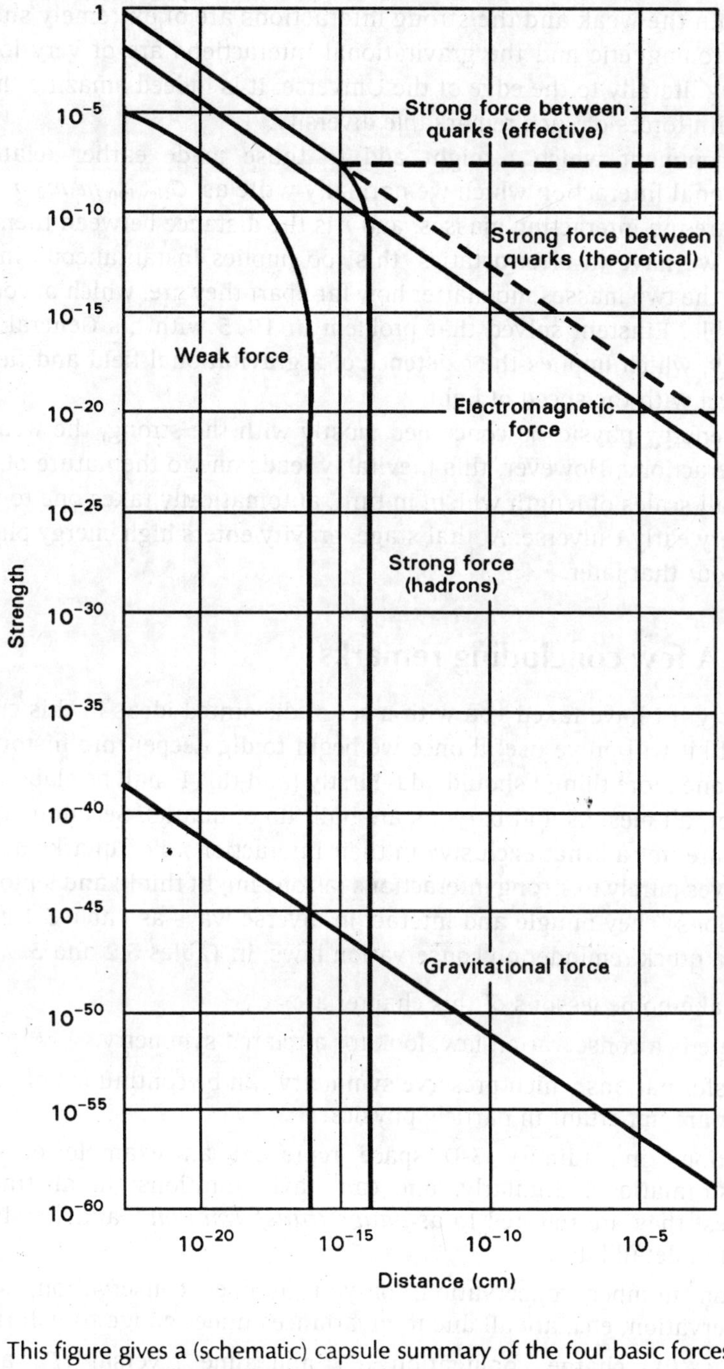

Fig. 3.11 This figure gives a (schematic) capsule summary of the four basic forces. Observe the wide disparity in strengths and the ranges. (The powers of ten obscure this fact!)

(This figure supplements Table A2.1.4 in the previous chapter.) Observe that while both the weak and the strong interactions are of extremely short range, the electromagnetic and the gravitational interactions are of very long range, stretching literally to the edge of the Universe. It is indeed amazing that Nature works with forces of such remarkable diversity.

One comment which I might add to those made earlier relates to the gravitational interaction which we normally write as $G_{\text{Newton}} m_1 m_2 /r$, where m_1 and m_2 are the interacting masses, and r is the distance between them. Like the formula we have in electrostatics, this too implies instantaneous interactions between the two masses, no matter how far apart they are, which of course is not permissible. Einstein solved that problem in 1915 with his General Theory of Relativity, which implies the existence of a gravitational field and field signals that travel with the speed of light.

High energy physics is concerned mostly with the strong, the weak and the e.m. interactions. However, this inevitably leads one to the nature of matter on very small scales of length which, in turn, automatically takes one to the nature of the very early Universe. At that stage, gravity enters high energy physics also; more about that later.

3.15 A few concluding remarks

I am sorry if I have taxed you with a set of disjointed ideas in this chapter, but this tool kit will prove useful once we begin to dig deeper into history. Perhaps there is one more thing I should add. Firstly (and this I shall be elaborating upon later too), all mesons and baryons are built up of quarks. Secondly, quarks and leptons are not all that exclusive in their interactions, i.e., quarks don't restrict themselves purely to strong interactions (as one might think) and leptons to weak interactions. They mingle and interact in diverse ways as Table 3.1 shows. And finally, a quick reminder on conservation laws, in Tables 3.2 and 3.3.

The take-home lessons of this chapter are:

- If there is a conservation law, look for a related symmetry.
- Transformations which preserve symmetry can be continuous or discrete; the latter are important in particle physics.
- Rotations in ordinary 3-D space represent an example of continuous transformations. Similarly, one can have 'rotations' in abstract internal spaces; they are referred to as *gauge transformations*, and we shall go into them in detail later.
- Lepton number conservation, baryon number conservation, strangeness conservation, etc., are all due to invariances under gauge transformations.
- Parity (P), charge conjugation (C) and time reversal (T) are discrete transformations.

- Continuous symmetries lead to additive quantum numbers while discrete symmetries lead to multiplicative quantum numbers.
- *P*, *C* and *T* may be individually violated in particle interactions, but *CPT* is always conserved. *CPT* invariance is called the Luders–Pauli theorem.
- All interactions are field-mediated. They can be represented by Feynman diagrams. The number of times the interaction vertex appears in the Feynman diagram indicates the order of the interaction process.

Table 3.1 Forces to which quarks and leptons respond

Species	Subspecies	Strong	e.m.	Weak
Quarks		Yes	Yes	Yes
Leptons	electron, mucn, tauon	No	Yes	Yes
	corresponding neutrinos	No	No	Yes

Remark

All particles having non-zero mass respond to the gravitational interaction; but in high energy physics, especially within the framework of the Standard Model, gravitational interaction is excluded as 'irrelevant'.

Table 3.2 Conservation laws and their validity with respect to the various interactions

Conservation of	Strong	e.m.	Weak
Energy	Yes	Yes	Yes
Momentum	Yes	Yes	Yes
Angular momentum (spin)	Yes	Yes	Yes
Electric charge	Yes	Yes	Yes
Isospin	Yes	No	No
Strangeness	Yes	Yes	No
Baryon number	Yes	Yes	Yes
Lepton number	Yes	Yes	Yes
P invariance	Yes	Yes	No
C invariance	Yes	Yes	No
T invariance	Yes	Yes	Yes

Remarks

1. *P* invariance is sometimes spoken of as parity conservation.

2. There is a special case in weak interactions (K decay) where *C* invariance does not hold. We shall discuss that in a later chapter.

Table 3.3 Conservation laws and related invariances

Symmetry/invariance	Quantity conserved
Time translation	Energy
Spatial translation	Momentum
Spatial rotation	Angular momentum
Gauge transformation	Electric charge
Rotation in isospin space	Isospin
Gauge transformation	Baryon number
Gauge transformation	Lepton number
Mirror reflection	Parity
Charge conjugation	Charge parity

Remarks

1. Once again let me remind you that while spatial translation, spatial rotation, time translation and gauge transformations are continuous transformations, *P*, *C* and *T* are discrete transformations.
2. Continuous transformations lead to additive conservation laws while discrete transformations lead to multiplicative conservation laws.
3. Gauge transformations represent rotations in abstract 'inner' space. Invariance under gauge transformations reflects internal symmetries.
4. Notice that time reversal does not figure in the above. Try and find out why!

4

From the Thirties to the Sixties

High energy physics really took off only from the late fifties to the early sixties. What happened before that could be described as a 'warm up'. This chapter reviews that period and is thus in the nature of a prologue to the drama that commenced in the sixties and is still going on.

4.1 Introduction

What we nowadays call particle physics or high energy physics, came into its own in the latter part of the fifties, but the foundations were laid much earlier. Between the thirties and the sixties the stage was being set, and it was rather like what happens before the curtain goes up at the start of a play — actors getting dressed, electricians fixing lights, carpenters doing last-minute hammering, stage hands moving props, etc. A lot of apparent confusion and disjointed activity all round, but everything becoming relevant and acquiring meaning once the play starts.

The period with which this chapter is concerned was a prelude of a similar nature. Though particle physics had not yet acquired a distinct identity of its own, the embryo was, nevertheless, definitely taking shape. But after the sixties everything changed most dramatically, as we shall see in later chapters. In fact, after the sixties, particle physics became not only one of the most active areas of physics but also the one which produced the maximum number of remarkable discoveries; it also emerged as the one consuming the maximum amount of research funds!

4.2 How it all began

So what happened during the prologue period, and how did it all get started? Hans Bethe and Victor Weisskopf are two of the giants whose careers began in the thirties and who were actively involved in much of what happened later. Suppose we were to corner them and ask, "Did you foresee the present explosive growth of particle physics?", they would probably respond, "My God, no! We did not have the faintest idea that it would grow so big, more or less pushing all other branches of physics to the sidelines. In those days, atomic and nuclear physics were the frontier areas, and some of the things we were nibbling at were merely called high energy nuclear physics." Well, that's how it all began!

Let us take a closer look. A good starting point would be the famous Dirac equation which, as Dyson once declared, succeeded in "weaving together Special Relativity and Quantum Mechanics in a seamless fabric."

The Dirac equation was a great pointer. Firstly, it drew attention to the fact that serious physics, at least where particles are concerned, must start by acknowledging special relativity. Next, beginning with the positron, it gave strong hints about the existence of anti-particles. Thirdly, it paved the way for the spectacularly successful quantum electrodynamics (what I call QED-III in ***QR-II***) which subsequently acted as a model for how one might approach the interactions between particles and fields.

4.3 The gate opens for particles

The *positron* (e^+) was the first among the new crop of elementary particles to be discovered, and that was just the beginning. The next whose existence was postulated was the *neutrino* (ν). Around the same time came the neutron, whose discovery went successively through suspicion, search and finally detection. Soon after Rutherford discovered the atomic nucleus (in 1911), the question arose, "What is the nucleus made up of ?" In those days it was believed that a nucleus (A, Z) with mass A and charge Ze consisted of A protons (p) and $(A-Z)$ electrons (e^-).

$$\text{Nucleus } (A, Z) = A\text{p} + (A-Z)\text{e}^-.$$

This implies that electrons exist in the nucleus. However, as the years went by, there was increasing evidence that electrons *cannot* exist in the nucleus. How then does one explain the mass and the charge of the nucleus? Rutherford thought that hiding in the nucleus was a neutral entity made up of a proton and an electron, but much more tightly bound than a hydrogen atom. In due course, this was named the *neutron* (n) and Rutherford set his team to search for it. Years passed and finally in 1932 Chadwick tracked the neutron down. It then became clear that

$$\text{Nucleus } (A, Z) = Z\text{p} + (A-Z)\text{n}.$$

This is the proton–neutron model of the nucleus, which replaced the earlier proton–electron model. The proton–neutron model continues to hold sway. Of course, the neutron was not right away declared to be an elementary particle. People kept wondering, and four months after the discovery, Chadwick hesitantly remarked, "It is, of course, possible to suppose that the neutron may be an elementary particle." It was left to Iwanenko to firmly declare, "We do not consider the neutron to be built of an electron and a proton, but as an elementary particle."

Thus around 1934 the count stood at: electron, proton, positron, neutrino and the neutron; and besides these there was also the photon, if one wanted to include it in the club. At that time, all except the neutrino had been physically detected. We now pick up threads of the story on the next particle to be admitted to the group.

4.5 Nuclear forces — phenomenological beginnings

We saw that the nucleus is made up of protons and neutrons. But what holds them together? Protons repel each other due to electrostatic forces, and inside the nucleus which is only $\sim 10^{-13}$ cm, this repulsion should be very high. So how come the protons still stay together? And then, what about the neutron? Being electrically neutral, why does it not leak away from the nucleus? What keeps it inside?

In 1932, Heisenberg applied his mind to these questions and asked, "What is the interaction between a proton and another proton, between a proton and a neutron, and finally between a neutron and another neutron?" He then came to the conclusion that there must exist within the nucleus a new and special kind of force, strong enough to keep all the particles together and bind the nucleus into a composite. Heisenberg also conjectured that this new force must be of short range. He offered a model for this nuclear force, and soon Wigner, Majorana and Bartlett offered their own versions; all of these represent the very first attempts to come to grips with what we now call *strong interactions* (S.I.).

The next major advance came in 1934 from Hideki Yukawa of Japan. Yukawa started his professional career in 1932, the very year that Heisenberg did his work on nuclear forces. Yukawa felt that the strong force was a fundamental force of Nature, rather like the electromagnetic and gravitational forces. In that case, should there not be an associated field like the electromagnetic and the gravitational fields? Now the electromagnetic field is a quantum field, with the 'particle' photon associated with it . In the same way, shouldn't the nuclear field also be quantum in nature with an associated particle? And just as two electrons exert a force on each other by exchanging photons, shouldn't two nucleons similarly interact by exchanging quanta of this nuclear field? Using the Feynman diagrams, we can make this comparison easier as in Fig. 4.1, but way back in 1934, Yukawa did not have that advantage; even then, he was on the right track. In fact, Yukawa went even further, arguing thus: "True, one can visualize the nucleon–nucleon potential the same way one visualizes the Coulomb potential, i.e., in terms of the exchange of a mediator or messenger particle. But there is also an enormous difference between the two situations as regards the range. How does one deal with this?" The answer came immediately—insert a 'switch' in the $(1/r)$ potential that would convert it from a long-range potential to a short-range potential. Yukawa discovered that the switch should be of the form $\exp(-\kappa r)$, where κ is a suitable constant. Thus the potential postulated by Yukawa has the form $[\exp(-\kappa r)/r]$, and is known as the *Yukawa potential* (see Fig. 4.2).

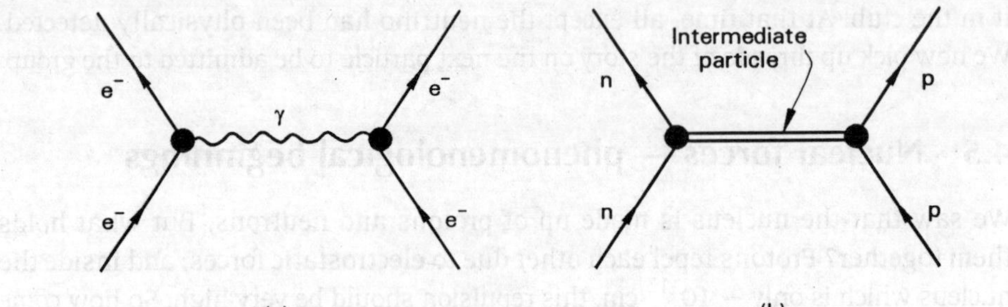

Fig. 4.1 Shown here are two Feynman diagrams, which illustrate how the strong interaction can be understood by analogy with the electromagnetic interaction. Just as the e.m. interaction is mediated by the photon [see (a)], Yukawa postulated that the strong interaction binding the nucleons within a nucleus was mediated by a new particle [see (b)] which, unlike the photon, had non-zero mass. Why non-zero mass, will soon become clear.

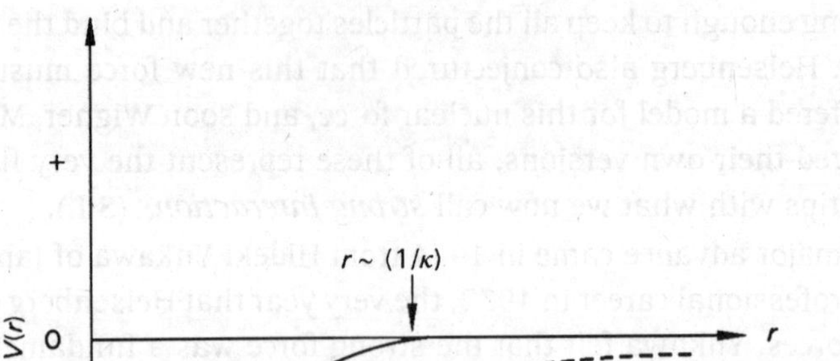

Fig. 4.2 This figure shows schematic plots of the $(1/r)$ attractive Coulomb potential and the attractive Yukawa potential which has the form $[\exp(-\kappa r)]/r$. This form is often called the screened Coulomb potential (because of the attractive nature of the two potentials, the value of the potential is negative). Notice that the Yukawa potential cuts off rapidly. This is due to the switch described in the text. It is this which makes the potential of short range, as it must be, if it is to describe nuclear forces.

"How does one fix κ?" This is no problem because $\kappa\sim$ (1/range). Thus, once one decides the range, the value of κ is automatically determined. Yukawa noted that dimensionally,

$$\kappa \sim (\text{mass} \times c)/\hbar \, . \qquad [4.1]$$

If the range is fixed as $\sim 10^{-13}$ cm, then the mass in Eqn. [4.1] works out to be several hundred electron masses. This mass is associated with the particle that acts as the mediator of the force, just as the photon acts as the mediator of the e.m. force. Notice that in the case of the e.m. interaction the range is infinite; therefore the mass of the photon emerges as zero and thanks to relativity, the photon must necessarily travel with the velocity c. In brief, Yukawa postulated that strong forces are mediated by the exchange of what he called 'heavy particles'. Later these were referred to as 'yukons', then mesotrons and finally mesons (for that story, see ***Bhabha***). In a nutshell:

Maxwell's theory: a) Coulomb potential
b) Infinite range
c) $(1/r)$ form
d) Mediator—photon, zero mass

Yukawa's theory: a) Yukawa potential
b) Short range
c) $\{\exp(-\kappa r)/r\}$ form
d) Mediator—meson, finite mass

Much later, Yukawa recalled:

> I decided to carry the theory (of Heisenberg) one step further .... Is this new nuclear force a primary one?... It seemed likely that it was a third force, unrelated to gravity and electromagnetism. If one visualizes the force field as a game of 'catch' between protons and neutrons (see Fig. 4.1), the crux of the problem was the nature of the 'ball' or the 'particle'.

4.5 The hunt for the meson

In those days, experiment usually preceded theory, a fact which pleased old Rutherford no end. But here at last was a case where theory had seized the lead and proclaimed to the experimenters, "Go and look for a particle with a mass around 300 electron masses."

How does one hunt for such a particle? Where does one look for it and how? Since high energy accelerators did not exist in those days, the best bet was to look for it in cosmic rays. As described in the companion volume ***Bhabha***, Yukawa himself suggested that the best way of looking for this particle was to study cosmic ray events.

Meanwhile, around 1934, Carl Anderson (the discoverer of the positron) and his graduate student Seth Neddermeyer had observed several cloud chamber tracks (for more about cloud chambers, see ***Relativity**** and ***Bhabha***) which Anderson strongly believed were due to 'particles of a new type' as he put it. These particles were, according to Anderson, heavier than an electron but lighter than a proton. Anderson knew nothing about Yukawa's prediction and he showed the pictures to Oppenheimer (later to become famous as the man behind the Atom Bomb Project; for more about him see ***QR-II***) seeking his opinion. Oppenheimer promptly dismissed these tracks as due to fast electrons. Thereupon Anderson gave up the idea of new particles but published his pictures nevertheless, with muted comments. The paper appeared in the *Physical Review* of August 1934.

Across the Pacific in Japan, Anderson's pictures produced great excitement, especially in Yukawa. Already the Japanese were struggling to detect the meson but here was an American who had actually seen the meson and yet missed it! How lucky for them!! However, the Japanese physicists faced a major handicap. They did not have a sufficiently powerful electromagnet to use along with their cloud chamber. The magnet was required to bend the particle, for that was the only way the mass of the particle could be established. The Japanese scientist Nishina (of Klein–Nishina formula fame) approached the Japanese navy and they gave the necessary help. Finally, Nishina obtained the required photograph giving the vital evidence, wrote the paper and mailed it to the *Physical Review* for publication.

Meanwhile, Anderson and Neddermeyer visited MIT where Street and Stevenson had also seen this particle and were about to publish their discovery. Anderson cursed Oppenheimer for misleading him, rushed back to his office and dashed off a paper to the *Physical Review*, this time firmly announcing the discovery. Just around this time, the Japanese paper was also received but unfortunately for them, the Editor decided that the paper was too long and returned it to the authors for shortening — valuable time was thus lost. And into this gap squeezed Street and Stevenson; eventually, the Japanese came a late third.

4.6 Was this the right fellow?

So the experimentalists had discovered a particle with a mass intermediate between that of the electron and the proton. But was this the particle that Yukawa had predicted? No one was sure. And then, Oppenheimer and Serber hesitantly said that maybe this was the Yukawa meson. When Yukawa heard of the discovery by Anderson, he also felt that it was the meson predicted by him. Ten years were to pass before it was finally established that the Yukawa meson was *not* the same as the one Anderson and Neddermeyer had seen. No wonder

* Throughout this book, the title ***At the Speed of Light*** has been referred to as ***Relativity***.

Oppenheimer later described this ten-year delay as a 'ten-year joke' played by Nature!

Meanwhile, experiments on cosmic-ray-produced mesons continued. Yukawa's meson was supposed to interact strongly with nuclear matter; it should therefore be easily absorbed and have a short lifetime. However, experiments seemed to suggest that the cosmic ray meson enjoyed a much longer lifetime.

The scene now shifts to Italy, and the year is 1943. Physics in Italy had bloomed during the thirties, due to the inspiring leadership of the incomparable Enrico Fermi. But then, like Germany, Italy also came under the fascist regime and its leader Mussolini passed, in 1938, laws that prohibited Jews from holding government jobs, university appointments, etc. Many eminent scientists, like Bruno Rossi for example, left Italy. Fermi was not a Jew but he knew that he was on the hit list as his wife was Jewish. He was therefore waiting for a chance to get away and it came in 1938 when he was awarded the Nobel Prize. Fermi went to Stockholm to receive the prize, and from there he fled to America.

Once Fermi departed, little was left of Italian physics barring a few brave souls like Amaldi, Conversi and Picconi. The latter two were young and so obsessed with physics that despite all the handicaps of war (including air raid attacks), they continued their experiments. And in June 1944, they discovered that the cosmic-ray-produced meson had a lifetime of around 2.2×10^{-6} sec. To us this may seem an incredibly short time, but by nuclear standards it is inordinately long. In fact, if you recall Fig. 2.A1.1, you will realise that the decay is through W.I. Anyway, getting back to the story, Conversi and Picconi now knew for sure that the Anderson meson was *not* the same as the meson predicted by Yukawa.

At this stage, these two enterprising Italians were joined by another young man, Ettore Pancini. Together they planned a new experiment. If the cosmic ray meson was of the Yukawa variety, then the positive and the negative species would behave differently. While the positive mesons would be repelled by the positive charge of the nucleus and therefore manage to live longer, the negative mesons should plummet into the nucleus and disappear quickly. Picconi and his collaborators found no such asymmetry — both the positive and negative species had the same long lifetime, too long for comfort. It was now quite clear that the cosmic ray meson was *not* the Yukawa meson but some other interloper. The year was 1946 and the war was just over.

Actually, even before the clandestine experiments of the Italians, the Japanese already suspected that there existed two types of mesons. Shoichi Sakata, a young colleague of Yukawa boldly made this proposal in June 1942 but with a war on, Sakata's proposal remained locked up in Japan.

In June 1947, a group of active young American physicists gathered at Shelter Island, New York for picking up the pieces and getting ready for serious post-war physics research. As described in *QR-II*, the main thrust of the Shelter Island conference was to understand Lamb's discovery which later spurred the development of modern QED. But some attention was also paid to other matters, including the discovery of Picconi and his co-workers. At this stage, Robert Marshak of Rochester suggested that perhaps there were two mesons, and argued as follows: one of the two mesons is produced by primary cosmic rays in the upper atmosphere during collisions with nuclei. This meson is the one predicted by Yukawa. This Yukawa meson then interacts strongly with nuclear matter (as Yukawa predicted) and is therefore quickly absorbed. Those that escape absorption decay quickly into another species which filters down to lower altitudes where they are easily observed (as did Anderson). This second species does not interact strongly with matter and is therefore able to enjoy a long lifetime.

Marshak was pleased with the reception his idea got at Shelter Island but when he went back to Rochester, he was in for a surprise. The May 24, 1947 issue of *Nature* carried a brief report by C.F. Powell (of Bristol) and his group, presenting evidence for the existence of two mesons. In October, Powell not only confirmed this evidence but also named the two species as π and μ. The names have stuck; only, today we refer to the two particles as pion and muon respectively (for more about them, see Table 4.1).

Table 4.1 Comparison of the pion and the muon.

Species	Symbol	Family	Charge	Mass (MeV)	Decay	Lifetime
Pion	π^0	Meson	0	135	2γ	$\sim 10^{-16}$ s
	π^+		+1	140	$\mu^+ + \nu$	2.6×10^{-8} s
	π^-		−1	140	$\mu^- + \bar{\nu}$	2.6×10^{-8} s
Muon	μ^+	Lepton	+1	106	$e^+ + \nu + \bar{\nu}$	2.2×10^{-6} s
	μ^-		−1	106	$e^- + \nu + \bar{\nu}$	2.2×10^{-6} s

Powell's discovery created a stir. One could understand the pion; it was forecast by Yukawa and had reason to exist. But what about the muon? Nobody knew why it was there and what purpose it served, and the great puzzlement of the period was echoed by Rabi who exclaimed, "Who ordered that?" Many years would pass before one got a better glimpse of the role ordained for the muon by Nature, in her Grand Design.

4.7 An aside and also an overview

So far I have been tracing specific events and the evolution of some of the specific ideas underlying these events. But it also helps to stand apart and take a global view of this crucial period because it was then that the style of physics itself changed.

Unquestionably, the single dominant event of this period was the Second World War. The war lasted close to six years in which time over twenty million people perished, not all of them at the battlefront. For the first time in human history, hundreds, if not thousands, of scientists played a key role in the war effort, the high point being the development of the atom bomb. A terrible weapon no doubt, but strangely, as Fermi once noted, the atom bomb was also good physics! What a galaxy of stars gathered together to create this most awesome weapon!! Many books have been written on the subject and I urge you to read at least one of them in order to get the flavour of the times.

A few years before the war, Hitler unleashed his gruesome witch-hunt against the Jews, driving out a large number of brilliant scientists of Jewish origin. While Germany (unknowingly) suffered the loss, America welcomed the refugees with open arms and thereby became immeasurably richer in scientific talent and capability. What a boon it was to receive Einstein, Wigner, Slizard, Fermi, Bethe, von Neumann and Weisskopf, to name just a few!

War always gives a boost to technology and this one yielded the radar (and with it, the whole gamut of microwave technology), the rocket, the jet engine, nuclear power and the digital computer. The American public realised and greatly appreciated the contribution that their scientists had made towards the victory of the Allies, and you have only to recall a story I have narrated in ***QR-II*** about the escort given to physicists as they were proceeding to the Shelter Island for that momentous conference. Public appreciation did not stop with mere thanks. On the contrary, the various agencies of the US Government vied with one another in pumping money into Science. Never was there such a windfall — one had merely to ask, and it was given. A glorious honeymoon which enjoyed a long innings (and ended only recently).

The Atom Bomb Project showed that a great thrust forward could be made by massive cooperative effort. This style percolated into experimental high energy physics, and physicists soon began to think of big machines and mega projects which involved millions of dollars in expenditure and hundreds of scientists working together as a team to execute the project. The era of big science had arrived and gone were the days of breakthroughs by lonely experimenters in Rome (Fermi) and Paris (Joliot Curie). To cap it all, old-world courtesies were replaced by brashness and aggressive competition — the rat race had begun and the slogan was: Publish or perish.

Big science revolved around high energy accelerators or machines as they were called, and they became possible as the result of war-time technology on the one hand, and generous funding on the other. No energy was high enough, and the labs raced against each other to build bigger, more powerful and better machines. When the war started, the most powerful accelerator boasted of an energy of about 200 MeV but by the mid-fifties (the war ended in 1945) the energy had already climbed to 6 GeV. The appetite for energy never abated. By the way, upto the sixties, most of the research was in America; thereafter, it spilled outside, especially to Western Europe and Russia.

It was really the big machines that completely changed the character of high energy physics. In 1955 when the Bevatron at Berkeley crossed 6 GeV, enough energy was available to create an anti-proton (remember, $E = mc^2$). Emilo Segre, earlier an associate of Fermi, exploited this, systematically went about hunting for the anti-proton, found it, and then won the Nobel Prize. Segre demonstrated that one could focus on what one wants to study, build the appropriate machine, back it up with all the necessary apparatus for detection, analysis, etc., and Presto! one would have arrived. By and large, this has been the style of high energy physics from then on.

4.8 Theorists also gear up

It was not merely the experimentalists who changed their style; the theorists too underwent a transformation — they became far more professional and businesslike.

It was the Dirac equation which ushered in a new era. Theorists realised that relativity must not be introduced as an afterthought but be built-in right from the beginning. This, as I have explained in *Bhabha*, is done quite systematically by demanding that the genetic equations satisfy what is called the Lorentz covariance. The role of symmetry was recognised, and it was also realised that one must go beyond space–time symmetries to accommodate internal symmetries (like isospin) revealed by experiments. And finally, the successful development of QED by Tomanaga, Schwinger and Feynman (see *QR-II*) strongly suggested that theories of matter (especially nuclear matter) are best formulated in the language of relativistic quantum field theory, notwithstanding possible hazards lurking in the form of infinities. With luck, these infinities could be tamed.

Relativistic quantum mechanics became an essential tool kit if one intended to do serious theoretical physics. And a new generation of physicists took it all in their stride without becoming dazed as their predecessors were, when first introduced to quantum mechanics.

4.9 Cosmic rays take a last bow

Since the thirties, the population of elementary particles has increased enormously, much of the increase occuring after the advent of the big machines. Few therefore realise that it was really cosmic rays which in those days provided the happy hunting ground. First there was the positron which was detected in cosmic ray studies; then came the pion and the muon, which between them played the 'ten-year joke'. But it did not stop with just these particles. Most importantly, cosmic rays also gave us our first introduction to what used to be called strange particles.

In their experiments, cosmic ray physicists basically resorted to two strategies. One was to employ the good old cloud chamber as Anderson and others did before the war. Only, the chambers kept growing bigger and better; also, all kinds of electronic controls were built in so that the chamber would respond only to a select category of events. Cloud chambers were convenient because they could be located in the lab, operated round-the-clock, and put on mountain tops if one wanted to see events at moderate altitudes.

The second approach was to catch the particles in thick and specially prepared photographic emulsions, wherein the particles left tell-tale tracks. The emulsion technique was developed after the war, by Powell of England (for which he was later rewarded with the Nobel Prize). Essentially, one had a stack of films and sent them up to very high altitudes using a balloon (see ***Bhabha***). However, the exposure would last only a few hours because after that the balloon would burst and the payload would come down. There was of course the risk of losing the payload if it fell into an inaccessible place as it sometimes did. (For example, when balloon flights were conducted in Chennai (Madras) around 1951, several of the film packages fell into the sea and were lost.) Even so, films played their own part in revealing the existence of many strange particles.

The story of these so-called strange particles really starts in 1947 when Rochester and Butler (of Manchester University) published two cloud chamber pictures of unusual events, one of which is sketched in Fig. 4.3. We see here a V-like fork appearing out of nowhere. Rochester and Butler surmised that the V-shaped event was caused by the decay of an unknown neutral particle of mass $\sim 1000m_e$. For a while this particle was called the V-particle. Being neutral, it did not leave a track but the two decay products manifested themselves via the prongs—obviously the decay products must have been charged particles.

For the next two years Rochester and Butler worked hard to obtain other examples of V-particles to confirm the two seen earlier, but with no luck. Later Rochester said, "Unfortunately, although the dear Lord was very kind to us in sending two of these, on the other hand He was extremely unkind in sending no more than two!" Around 1949 Rochester went to America where he spoke about the V-particle to Carl Anderson. The latter was extremely interested and straightaway diverted his best cloud chamber to the problem. Anderson's advantage was that his cloud chamber was on a mountain top (unlike that of Rochester). V-particles, if they existed, should be more evident at higher altitudes because cosmic rays are more plentiful there. And sure enough, in just a few months, Anderson picked up 34 events!

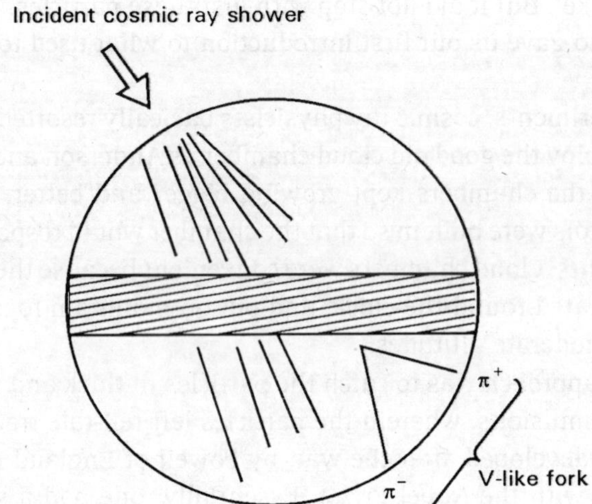

Fig. 4.3 This is a sketch of the famous cloud chamber photo of Rochester and Butler which revealed for the first time the existence of strange particles. On the basis of today's knowledge, the particle in this case can be identified as K^0 which, after a brief existence decayed into a pair of pions, i.e., π^+ and π^-. Notice that since K^0 is neutral, it does not leave any tracks. The V-like prong is produced by the decay products, i.e., the pions, because they are charged. On account of their characteristic V-like signature, strange particles were for a while known as V-particles.

All kinds of strange particles were then observed and as their tribe increased, there arose the problem of systematically classifying and naming them. Sometimes these particles were charged and sometimes they were not. A variety of symbols and alphabets were assigned to these particles, like: Λ (Lambda), Ξ (Cascade), K, τ, etc. One common but puzzling feature was that although these particles were born in strong interactions, they did not decay as rapidly as one would expect. At that time no one knew why, but later, remarkable physics came out of it all. One should therefore not forget the debt owed to cosmic ray studies which first revealed the existence of many exotic particles unknown till then.

In those days, cosmic rays was the exciting frontier where one could catch new particles and deal with unheard-of energies. Few people were interested in elementary particles. Periodic conferences were held to review progress and in 1950, the venue was Mumbai (Bombay). The last of these conferences was a memorable one held in 1953 at Bagneres in France, quite close to a mountain where there was a cosmic ray laboratory. The meeting was lively but everyone sensed that the days of cosmic rays were over. Leprice Ringuet explained why. He said, "We are being pursued — pursued by the machines." Delivering the conference summary, Powell observed similarly, "Gentlemen, we have been invaded . . . the accelerators are here."

Yes, the machines had arrived and the days of random fishing were over. Instead, one could now design the experiment to look precisely for what one wanted and perform the experiment under controlled conditions. The machines took over completely, and have been in command ever since. But those pioneering days are not without charm, even though one knew so little. As Abraham Pais recalled:

> It was a wonderful mess at that time. Wonderful! Just great! It was so confusing — physics at its best when everything is confused and you know something important lies around the corner.

In the next chapter, we shall start going round that corner.

4.10 Chapter summary

Since we have embarked on some kind of a random survey, it would perhaps be useful to take stock.

- This chapter deals with the period 1930 to about 1960, of which the period 1930–1935 (roughly) was something like a curtain raiser.
- The Dirac equation set the stage for anti-particles by heralding the positron. Further, it highlighted the importance of relativity.
- A beginning was made in understanding the strong force. Yukawa in particular, introduced the idea of mediation via a particle with a non-zero mass.
- Another important concept to be born in this period was isotopic spin.
- It was also in this period that the idea of a weak force was born (that story will be told in Chapter 5).
- Many new particles were either postulated and/or discovered: the neutrino, the positron, the pion and the muon, and finally the strange particles.
- On the social side, scientists got deeply involved in war work between 1940 and 1945, and made monumental contributions. A grateful America rewarded them in the post-war era by generously pumping in money on a scale undreamt of before.
- Big science burst forth on the scene and accelerators made a strong appearance.
- A great triumph belonging to this era is the QED of Feynman, Schwinger and Tomanaga.
- The electron and the electromagnetic interaction were now fully understood but where strange particles were concerned, the plot was beginning to thicken.

Now, on to some serious physics!

5

A Weak Story

There are four major milestones in the history of weak interactions. They are: (i) Fermi's postulation of the weak interaction and his model for it, (ii) the discovery of the failure of parity in weak interactions, (iii) the discovery of the Universal Fermi Interaction, and finally (iv) the unification of the weak and the electromagnetic forces. This chapter discusses the first three of these events.

5.1 Beta decay: the beginning

Among the three forces that dominate the sub-atomic world, the so-called weak force is insignificant where raw strength alone is concerned. However, it has not only produced its quota of surprises, but has, in addition, also given us a glimpse of the remarkable unity that underlies Nature.

The story begins in 1896 when Becquerel discovered radioactivity. It was noticed then that some atoms spontaneously emit electrons. Later it was realised that these electrons actually come from the nucleus. This produced a dilemma since, using quantum arguments, one can show that electrons cannot reside in the nucleus (this was one of the problems that the great Enrico Fermi used to give his students; can you figure out the argument?). The difficulty was later resolved by declaring that the electron is born at the instant it is ejected!

Three basic types of radioactivity have been identified; in all of these, the nucleus transforms from one species to another. Symbolically, the processes are:

a) $(A, Z) \longrightarrow (A, Z+1) + e^- + ...$ (electron emission),

b) $(A, Z) \longrightarrow (A, Z-1) + e^+ + ...$ (positron emission),

c) $(A, Z) + e^- \longrightarrow (A, Z-1) + ...$ (electron capture).

In all cases the mass of the nucleus remains more or less the same but the charge alters during the transformation. Are you wondering about the blank .... in each of the above equations? There lies a mystery about which you will soon hear.

5.2 The puzzling continuous β-spectrum

By the early part of this century, nuclear transformation was no surprise because most of the elements at the tail end of the periodic table were observed to undergo radioactive decay, usually by the emission of an alpha particle (helium nucleus). Invariably, the emitted α-particles had well-defined energies. Consider the reaction

$$X \longrightarrow Y + \alpha,$$

i.e., the nucleus X emits an alpha particle and transforms to the species Y. The energetics of the transformation are sketched in Fig. 5.1 and one would expect, based on the principle of conservation of energy, that the emitted α-particles would have the energies $(E_{X1} - E_Y)$, $(E_{X2} - E_Y)$, etc., depending upon from which level of X the decay occurred. In other words, in α decay the energy spectrum of the emitted particles is discrete.

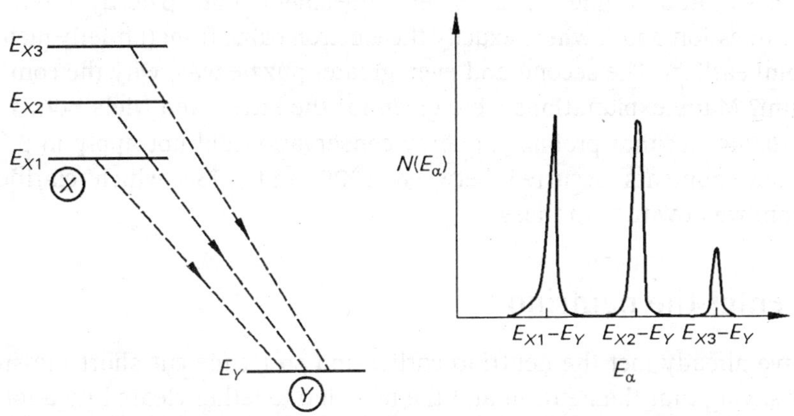

Fig. 5.1 Typical α decay scheme in which nucleus X decays to nucleus Y by emitting an alpha particle. Since the nucleus X can be in different states of excitation prior to the decay, the emitted α-particle can have different energies. On the right is shown the schematic of the α-spectrum. Observe its discrete nature.

The energy spectrum of the electrons emitted in β decay was found to be in marked contrast to that in α decay — instead of being discrete, it was a continuum as sketched in Fig. 5.2. This was a total surprise because it seemed to imply that energy conservation, a sacred principle of physics, had failed (can you explain why people thought so?) and so, at first, people suspected the experimental results. However, even after careful experiments the continuous spectrum refused to go away; it was a fact of life and had to be explained.

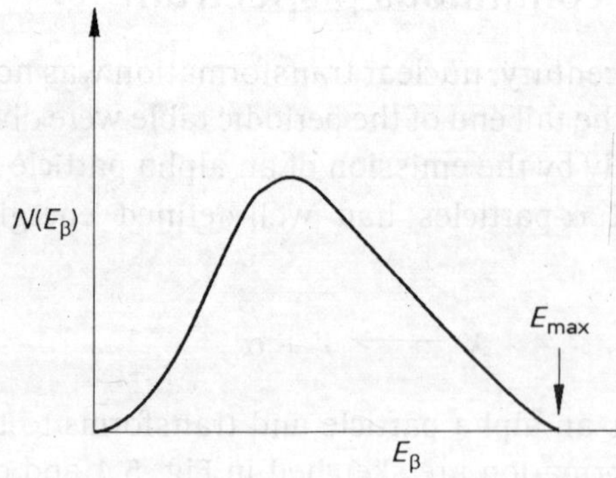

Fig. 5.2 Schematic diagram of the energy spectrum of the electrons emitted in β decay. Observe that in contrast to the discrete spectrum that one has in the case of α decay, the spectrum in this case is continuous. Also notice the existence of the energy cutoff.

There was, at that time, a two-fold embarassment about β decay. Firstly, there was the question about where exactly the electron came from (I briefly mentioned this point earlier). The second and even greater puzzle was, why the continuous spectrum? Many explanations were given for the latter, and Niels Bohr went so far as to suggest that probably, energy conservation did not apply in β decay. Three developments occurred between 1929 and 1934 which clarified the situation; we now turn to these.

5.3 Enter the neutrino

You have already met the neutrino earlier and so let me cut short the story by merely saying that the neutron and the neutrino together cleared up a lot of the prevailing mystery.

The neutrino story is quite interesting. It all began with the continuous spectrum. As I just mentioned, Bohr tried to explain it away by suspending the sacred principle of conservation of energy, but Pauli would have none of that. Instead he invoked a new particle that rescued the energy conservation principle and also neatly explained the mystery. This new particle proposed by Pauli had zero electric charge, nearly zero mass and spin 1/2; Pauli called it 'neutron'. Later, it was renamed by Fermi as neutrino. For the moment, we shall continue to call it the neutron, adding (P) to denote that we are referring to Pauli's hypothetical rescue particle.

Pauli did not publish his idea in a formal paper. Instead, he casually mentioned it in a letter. In December 1930 there was a meeting in Tubingen, Germany, to discuss radioactivity, and Pauli was one of the invitees. Unable to attend he wrote a letter of regret in which he said:

> I have come upon a desperate way out (note the words!) regarding the .... continuous β–spectrum ...... There could exist in the nucleus elementary neutral particles, which I shall call neutrons ....
>
> I admit that my way out may not seem very probable *a priori* since one would probably have seen the neutrons a long time ago if they exist. But only one who dares wins ... One must therefore discuss seriously every road to salvation. Thus, dear radioactive ones, examine and judge ...

Pauli did not offer a theory but merely tossed an idea that was very daring and one which later provided the solution. Pauli did not stop with merely writing a letter. He kept on asking why, if the nucleus contained neutrons (P), they had not been spotted so far. He then made an important suggestion, the principle of which is illustrated in Fig. 5.3. The essence is the following: Why not use momentum conservation to check out if there is an elusive third particle? But the proposed experiment seemed very difficult, and it was back to square one.

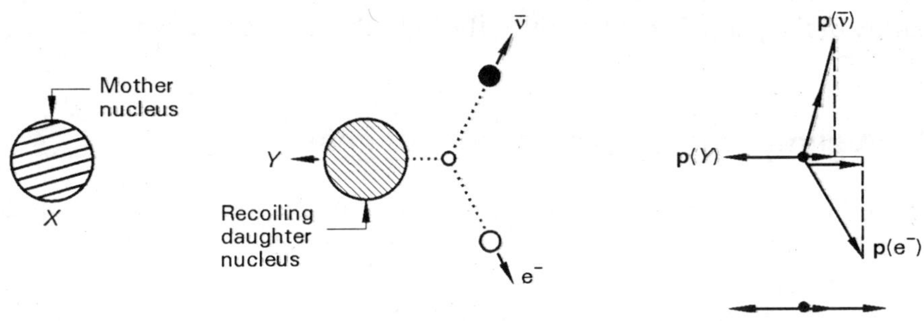

Fig. 5.3 This figure discusses momentum conservation as applied to β decay, and how one may use this principle to obtain indirect evidence for the existence of the neutrino. Consider the decay $X \rightarrow Y + e^- + \bar{\nu}$. X disappears but three other particles appear; of these, the anti-neutrino escapes detection. Since X is stationary, the intial momentum is zero; therefore, the final momentum also must be zero, i.e., $\mathbf{p}(\bar{\nu}) + \mathbf{p}(e^-) + \mathbf{p}(Y) = 0$; or, $\mathbf{p}(e^-) + \mathbf{p}(Y) \neq 0$. Pauli argued that if one measured $\mathbf{p}(e^-) + \mathbf{p}(Y)$ and showed that this sum is not equal to zero, then it implied the existence of the third particle, i.e., the anti-neutrino. Electron momentum is easy to measure but not the recoil of Y, i.e., $\mathbf{p}(Y)$.

5.4 Fermi baptizes the neutrino

Pauli came up with his unusual idea in 1930. Two years later, Chadwick discovered a particle which he named the neutron. Clearly, Chadwick's neutron was not the same as the neutron Pauli was talking about. Like Chadwick's neutron,

Pauli's particle also was electrically neutral and had a spin of 1/2. But while Chadwick's neutron had a mass almost equal to that of the proton, Pauli's particle had a mass almost equal to zero; so it was really a different particle. You may say that Chadwick hijacked the name 'neutron', thereby rendering Pauli's particle nameless! However, that situation did not last long.

In 1933, Fermi put all the pieces together and declared that radioactive decay was caused by a new force, viz., the weak force. Thus it was Fermi who formally recognised the existence of the weak force. He then showed how this weak force caused neutrons to emit electrons and positrons, leading to radioactivity. Energy and momentum conservation were ensured by including Pauli's now nameless particle, which Fermi renamed as *neutrino* (in Italian, the suffix *no* means 'small one'; thus neutrino is the small neutral particle).

Fermi wrote a paper describing his theory; it was brief, clear and matter of fact, as always. Fermi sent his paper to the prestigious journal *Nature* in England as that was where important papers were published. However, making one of its greatest blunders, *Nature* returned the paper to the author rejecting it! Fermi then submitted the paper to an Italian journal where it should have got lost for eternity. But it did not, because Fermi's theory was really a breakthrough and even the obscurity of the journal could not hide the light that shone through.

5.5 A glimpse into Fermi's work of genius

How did Fermi explain radioactivity in the light of the neutron as well as the neutrino? Simple. He said that a nucleus emits an electron when one of its neutrons transforms according to the reaction

$$n \longrightarrow p + e^{-} + \nu.$$

The reaction, said Fermi, is driven by the weak force. Similarly, positron emission arises when a proton in the nucleus transforms according to

$$p \longrightarrow n + e^{+} + \nu.$$

According to Fermi, the electron or the positron was expelled the moment it was created and therefore there was no need to imagine their pre-existence inside the nucleus. The neutrino provides both energy and momentum conservation. In the light of the above picture, the radioactive decay of an actual nucleus, e.g., tritium (^{3}H, isotope of hydrogen) can be visualised as

$$^{3}\text{H} \longrightarrow {}^{3}\text{He} + e^{-} + \nu.$$

For the moment, I am avoiding the neutrino–anti-neutrino complication; that will be taken up later.

The basic idea behind Fermi's theory is simple. The essential facts are:

(a) The nucleon is transformed, accompanied by the emission of light particles — what we would call leptons.

(b) Both energy and momentum are conserved.

(c) There is participation by the neutrino.

What is the force that drives the reaction? This is where Fermi's genius is evident. He simply said, "Why not model it after the e.m. interaction?" Today this might seem natural and obvious, but sixty years ago when one did not even know for certain that a weak force existed, such a guess was nothing short of pure genius. It was a daring and imaginative move.

Consider a charge distribution ρ and an electrostatic (i.e., scalar) potential φ (such as is produced by static charges). The energy of interaction is $\sim\rho\varphi$. In the rather more general case where one has a current **j** and a vector potential **A**, the interaction is described by (**j.A**) — this is the famous "jay dot **A**" form of the interaction. Now the vector potential can be expressed in terms of another current, so that (**j.A**) ends up as (**j.j**′), which is referred to as the current–current interaction. More explicitly, if one has two currents $\mathbf{j}(\mathbf{r})$ and $\mathbf{j}'(\mathbf{r}')$, the interaction energy is given by:

$$H_{\text{int}} \sim \int d\mathbf{r}\, d\mathbf{r}' \mathbf{j}(\mathbf{r}) \cdot \mathbf{j}'(\mathbf{r}') \frac{1}{|\mathbf{r}-\mathbf{r}'|}\,.$$

Fermi basically used this idea. Consider for example, the decay of the neutron (recall the decay equation given earlier). In this decay, one has the nucleon on the one hand, and the light particles on the other. Thus, said Fermi, one can think of two currents, one associated with the nucleon and the other associated with the light particles; β decay can thus be considered in terms of an interaction Hamiltonian of the form of 'nucleon current dot light particle current' — that is the general idea.

There is an elegant way of representing Fermi's ideas using Feynman diagrams which is done in Fig. 5.4. Now you might wonder, "In the diagrams one draws in the case of QED, one has the photon. Is there not an analogous particle in the case of weak interactions? Why is it not appearing in Fermi's theory?" Good question. Fermi perhaps realised this himself, but did not care to speculate along those lines. In the 'j dot A' language, what Fermi did was to assume that:

$$H_{\text{int}} \sim \int d\mathbf{r}\, \mathbf{j}(\mathbf{r}) \cdot \mathbf{j}'(\mathbf{r})\,.$$

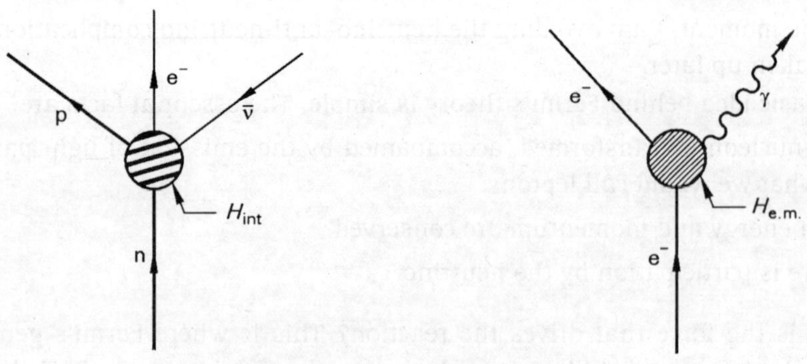

Fig. 5.4 Feynman diagram to illustrate the basic process considered by Fermi. Also shown for comparison is the simplest module of the electromagnetic interaction.

This is called *contact interaction*, since the current **j** interacts with the current **j**′ at the same point **r**. In the Feynman language, this implies that there is no intermediate particle. It is also like using electrostatics instead of the full electromagnetic equations.

Even though to us today Fermi's theory might seem naive, people got a lot of mileage out of it for nearly thirty years. I should mention here that in his theory, Fermi explicitly assumed that parity was conserved in weak interactions — at that time, there was no reason to believe otherwise.

5.6 The anti-particle of the neutrino

I had mentioned earlier that for each particle there is an anti-particle but so far in this chapter, I have avoided referring to the anti-particle of the neutrino. Such an anti-particle does exist, and it is the anti-neutrino. Neutrino is the particle that is emitted when a proton turns into a neutron, i.e.,

$$p \longrightarrow n + e^{+} + \nu.$$

But when a neutron turns into a proton, it is the anti-neutrino that is emitted along with the electron, i.e.,

$$n \longrightarrow p + e^{-} + \bar{\nu}.$$

How do we remember what particle is emitted when? This is where the lepton number is useful. I shall illustrate this with an example. Recall the following:

(a) The lepton number for the proton and the neutron is zero.

(b) The lepton number for the electron is +1 and for the positron is –1.

(c) The lepton number for the neutrino is +1 and the anti-neutrino is –1 (the anti-particle has –1).

(d) Lepton number is conserved.

Now consider neutron decay. Write $n \rightarrow p + e^- + \nu$. Apply lepton number conservation. We have $0 = 0 + 1 + 1$, which is clearly wrong. So, instead of the neutrino on the right-hand side, we must have the anti-neutrino there.

5.7 A puzzle and a suspicion

The Fermi theory was bread and butter to experimentalists who kept themselves busy by collecting a lot of data and checking out various aspects of the theory. What caused the next major flutter and advance in the theory of weak interactions was not this routine activity but a puzzle from a different quarter.

I had earlier mentioned the so-called strange particles and how they were first spotted in cosmic rays. Two strange particles that attracted special attention were the τ and the θ. Although initially they were thought to be different from each other, many facts strongly suggested that they were actually one and the same particle (now called the K^+). For example, both particles had the same mass (~ 966 times the electron mass); their lifetimes were also the same (~ 1×10^{-8} sec). But there was one fly in the ointment, and that showed up in the decay modes. Specifically, the θ^+ and the τ^+ decay according to:

$$\theta^+ \longrightarrow \pi^+ + \pi^0,$$

$$\tau^+ \longrightarrow \pi^+ + \pi^+ + \pi^-.$$

So what's the problem here? The problem shows up if one considers the parity on both sides in the two decays.

First, let me remind you that every particle has assigned to it an intrinsic parity which is either + or – . What does that mean? Simply this: Let ψ be the wave function of the particle. On the application of the parity transformation P, let $\psi \rightarrow \psi'$. Now either $\psi' = +\psi$ or $-\psi$. In the former case the parity is said to be + and in the latter case it is said to be –.

Now let us consider $\theta^+ \rightarrow \pi^+ + \pi^0$. The intrinsic parity of the pion is –. Ignoring complications due to relative motions, we can write

$$\text{Parity of } \theta^+ = (-) \cdot (-) = +.$$

Similarly,

$$\text{Parity of } \tau^+ = (-) \cdot (-) \cdot (-) = -.$$

And this precisely is where the problem is. If the two particles were one and the same, how come it has positive parity in one case and negative parity in the other? Clearly, something was wrong somewhere. What, and where?

The parity problem first surfaced in 1956 during a conference at Rochester organised by Marshak. In the course of the discussions, Feynman appears to have remarked, "Could it be that the τ and the θ are different parity states of the same particle which has no definite parity, i.e., is it that parity is not conserved?" No one knew.

One of the participants at the Rochester conference was C.N.Yang, then working at the Brookhaven National Laboratory (near New York). He travelled back to New York by train along with Wheeler, and during the ride Yang bet Wheeler one dollar that the τ and θ are really different particles. What happened to that bet you will hear about shortly.

Back in New York, Yang called on his friend and classmate T.D. Lee (then at the Columbia University) and together they began to think about the τ–θ puzzle. If the two particles were one and the same — and it seemed they were — the only resolution to the parity clash was to say that parity is not conserved in weak interactions which was the process driving the decay. How does one know it is the weak interaction that is responsible for the decay? From the lifetime. The lifetime of 10^{-8} sec is too long for the decay to be by any process other than weak interactions. So the question was, "Is parity violated in W.I.?" How could that be? Parity is never violated. Who says so? Fermi of course. But is there convincing experimental proof? Yang and Lee did not know and so they went to Madame Wu of Columbia, an expert in β decay and asked her point blank whether she was aware of any *firm* experimental evidence in favour of parity conservation in W.I. Wu said that everyone *believed* in parity conservation but there was *no* explicit experimental proof.

Yang and Lee now decided to throw their hats into the ring. They wrote a paper entitled "Is Parity Conserved in Weak Interactions?" and submitted it to the *Physical Review*. The Editor did not like the title to be in the form of a question and changed it to: "Question of Parity in Weak Interactions". Yang and Lee said:

> One way out of the difficulty (regarding τ and θ) is to assume that parity is not strictly conserved so that τ and θ are two different decay modes of the same particle, which necessarily has a single mass value and a single lifetime. We wish to analyse this possibility in this paper against the background of existing experimental evidence for parity conservation.

Yang and Lee did not state that parity was violated. They simply pointed out that there was no experimental evidence available strongly in favour of parity conservation, and added:

> To decide unequivocally whether parity is conserved in weak interactions, one must perform an experiment to determine whether weak interactions differentiate the right from the left.

They naturally made several suggestions. The experiments were duly performed and soon it was realised that parity was violated in W.I., as Yang and Lee originally suspected. Quickly came the reward — the Nobel Prize. But Yang lost his bet with Wheeler!

5.8 The fall of parity

While they were thinking about the parity problem, Yang and Lee sounded out several experimenters about doing an experiment to verify whether parity was actually conserved or not. At that time, these two were not famous and therefore few people took them seriously. The experimenters merely raised their eyebrows, as if to say, "Don't you think I have better things to do?" But one person did take Yang and Lee seriously, and that was Madame Wu (who was already being consulted regularly). She instinctively knew that Yang and Lee were onto something hot, and decided to pitch in. Presently I shall describe the famous Wu experiment but here let me mention that Wu's plan called for very low temperatures, ~ 0.01 K (see ***A Hot Story*** for more about low temperatures). Such facilities were not available at Columbia University (where both Wu and Lee worked) but they were at the National Bureau of Standards (NBS) at Washington. So Wu worked out a collaboration — the deal was that the NBS would provide the low temperature while Wu would organise the parity checkout. Just around that time, Wu was in fact all set to visit China, her homeland, but she cancelled the trip at the last minute in view of the urgency of the experiment.

A few preliminaries, before we consider the principle of the Wu experiment. Recall that the parity transformation P changes vector $\mathbf{r}$ to $-\mathbf{r}$. One can ask what P does to other physical quantities like the electric charge, for example. Charge, of course, is not affected by P but many other physical quantities are, especially those based on $\mathbf{r}$ in some way. So in general we could ask:

$$\mathbf{r} \xrightarrow{P} -\mathbf{r},$$

$$X \xrightarrow{P} ?$$

i.e., what happens to a physical quantity X under the P operation? Suppose X is a scalar, i.e., a mere number; then obviously it will not change. But, if X is a vector like $\mathbf{r}$ it will change sign. These are easy to understand. Next we consider composite quantities like $(\mathbf{X}\cdot\mathbf{Y})$, $(\mathbf{X} \times \mathbf{Y})$, etc., where $\mathbf{X}$ and $\mathbf{Y}$ are vectors. Here we have to be careful. Orbital angular momentum $\mathbf{L}$ provides a good example. As you know, $\mathbf{L} = \mathbf{r} \times \mathbf{p}$, where $\mathbf{p}$ is the momentum. Now,

$$\mathbf{r} \xrightarrow{P} -\mathbf{r},$$

$$\mathbf{p} \xrightarrow{P} -\mathbf{p},$$

but, $\mathbf{L} \xrightarrow{P} \mathbf{L}.$

So, even though **L** is a vector, under reflection it does *not* change sign. It is therefore called an *axial vector*. Suppose we consider $(\mathbf{L}\cdot\mathbf{p})$, which is the scalar product of an ordinary vector (**p**) and an axial vector (**L**). We then have

$$\mathbf{r} \xrightarrow{P} -\mathbf{r}, \quad \text{and} \quad \mathbf{L} \xrightarrow{P} \mathbf{L};$$

$$\text{So,} \quad (\mathbf{L}\cdot\mathbf{p}) \xrightarrow{P} -(\mathbf{L}\cdot\mathbf{p}).$$

Contrary to the usual case, the scalar product here changes sign. Quantities like $(\mathbf{L}\cdot\mathbf{p})$ are therefore called *pseudoscalars*. A word of caution here. Popular books often discuss the parity business by drawing a mirror, showing reflections in the mirror, etc. In all this one must be very careful because the mirror that is referred to in the P operation is not the looking glass or the shaving mirror which merely reflects in the optical sense. The 'mirror' that P is supposed to create is, as Wigner emphasises, a physicist's mirror. Unlike the shaving mirror, the physicist's mirror deals properly with pseudoscalars and axial vectors. By the way, one does not actually have a physicist's mirror — all that one does in practice is to create a situation that looks as if a reflection has been produced in a physicist's mirror; always remember this.

After all this preamble and cautionary remarks, we are now ready to consider the Wu experiment. Wu decided to use ^{60}Co (radioactive cobalt, often used in cancer therapy) as the test vehicle. ^{60}Co decays according to

$$^{60}\text{Co} \longrightarrow {}^{60}\text{Ni} + e^- + \bar{\nu}.$$

^{60}Co has spin (**J**) equal to 1, while ^{60}Ni has spin zero (see Fig. 5.5). **J** is like **L** and therefore, $(\mathbf{J}\cdot\mathbf{p})$ is a pseudoscalar. Wu now argued as follows. Let the spin **J** of the cobalt nucleus point upwards (see Fig. 5.6(a)). Let the momentum of the emitted electron be **p**. We denote this situation symbolically as $(\mathbf{J}\uparrow, \mathbf{p}\uparrow)$. Let the counting rate in this case be I_1. Suppose the parity operator is applied to this state. We then get the state $(\mathbf{J}\uparrow, \mathbf{p}\downarrow)$, which is no different from $(\mathbf{J}\downarrow, \mathbf{p}\uparrow)$, since a simple rotation transforms the latter into the former (Fig. 5.6 (b)). Thus, $(\mathbf{J}\downarrow, \mathbf{p}\uparrow)$ corresponds to the state obtained from the previous one $(\mathbf{J}\uparrow, \mathbf{p}\uparrow)$ by P operation. If I_2 denotes the counting rate in the second case $(\mathbf{J}\downarrow, \mathbf{p}\uparrow)$, then clearly I_1 and I_2 would represent decay rates in the real and the mirror worlds. If $I_1 = I_2$, then parity is conserved; if $I_1 \neq I_2$, then parity is not conserved. This is the essence of the Wu argument. So it all boils down to the following steps:

(a) Measure I_1 corresponding to $(\mathbf{J}\uparrow, \mathbf{p}\uparrow)$.

(b) Measure I_2 corresponding to $(\mathbf{J}\downarrow, \mathbf{p}\uparrow)$.

(c) Compare I_1 and I_2.

(d) If $I_1 = I_2$, then parity is conserved; otherwise, parity is not conserved.

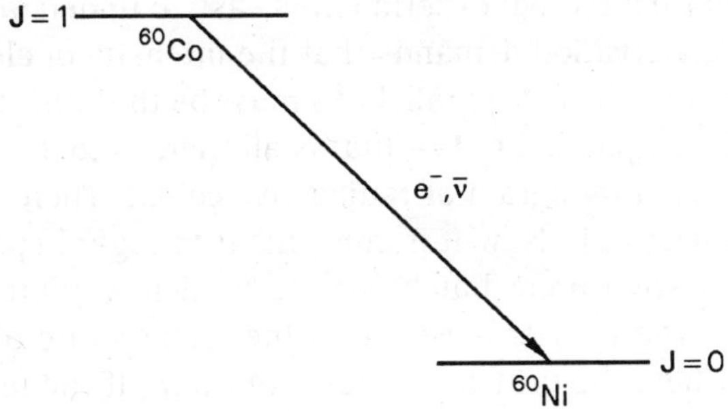

Fig. 5.5 Decay scheme for ^{60}Co, used by Wu in her famous parity experiment. Note that $J \neq 0$ for ^{60}Co. This is a must for the experiment.

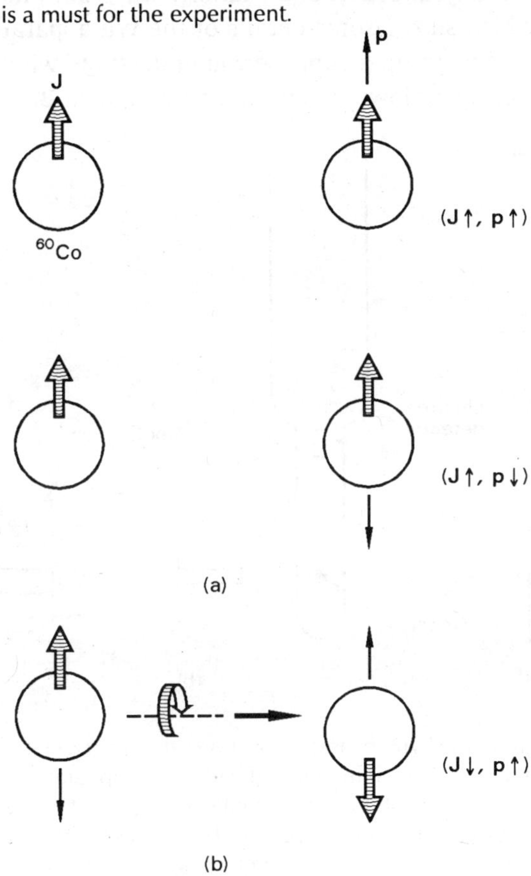

Fig. 5.6 This figure illustrates the logic behind the Wu experiment. If parity is conserved, then electrons must be emitted at the same rate parallel to **J** as anti-parallel to **J**, i.e., the counting rate I_1 for (**J**↑, **p**↑) must equal the counting rate I_2 for (**J**↑, **p**↓). But, as (b) shows, (**J**↑, **p**↓) is the same as (**J**↓, **p**↑), being related by a simple rotation. Depending upon the convenience, one can use either the scheme (**J**↑, **p**↓) or (**J**↓, **p**↑) for determining I_2. Wu compared counts for (**J**↑, **p**↑) with those for (**J**↓, **p**↑).

Once you grasp this, the Wu experiment is easy to understand. To repeat, Wu said that parity conservation demands that the intensity of electrons emitted in the radioactive decay of cobalt parallel to **J** must be the same as the intensity of electrons emitted anti-parallel to **J** — that is all there is to it.

Suppose you take a milligram of radioactive cobalt. There would be trillions and trillions of nuclei in it. Now it is convenient to regard spin as an arrow, in which case the question immediately arises, "Which way are these trillions of arrows pointing?" The answer depends on the temperature of the sample, and that precisely is where the 0.01 K business comes in. If the temperature is very low then the spins become disciplined and could, with the help of a small magnet, be made to point in any direction one wants. With this in mind, you may now consult Fig. 5.7 which shows some details of the Wu apparatus. In this set up, the direction of **p** is always up (I suppose you understand why), whereas **J** can be made to point either up or down, via the controlling field **B**.

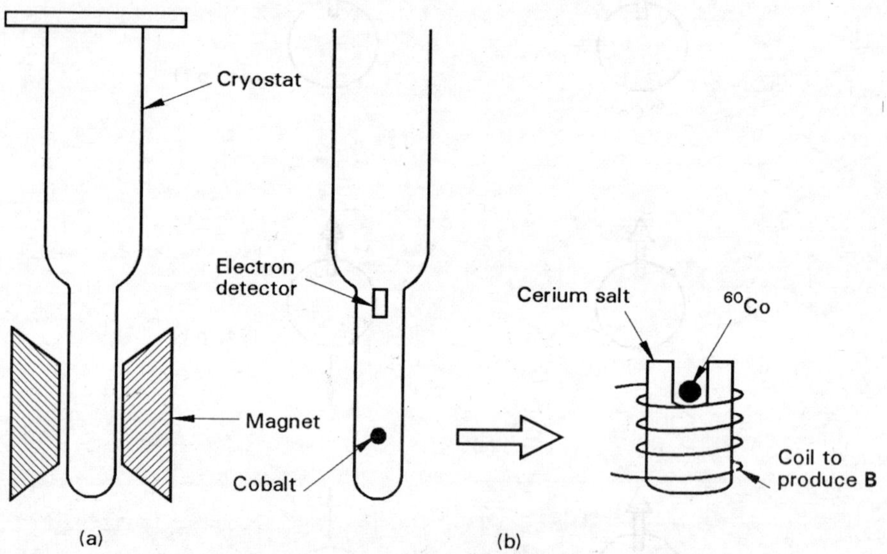

Fig. 5.7 Some details of the Wu apparatus. (a) shows the big magnet used for the adiabatic demagnetisation and the cryostat. Some more details of the apparatus are available in ***A Hot Story***. (b) shows the nuclear part of the experimental arrangement. Observe that the electron detector is above the radioactive source — see (b). Thus only electrons going up are detected, which means that **p** is always up in the experiment. However, **J** can be made to be up or down by switching the field **B** produced by the coil.

What Wu and her collaborators did was to cool the cobalt sample to the lowest temperature they could attain — they used the technique of adiabatic demagnetisation which is described in *A Hot Story* — and then 'let go', i.e., allowed the sample to slowly warm up. And as it did so, they measured I_1 and I_2 every few seconds by switching the field **B**. The results they obtained are shown in Fig. 5.8. As you can see, at start, I_1 was lower than I_2, the message being that more electrons were emitted anti-parallel to **J** than parallel to **J**; but with the passage of time, the two counting rates approached each other. This, of course, is expected, since with the passage of time the cobalt slowly warms up. Once that happens, nuclear discipline is lost, i.e., the spins point randomly; under such circumstances, the subtle difference in count rates produced by parity effects gets washed out. The message of the Wu experiment was loud and clear: Parity is *not* conserved in β decay.

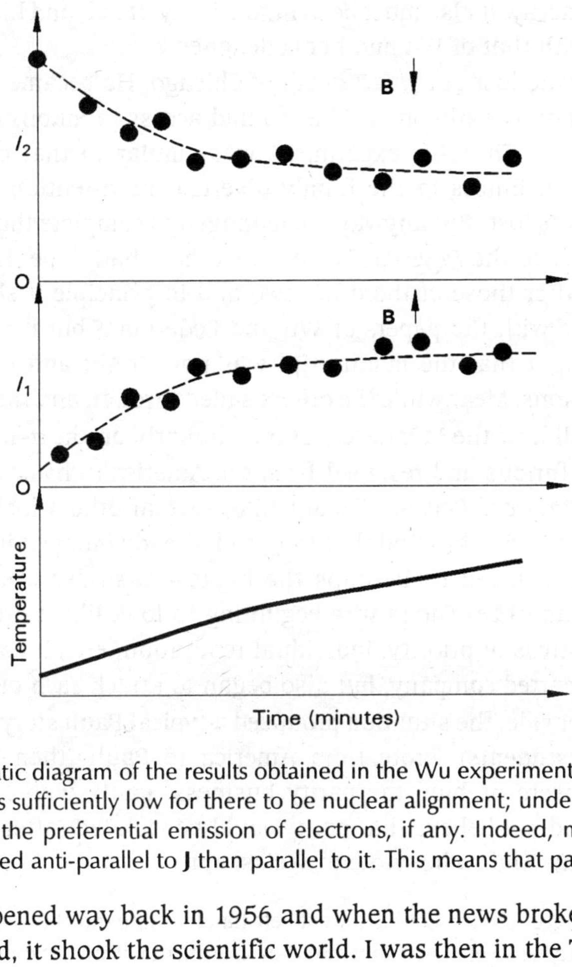

Fig. 5.8 Schematic diagram of the results obtained in the Wu experiment. In the beginning, the temperature is sufficiently low for there to be nuclear alignment; under these conditions, one can observe the preferential emission of electrons, if any. Indeed, more electrons are found to be emitted anti-parallel to **J** than parallel to it. This means that parity is violated.

All this happened way back in 1956 and when the news broke out that parity is not conserved, it shook the scientific world. I was then in the Tata Institute of Fundamental Research, Mumbai, and I still recall how nuclear physicists went about excitedly explaining the tau–theta puzzle and fall of parity to anyone who cared to listen. Today, as described in ***A Hot Story***, the Wu apparatus is routinely used as a low-temperature thermometer.

Wu was not the only one in the great parity race; Lederman of Columbia was another. Much earlier, when Lee had approached him, Lederman was not very keen on the parity business but later, when he heard that things were happening in Washington, he felt that he too must plunge into the race. In Columbia they had an accelerator which produced muons, and Lederman decided to examine parity conservation in muon decay:

$$\mu^- \longrightarrow e^- + \bar{\nu}.$$

Lederman's point was simple. The muon decayed by weak interactions. If parity is violated in β decay, it also must be in muon decay. It did, and his paper appeared back-to-back with that of Wu and her colleagues.

One person who lost out was Telegdi of Chicago. He became interested in the parity business quite early on and he too had access to muons (produced in the Chicago machine). Thus his experiment was similar to that of Lederman but unfortunately, an illness in the family diverted his attention for a while and valuable time was lost. But anyway, he managed to complete the experiment and rushed his paper to the *Physical Review*, as others had done. His paper arrived just two days after those of the other two and in principle it should have been published along with the papers of Wu and Lederman, but it was not, because the Editor thought that the manuscript was not alright and returned it to the author for revisions. Meanwhile the others sailed through and the excitement was soon over (recall how the Japanese lost out similarly on the π–μ difference).

Telegdi was furious and resigned from the American Physical Society (which publishes the *Physical Review*). Meanwhile, several others looked up their old data and were chagrined to find that they had already bumped into parity failure but had not recognised it! Perhaps the biggest casualty was the friendship between Yang and Lee. Theirs was beginning to look like a dream partnership but under the stress of priority, individual reputation, etc., it collapsed. Lee and Yang not only parted company, but also began to attack each other.

On the lighter side, the situation produced a typical Pauli story. Weisskopf (one of Pauli's old students) wrote from America to Pauli (then in Europe) that experimenters were probing the parity business. Pauli, who always held very strong views, did not believe that parity could be violated (after all, he belonged to the old guard). And so he wrote to Weisskopf:

> I do *not* believe that the Lord is a weak left-hander (note the play on words), and I am ready to bet a very large sum that the experiment will give symmetric results.

Ten days later he wrote to Weisskopf again but this time in a different tone:

> Now, after the first shock is over, I begin to collect myself again (as one says in Munich). Yes it was very dramatic. On Monday 21st, at 8.15 P.M. I was to

> give a lecture on 'Past and Present History' of the neutrino. At 5 P.M. the mail brought me three experimental papers ... Now where shall I start? It is good that I did not make a bet. It would have resulted in a very heavy loss of money (which I cannot afford); I did make a fool of myself, however (which I think I can afford to do) ..... incidentally, only in letters or orally and not in anything that was printed. But others now have the right to laugh at me. What shocks me is not the fact that "God is just left-handed" but the fact that in spite of this He exhibits Himself as left/right symmetric when He expresses Himself strongly (again note the play on words) .... How can the strength of an interaction produce or create symmetry groups, invariances or conservation laws? ... Many questions, no answers!

And finally, when Madame Wu stood up to speak at the 1957 International Conference on High Energy Physics, she began with the words:

> I am here on the strength of weak interactions.

A clever play on words!

5.9 The master criminal is caught

It is time now to return to the neutrino which Prof. Fraunfelder has described as a master criminal — first it escaped suspicion, and then detection for a long time. Pauli only gave a vague hint of its existence and it was really Fermi who gave the neutrino a firm place in β decay. Still, seeing is believing and many were hesitant about accepting the neutrino into the family of elementary particles. In 1938, Bethe and Peierls calculated that low energy neutrinos would traverse billions of kilometres of water without interacting with a single atom! If it was so elusive, then how does one catch it?

Meanwhile, there were others who said that elusive or not the neutrino and anti-neutrinos must be taken into account. There is one interesting point about the neutrino, to which we call attention in Fig. 5.9. We have here the positive and the negative energy states for the electron. You will recall from ***QR-II*** that the negative energy states relate to the anti-electron or the positron, and that there is an energy gap of $2m_e c^2$ which separates the positive and the negative energy states. Since the neutrino mass is zero (so it is believed), the energy gap is zero.

Getting back to the elusive 'master criminal', one person intent on capturing it was Frederick Reines of Los Alamos. That is the place where the first atom bomb was built and after the war they continued the bomb business to produce 'new, improved' bombs (like in the case of toothpaste!). Frequently, these new-model bombs were tested, and this gave Reines an idea, "Why not explode a bomb and catch a few of the trillions of anti-neutrinos let loose?" Reines discussed this idea with his friend Clyde Cowan and together they approached Fermi for his blessings. Fermi wryly declared that one couldn't set off an atom bomb everytime one wanted to do a neutrino experiment. Disappointed but not discouraged, Reines and Cowan thought again and said, "Why not design an experiment to catch the anti-neutrinos pouring out of a nuclear reactor?"

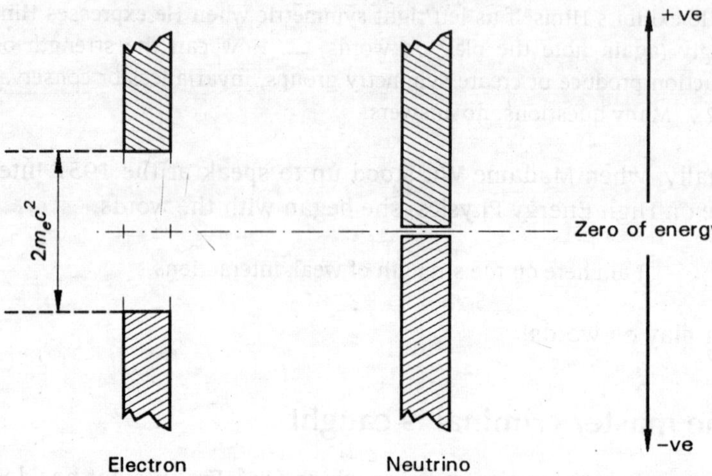

Fig. 5.9 Schematic representation of the positive and the negative states of the electron and the neutrino. In the latter case, the energy gap is zero because the neutrino is massless.

That is how they came up with the scheme illustrated in Fig. 5.10. Their basic idea was to exploit the reaction

$$\bar{\nu} + p \longrightarrow n + e^+.$$

If both the neutron and the positron born in the reaction are caught and tagged, then one would have evidence for the anti-neutrino. And once one accepts the anti-neutrino, one would have to accept the neutrino as well.

The scheme may be simple but the experiment wasn't, and Reines and Cowan toiled for three long years to plug all loopholes so that the evidence for the anti-neutrino was firm and convincing. The funny thing is that while low energy neutrinos (and anti-neutrinos) are so shy of interacting with matter, high energy neutrinos are more 'sociable'; experiments involving the interactions of high energy neutrinos are much easier, and thus it is that these days one hears of many neutrino experiments. If there is any moral in the neutrino tale it is that no criminal can escape detection for ever! In 1995, the Nobel Committee gave belated recognition to the discovery of the neutrino. Alas, Cowan was no more and Reines alone received the honour.

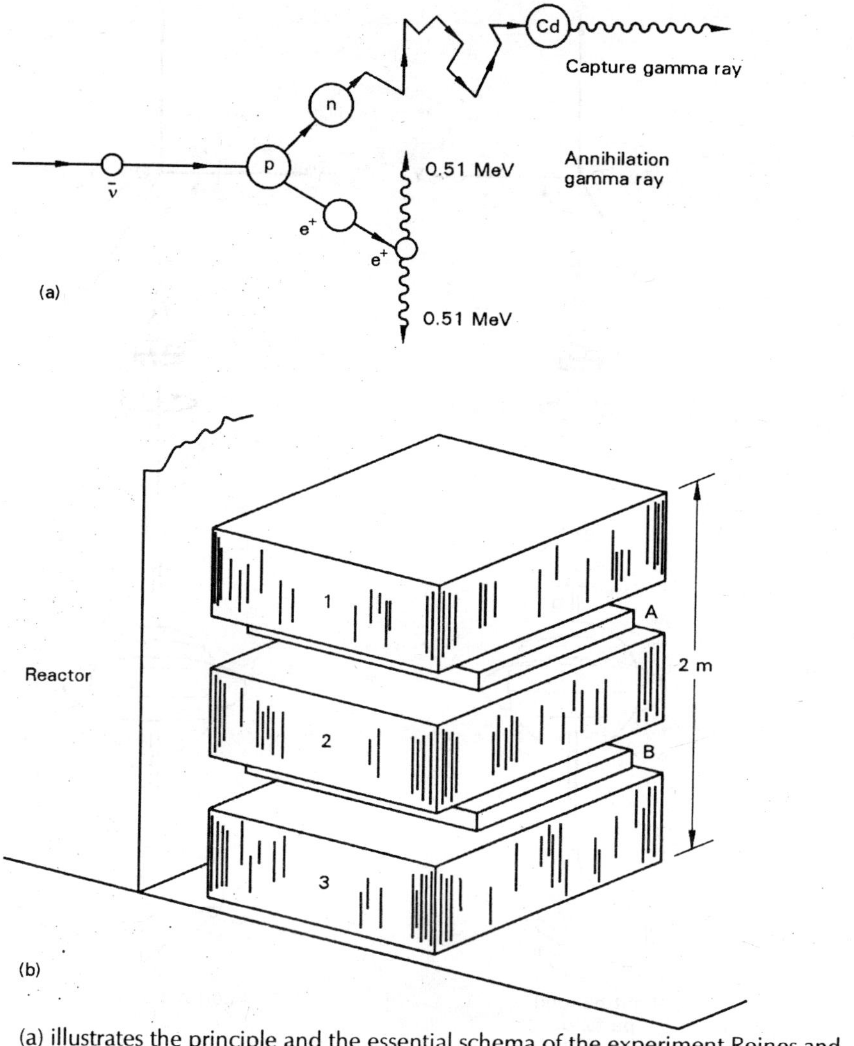

Fig. 5.10 (a) illustrates the principle and the essential schema of the experiment Reines and Cowan used to detect the anti-neutrino. The $\bar{\nu}$'s from the reactor interact with the protons in the water, causing the reaction $p + \bar{\nu} \rightarrow n + e^{+}$. The detection of the anti-neutrino is based on the successive detection of the neutron as well as the positron produced by the reaction. The positron annihilates almost as soon as it is produced, resulting in the emission of two photons, each of energy 0.51 MeV, which travel in opposite directions. The neutron suffers collisions while it is in motion, gets slowed down in the process, and is eventually captured by cadmium loaded into the water in the form of a soluble salt. When the neutron is captured, again a characteristic gamma ray is emitted; this happens about 10^{-5} sec after the positron is annihilated. Thus, clear signatures are available. (b) shows the detector arrangement. A and B are two targets, each containing 200 litres of $CdCl_2$ solution. 1, 2, 3 are tanks which contain liquid scintillators. They are meant to catch the gamma rays and produce light flashes. Each tank is viewed by 110 photomultipliers which pick up the light flashes produced by the different gamma rays.

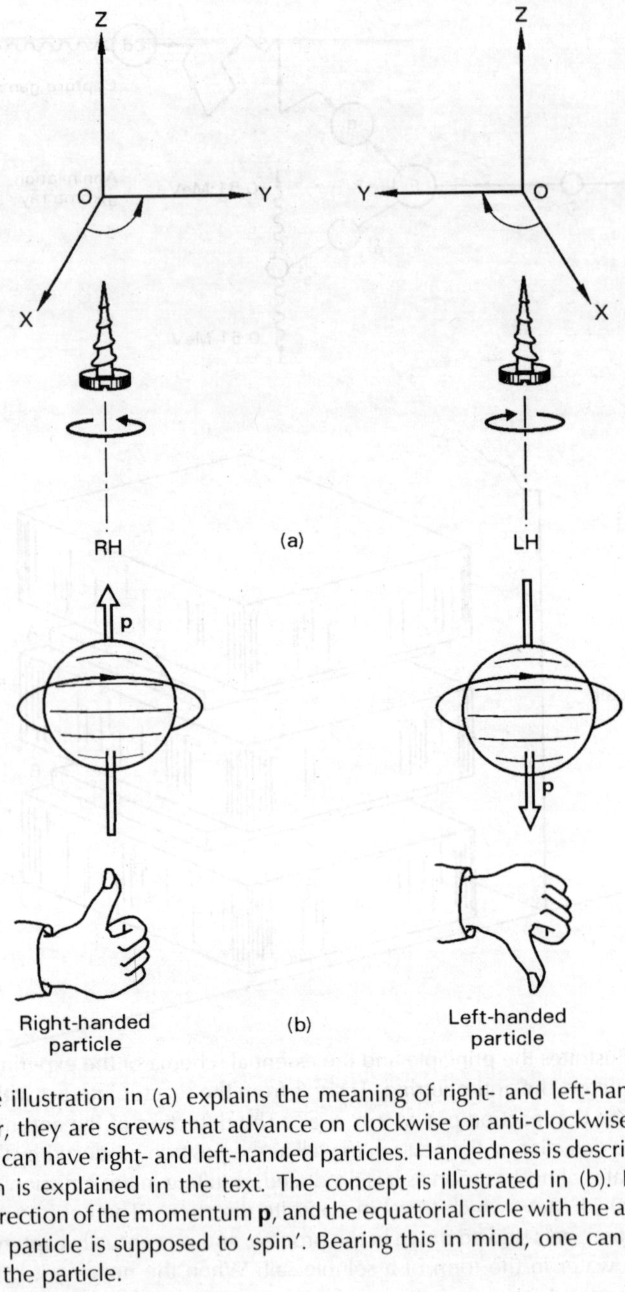

Fig. 5.11 The illustration in (a) explains the meaning of right- and left-handed screws. As should be clear, they are screws that advance on clockwise or anti-clockwise turning. In the same way, one can have right- and left-handed particles. Handedness is described by the term chirality, which is explained in the text. The concept is illustrated in (b). Here the arrow indicates the direction of the momentum **p**, and the equatorial circle with the arrow is symbolic of the way the particle is supposed to 'spin'. Bearing this in mind, one can understand the handedness of the particle.

5.10 Separating the twins

I am sure you must be quite uneasy about the neutrino–anti-neutrino pair. Suppose you encountered one of them. How do you tell whether it was a neutrino or an anti-neutrino? There is a way, and the trick is to use what is called *helicity*. Simply stated, helicity means handedness, and this is something familiar to us from screws (see Fig. 5.11). Handedness is alright with screws but handedness for particles? Yes, it is possible, and is called chirality or helicity. It is introduced via the definition

$$H \equiv (\mathbf{p}\cdot\sigma) / (|\mathbf{p}|\ |\sigma|),$$

where σ denotes the particle spin and **p** its momentum. Let us say that the spin can be either parallel or anti-parallel to the momentum. In that case, the helicity H is either $+1$ or -1. If it turns out that the neutrinos have one kind of helicity and the anti-neutrino the other kind, then we have a means of distinguishing the neutrino from the anti-neutrino.

This looks like a plausible scheme but are there any strings attached? Indeed there are, and the restriction is that the mass of the particle must be zero; it follows then, from the special theory of relativity, that the particle must travel at the speed of light c. Unless the particles travel at the speed of light, we can get confused about particle helicity; so we must have the particle travelling at speed c, and that automatically implies that the particle's rest mass must be zero.

This argument can be understood as follows. Consider a particle with spin and momentum as shown in Fig. 5.12(a). This is the view of a stationary observer O. For the moment, let us assume that the particle is travelling with a speed $v < c$. Suppose there is an observer O′ travelling in a rocket with a speed $v' > v$. To this

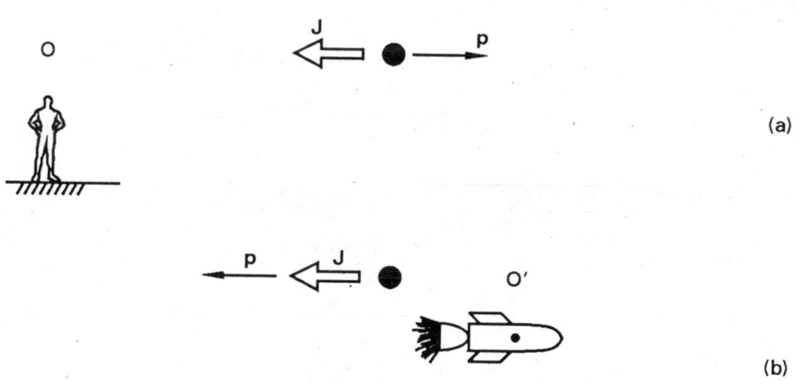

Fig. 5.12 One is able to assign a unique helicity to the neutrino because it has zero mass and therefore moves at the speed c. This figure illustrates what would happen if the velocity v of the particle is less than c. (a) shows a stationary observer O. For him, the spin angular momentum **J** points to the left and **p** to the right. (b) shows observer O′ travelling at a speed $v' > v$. For him, **J** points left but so does **p**. Thus, O and O′ obtain different results for the same particle, which is not allowed. However, if $v = c$, then the same result is obtained by the two observers. This implies that in order to have unique helicity, the particle must travel at the speed of light.

observer (who does not know that he is moving), the particle would appear to move from the right to the left instead of from the left to the right, as it did to observer O. So, **p** for observer O′ would be directed *opposite* to that of **p** for observer O. Thus the two observers would report different helicities for the particle. But that cannot be, which means that the particle *must* travel at the speed *c*.

All this however, does not completely solve the problem, for we now have two alternatives as below:

Alternative A	Alternative B
ν: $H = -1$ (left handed)	$\bar{\nu}$: $H = +1$ (right handed)
ν: $H = +1$ (right handed)	$\bar{\nu}$: $H = -1$ (left handed)

Which of these two alternatives has Nature chosen? For a while there was uncertainty till Goldhaber, Grodzins and Sunyar solved the problem.

5.11 Making the right choice

Goldhaber's experiment is very clever, which is why I am discussing it, particularly as the experiment is no longer remembered these days. Goldhaber and his collaborators investigated the decay of ^{152}Eu. This isotope decays by the capture of a K electron, and the decay occurs in two steps as follows (see also Fig. 5.13):

$$\text{(i)} \quad {}^{152}\text{Eu} + e^- \longrightarrow {}^{152}\text{Sm}^* + \nu,$$

$$\text{(ii)} \quad {}^{152}\text{Sm}^* \longrightarrow {}^{152}\text{Sm} + \gamma.$$

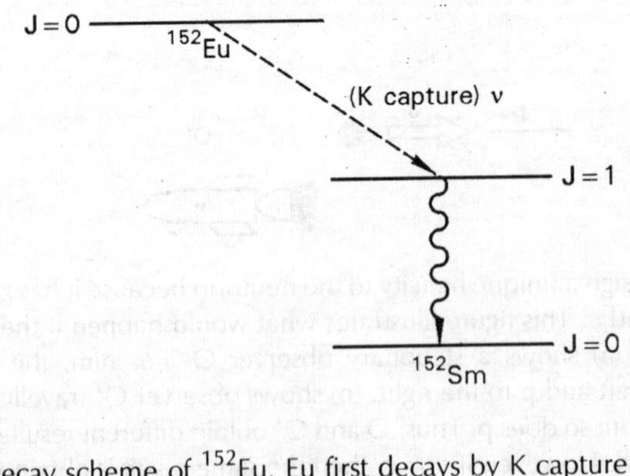

Fig. 5.13 Decay scheme of ^{152}Eu. Eu first decays by K capture to the excited state of Sm. This is followed by a transition from the excited to the ground state of Sm, with the emission of a gamma ray. Thus, the decay is to be visualized as occurring in two steps.

Although two steps are indicated, for all practical purposes, they are simultaneous. Notice that while the excited state of Sm has a spin of 1, the ground state has a spin of 0; clearly then, the emitted gamma ray has to carry away the angular momentum available. In practical terms what this means is that the emitted gamma ray would be circularly polarised like ordinary light can be.

Let us look at the scenario more closely. Three entities are involved, viz., e^-, ν and γ. We denote their angular momenta (in units of $\hbar$) as: $J_z(e)$, $J_z(\nu)$ and $J_z(\gamma)$. Since J for ^{152}Eu is zero, we must obviously have

$$J_z(e) + J_z(\nu) + J_z(\gamma) = 0.$$

Putting in without discrimination all the possible values, leads to

$$(1/2) \pm (1/2) \pm 1 = 0.$$

Clearly, all the possible combinations we have put in are not allowed. The permitted ones are:

$$+ (1/2) + (1/2) - 1 = 0,$$

which implies

$$J_z(\nu) = 1/2 \quad \text{and} \quad J_z(\gamma) = -1,$$

or,

$$- (1/2) - (1/2) + 1 = 0,$$

corresponding to which we have

$$J_z(\nu) = - 1/2, \text{ and } J_z(\gamma) = +1.$$

Notice that the two J_z's have opposite signs. If $J_z(\gamma)$ can be measured, then the above analysis can be used to determine $J_z(\nu)$. $J_z(\gamma)$ was determined by measuring the polarization of the gamma ray.

So we know $J_z(\gamma)$ and that tells us how the spin of the neutrino is pointing. What about the momentum of the neutrino? We can't detect it and so how do we know in which direction the neutrino is going? This is where the smartness of the experimenters shows. Now to start with, the ^{152}Eu nucleus is at rest and so the initial momentum is zero. During decay one has the γ, the ν and the recoiling Sm nucleus; the combined momentum of all the three must add up to zero. You will agree that in some cases, the Sm might not recoil at all and therefore have zero momentum, in which case the momenta of the neutrino and the gamma ray must add up to zero, or in other words, in such cases, the gamma ray will be travelling in a direction opposite to that of the neutrino. Goldhaber, Grodzins and Sunyar used a clever means of selecting only such cases where γ and ν went in opposite directions.

Now consult Fig. 5.14 which shows two basic possible outcomes of the experiment. Goldhaber's experiment showed that Nature chooses the option in Fig. 5.14(b ii), in turn implying that the helicity of the neutrino is –1, i.e., it is *left-handed*.

You should note here that if an electron is reflected in a mirror, i.e., if the parity operation is applied, it remains an electron. This is true of almost all particles, except the neutrino. If a neutrino is reflected in a mirror, the direction of spin is not reversed but the **p** is; in other words, its helicity is reversed. Once that happens, the particle ceases to be a neutrino; it becomes an anti-neutrino. In other words, the particle in the mirror is different from the particle in the real world, which is a case of parity violation. Thus, *the idea of neutrino helicity is central to the non-conservation of parity in weak interactions*.

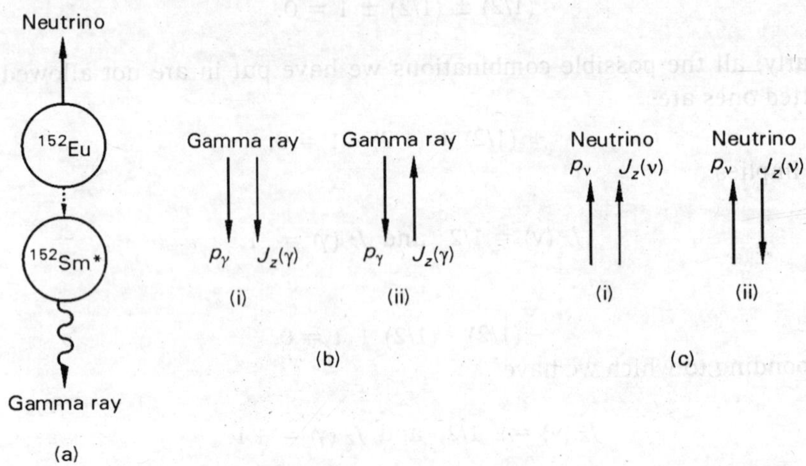

Fig. 5.14 This figure shows two possible outcomes in the Goldhaber experiment. In both cases, the gamma ray goes one way and the neutrino in the opposite direction. (This is ensured by a clever trick employed by the experimenters.) As far as the gamma ray is concerned, there are two possibilities, which are illustrated in (b). If these possibilities for the spins are combined with the momentum information in (a), then two possibilities emerge for the helicity as shown in (c). Experiments showed that the gamma ray emitted was according to option b(ii). This shows that the neutrino prefers option c(ii), i.e., it is left-handed.

5.12 The Universal Fermi Interaction

The weak force, people realised, does not bind; rather, it presides over disintegration or decay. Fermi succeeded in showing that the form of the weak interaction proposed by him could neatly tie up diverse facts pertaining to β decay. Then came the μ meson which was first mistaken to be the particle proposed by Yukawa but later recognised to be a different one. You will recall that at that time, people wondered what on earth the muon was supposed to do. But slowly it was realised that the muon also had features that smacked of weak interactions. In short, by the late forties, attention had narrowed down to the following reactions:

$$\text{(a)}\ \ n \longrightarrow p + e^- + \bar{\nu}\ \ (\beta \text{ decay}),$$

$$\text{(b)}\ \ \mu^+ + n \longrightarrow p + \nu\ \ (\mu \text{ capture}),$$

$$\text{(c)}\ \ \mu^\pm \longrightarrow e^\pm + \nu + \bar{\nu}\ \ (\mu \text{ decay}).$$

Everyone agreed that reaction (a) was driven by weak interactions. But what about the other two? In other words, the question boiled down to this, "Is there a Universal Fermi Interaction, and if so, what is its form?" By Universal Fermi Interaction, I mean one which drove reactions (a), (b) and (c) above, and not merely conventional beta radioactive decay.

In the above, by form I mean the mathematical form. Earlier we saw that Fermi wrote the weak interaction as a current–current interaction. Fermi assumed a scalar form for the interaction but a short while later, Gamow and Teller, analysing some experimental data, concluded that an axial-vector form is needed to handle some cases; thus, in those days, one spoke of the nuclear transitions leading to β decay as being of the Fermi or Gamow–Teller type. Around 1929, von Neumann showed, using Dirac's relativistic theory of the electron, that interactions in field theory must necessarily have one of the following forms: Scalar (S), vector (V), tensor (T), axial vector (A) and pseudoscalar (P). On this basis, allowing for all possibilities, one could write the weak interaction Hamiltonian in the general form

$$H_{\text{int}} = g_S H_S + g_V H_V + g_T H_T + g_A H_A + g_P H_P$$

where g_S, ..., etc., indicate the strengths, and the meanings of H_S, etc., should be self-evident. The form given above is the most general one possible but does Nature really need all the components posited above?

Now let's move to the conference on Particle Physics held at Rochester in 1957. I must mention here that even as the cosmic ray conferences were dying out, Robert Marshak of Rochester University took over and organised periodic meetings bringing particle physicists together. These high energy physics conferences continue to be held once in three years and they are always referred to as 'Rochester Conference', no matter what the venue is! Back in 1957, parity had just been overthrown, and all the bigwigs — Lee, Yang, Wu, etc., — were there in full force. One of the issues discussed was the Universal Fermi Interaction — was it there or not? Lee emphatically said *no*! Wu made it worse by citing various pieces of experimental evidence, all of which seemed to refute the idea of a Universal Fermi Interaction. The audience was thrown into confusion.

Marshak, in particular, was rattled because he and his student Sudarshan (see the Appendix to this chapter) had been quietly sifting the experimental evidence and were veering towards the view that

(a) There *was* a universal interaction,

(b) it involved only V and A,

(c) in fact, the V and A acted in opposition, i.e., it was $(V-A)$ interaction. Later, people would refer to it as the 'V minus A' interaction.

True, some of the then-available experimental evidence did not support the $V-A$ theory of Sudarshan and Marshak, but Sudarshan strongly felt that these experiments themselves were in doubt. Actually, Marshak wanted to unveil the $V-A$ theory during the conference but here were the experts decrying the Universal Fermi Interaction. Even as Marshak was wondering what course of action to take, circumstances conspired such that neither Marshak nor Sudarshan could speak on the subject (see the Appendix to this chapter).

A couple of months later, Marshak and Sudarshan were in California where they met Gell-Mann (more about him soon) and Felix Boehm. They lunched together and at that time, Marshak and Sudarshan explained their $V-A$ theory and asked Boehm (an experimenter) what the latest results were like. Boehm replied that the recent results ruled out (V, T) but were compatible with $(V-A)$. Marshak and Sudarshan were delighted.

In August, Marshak went to Italy to attend a conference in Padua where he categorically asserted that Universal Fermi Interaction did exist and that it was $(V-A)$. Later he recalled, "I said that it *had* to be $(V-A)$ for a universal interaction, and to be so will require that four experiments be murdered — I used the word *murdered* — including the ^{6}He experiment." (This ^{6}He experiment was a crucial one and the result obtained from it was not favourable to the $V-A$ theory.) What Marshak meant was that many of the previous experimental results were suspect, and that those experiments would have to be repeated. At that time people said that Marshak was crazy but the suspect experiments were in fact later repeated and *every one* of the new experiments pointed to $(V-A)$. So, did Marshak and Sudarshan become famous? No, and thereby hangs a tale!

Remember the California luncheon meeting that Marshak and Sudarshan had? Gell-Mann who was present at that meeting was quietly listening with interest, because he too was wondering about a Universal Fermi Interaction. Hearing that Marshak and Sudarshan were also on the same track, he decided to move quickly. Together with Feynman, he gave the theory shape and raced into print. Later, Feynman recalled:

> Finally they got all this stuff into me, and they say, "The situation is so mixed up that even some of the things they've established for years are being questioned — such as the beta decay of the neutron is S and T. Murray

> (Gell-Mann) says it might even be V and A, it is so messed up." I jump up from the stool and say, "Then I understand EVVVVERYTHING!" They thought I was joking. But the thing I had trouble with at the Rochester meeting — the neutron and the proton disintegration — everything fit *but* that, and if it was V and A instead of S and T, *that* would fit too. Therefore, I had the whole theory!
>
> I went on and checked some other things, which fit, and new things fit, and I was very excited. It was the first time, and the only time, in my career that I knew a law of Nature that nobody else knew. The other things I had done before were to take somebody else's theory and improve the method of calculating (an obvious reference to QED) — I thought about Dirac, who had his equation for a while — a new equation which told how an electron behaved — and I had this new equation for beta decay, which wasn't as vital as the Dirac equation, but it was good. It was the only time I ever discovered a new law.

Sudarshan and Marshak had it earlier but Gell-Mann and Feynman scooped them. Well, that's the way it sometimes happens!

5.13 The two-component neutrino theory

Way back in 1929, Herman Weyl, who inspired many to consider the symmetry aspects in physics, was playing with the Dirac equation, just then revealed to the world. Dirac's theory, you will recall, embraces both electrons and positrons (or anti-electrons if you wish), and since both the electron and the positron have two spin orientations possible, there are in all four components to the wave function; correspondingly one has positive and negative energy states with a gap in between.

Much before the discovery of the positron, there was a tendency to regard the proton as some kind of an anti-particle of the electron. If one believes in this, then since the proton is so different from the electron one must somehow restrict the four components of the electron to just two. Weyl found that if parity violation is assumed, then such a thing happens; the electron gets two components and the proton possibly the other two. Today we know that that is not the way things happen in Nature, but back in the late twenties, Weyl did the best he could to explain at one stroke both the electron and the proton. Weyl's ideas were promptly rejected because it involved the violation of parity, which nobody was ready for then.

We now move to around September 1956 and focus on a young Pakistani scientist named Abdus Salam. More will be said about him later but here we consider the fascinating story of how he came close to fame in the parity non-conservation business but was cheated out of his chance. It all happened as follows.

Salam was at that time working in the Cambridge University, England, and had become interested in elementary particle physics. He managed to attend a conference in Seattle, America (that is where the Boeing aircrafts are made), and was returning to England. In those days, many scientists took advantage of free plane rides on transport aircraft, made available by the U.S. Air Force which constantly had huge planes flying across continents. The rides were free but they took a long time and there were other inconveniences besides; but Salam, like many a young scientist, did not mind, for the ride was free; otherwise, how could they attend conferences?

At the conference, he had heard a talk by Yang, who had discussed the tau–theta puzzle. Parity breakdown was perhaps not explicitly mentioned by Yang, but it was in the air, so to speak. During the long journey back from Seattle to London, Salam kept thinking about Yang's lecture. Later he recalled:

> I could not sleep. I kept reflecting why Nature should violate left–right symmetry in weak interactions. Now the hallmark of most weak interactions was Pauli's neutrino. While crossing the Atlantic, there came back to me a deeply perceptive question which Rudolf Peierls had asked me at my Ph.D. examination a few years before — "Why is the mass of the neutrino zero?"

At the time of the Ph.D. viva, Salam did not know the answer; nor did Peierls! But the examiner graciously said that if Salam ever found the answer, he should come back and report. The answer now suddenly came to Salam, flying ten thousand metres above the Atlantic. In his cramped seat, Salam wrote down the Dirac equation with mass zero. He immediately saw that

(a) the four components reduced to two, i.e., the neutrino needed only two components instead of four,

(b) parity was of course automatically violated, and,

(c) the neutrino had to travel at the speed of light, which in turn meant that it had to have zero mass.

The answer to Peierls' tricky question was now at hand; more important, he had a theory which gave the foundation of parity violation in weak interactions. Salam realised that the zero mass neutrino was like a miniscule corkscrew, drilling its way at the speed of light.

The following morning, a highly elated Salam stepped off the plane in London, rushed to Cambridge, frantically gave his theory proper shape and, clutching the papers, hopped onto a train going to Birmingham to report his findings to Peierls. The former examiner gave a patient hearing but his verdict was disappointing, "I do not believe that the left–right symmetry is violated in weak interactions at all". Remember, this was several months before the Lee and Yang paper, Wu experiment, etc. At that time, only a mad person could talk about parity violation!

Peierls might not have been convinced but Salam was; he felt it in his bones that parity violation was true and tied to the existence of the two-component neutrino. If Peierls did not approve, it did not matter. Why not approach the Oracle himself, i.e., Pauli? And so, Salam gave his neutrino paper to a young physicist going to Zurich, where Pauli lived. Pauli was scornful. "Give my regards to Salam and tell him to think of something better" was the message from the Oracle! Meanwhile, Salam went on to claim the existence of the Universal Fermi Interaction, and was treated to a second blast from Pauli. This time the message was, "Withdraw your paper!"

Salam did eventually withdraw the paper, but under different circumstances, and only to deeply regret it later. Salam wrote his paper which had the neutrino business, Universal Fermi Interaction and all that, and sent it to the *Physical Review*. In between he went to Rochester to attend the conference organised by Marshak. That was where Wu gave a report which threw cold water on the Universal Fermi Interaction — I have mentioned this earlier. Alarmed, Salam stopped at the office of the *Physical Review* and withdrew his paper. Months later, the parity business exploded; Lee and Yang had their paper, Wu had hers and it was all over. Salam who could have covered himself with glory if only he had not acted impulsively (and possibly under pressure from Pauli), was now left to bitterly rue his fate. But Fate, as it turned out, was kind to Salam, and eventually he won the coveted Nobel Prize, for another and an even more important discovery.

5.14 Chapter summary

This is an important chapter, where I have mentioned many things. I am sure you would welcome a summary.

- Compared to alpha decay, beta decay represented a major puzzle on account of the continuous spectrum exhibited by the emitted electrons.
- Bohr thought that the continuous spectrum signalled the breakdown of the conservation of energy.
- Pauli said no, and added that it called for one more particle which was neutral, had a spin of (1/2), and had zero mass. Today, thanks to Fermi, we call this particle the neutrino.
- Using the neutron and the neutrino, Fermi developed a theory of beta decay. It respected both energy and momentum conservation.
- A unique feature of Fermi's theory was that it required beta decay to be mediated by a new force, the weak force. Fermi modelled this weak force by analogy with the e.m. force, and expressed it in the form of a current–current interaction (**j** dot **A** form). However, the interaction was of the contact type — in modern language it means that Fermi's version of the weak force did not call for any mediating particles.
- Fermi's theory explicitly assumed the conservation of parity.

- Fermi, and independently Gamow and Teller, examined allowed transitions in beta decay. Together they provided room for *S*, *V*, *A* and *T* interactions.
- From very general considerations, von Neumann showed that the only possible interactions are *S*, *V*, *A*, *T*, and *P*.
- After the war came the tau–theta puzzle. Were the tau and theta two different particles or were they one and the same?
- Yang and Lee said that if they were one and the same, then parity must break down in W.I. No supporting experimental evidence was available at that time, and so they proposed new experiments.
- All experiments voted for parity breakdown in W.I.
- Parity breakdown implies that where W.I. is involved, the left can be distinguished from the right, or the real world from the mirror world.
- In turn this implies that the neutrinos have a screw sense, i.e., helicity. Goldhaber further determined that neutrinos are left-handed.
- Sudarshan and Marshak, and independently Gell-Mann and Feynman, proposed the existence of the Universal Fermi Interaction, and declared that its form was $(V - A)$.
- Salam had a theory of the two-component neutrino but, misguided by wrong experimental evidence (and possibly objections from Pauli as well as Peierls), failed to publish it! The two-component theory later emerged from the work of Lee and Yang.

A tremendous amount of work and progress you would think — yes indeed, but, compared to what came later, all this represents merely the tip of the iceberg!

Appendix to Chapter 5

All scientists look forward to making profound discoveries, for there is no greater thrill. After making the discovery and savouring its beauty, they then like to bask in the glory of praise, recognition, etc. Sometimes it happens that the discoverer is completely overlooked. There could be several reasons for this but no matter what they are, the discoverer feels dejected and even cheated. There is a tragic episode of this nature associated with the $V - A$ interaction, and the affected person was Prof. E.C.G. Sudarshan of Texas University.

Born in Kerala in 1931, Sudarshan worked for a while as a Research Assistant at the Tata Institute of Fundamental Research, Bombay (now Mumbai). In 1955 he left for America to work for his doctorate degree under Prof. Robert E. Marshak of Rochester University. The big excitement at that time was weak interactions, and on Marshak's advice, Sudarshan plunged into that subject. He recalls:

> It became clear to us from muon decay that the $V - A$ interaction was the natural choice; muon capture had to have an A (or P) interaction, otherwise the muon would not decay! So we set about examining in detail β decay data to see if we could challenge the β decay assignment of $S - T$ which was generally accepted.

As mentioned in the text, the ultimate arbiter is always the experiment, and there were a few experimental results which seemed to be in conflict with the $V-A$ hypothesis. However, Sudarshan firmly believed that it was $V-A$ and nothing else; so convinced did he feel that to him, if experiments said something different, it simply meant that those experiments were wrong! Let us hear him again reminiscing:

> We (Marshak and Sudarshan) were quite convinced it was $V - A$. Then came the VIIth Rochester Conference at which I was to present this theory in outline. But a few days before the conference, Marshak realised that as a graduate student I was not a delegate to the conference and felt that it was therefore not possible to make a presentation. Marshak, the Chairman of the Conference, was giving a principal talk on nucleon–nucleon interactions and could not present another paper in which he was the co-author. So Marshak and I jointly requested Paul Matthews, then a Visiting Professor at Rochester, to make the presentation. Matthews never did, and *I could not get to speak*. It was only much later that I realized that I would have to wait approximatley 28 years before I got a chance to talk about it.

Sudarshan is obviously bitter but there are reasons, as you will presently see. If only someone had said a word at the Rochester Conference, things might have been different. But as it transpired, for Sudarshan, the conference was a lost opportunity.

Around August 1957, there was a conference due to be held in Padua, Italy. Sudarshan was not invited but Marshak was and so Sudarshan prepared a paper on their ($V - A$) work; here was another opportunity to unveil it. Marshak did speak about the theory but even so, luck was not on Sudarshan's side. Meanwhile, Sudarshan completed his Ph.D. formalities and went to Harvard.

By this time, the $V-A$ idea had become a hot topic and word was going round that Gell-Mann and Feynman had done some important work on the subject. (Recall the California luncheon meeting when Marshak and Sudarshan explained their work and Gell-Mann was a silent listener.) News about the breakthrough of Gell-Mann and Feynman reached Harvard. Over now to Sudarshan:

> It was amazing that at that place no one would be bothered that there was a new theory of weak interactions and that one of the authors of the theory was at Harvard. The few who knew about the theory conveniently forgot that Marshak and I were the authors. C.N. Yang mentioned the ideas he had heard from Feynman. I looked up, down, sideways at the Harvard audience and finally decided that this humble foreigner should speak up and I did. Not that it made any difference.

Soon after this Sudarshan heard that Feynman and Gell-Mann were preparing to publish a paper on the $V-A$ theory, and now he became really anxious. He had made the discovery earlier and did not wish to be scooped. And so, what did he do? In his words:

> I telephoned Marshak to ask about the status of our paper (Sudarshan wanted a comprehensive version of the Padua conference paper, prepared for a journal); he (Marshak) said that it was already presented at Padua, Venice, and presented as a Rochester preprint. He assured me that it was a sufficient way to inform all around about our work.

Marshak was taking the old-world view that one can't publish the same thing over again whereas Sudarshan's contention was that a conference paper was not really a publication (indeed, that is the convention today). Feynman and Gell-Mann who discovered the $V-A$ interaction after him, were now going ahead with their paper. And if they published first, gone was the credit due to him as the original discoverer.

In the end, that is precisely what happened. The Feynman–Gell-Mann paper appeared in the *Physical Review* and Sudarshan was left high and dry. As a sop, Marshak agreed to publishing a short paper on their work in the same journal, but by then it was too late. A pained Sudarshan recalls:

> Unfortunately, years later it became usual for people to make no reference at all to our work — It puzzled me a lot how anyone who had read our paper and

seen the analysis of experimental data and the subsequent outline of the chiral theory could ignore it.

Sudarshan is justified in his complaint. A recent book entitled *Electroweak Interactions* by Peter Renton and published by the Cambridge University Press in 1990 says:

> The framework of the phenomenological $V-A$ theory was first published by Feynman and Gell-Mann (1958).

There is no reference whatsoever to Sudarshan and Marshak; and so it is in many other cases. The story does not quite end here. Once there was a conference at Gatlinberg, USA, on weak interactions, particularly the $V-A$ business. Sudarshan again:

> I was invited to chair a session ... I was flattered but was not invited to talk at this conference even though our work was the reason the conference was held at all! Marshak was too ill to attend. One was like a witness in a court not allowed to speak, even when what one had to say was important. But worse was to come. Several years later the Argonne National Laboratory held a conference on weak interactions in which neither Marshak nor I had even a ceremonial role, not to mention an invited talk at the meeting. When I asked my friend Kameshwar Wali (author of *Chandra*, the biography of Chandrasekhar), a kingpin of the conference, "How come?", his response was, "George, you make it difficult for me to be your friend." At Gatlinberg, Richard Feynman gave a leisurely evening lecture on β decay .... The next evening Dick (Feynman) and I sat next to each other; he very graciously said that he had been told that I had the essential $V-A$ ideas much earlier than anyone else.

We skip years and jump to 1974. The scene is the University of Pennsylvania, and Feynman is giving a talk. What does he say?

> We have a conventional theory of weak interactions, invented by Sudarshan and Marshak, published by Feynman and Gell-Mann, and completed by Cabibbo ....

In 1985, Feynman met Marshak at a conference in Kyoto, Japan, arranged to celebrate the fiftieth anniversary of Yukawa's proposal about the meson. Marshak mentioned to Feynman about Sudarshan's unhappiness in not receiving proper recognition for his discovery. Shortly thereafter, Feynman wrote Marshak a letter in which he said:

> It was great seeing you and talking. I hope someday we can get this straightened out and give Sudarshan the credit for priority that he justly

> deserves ... . I wish I had not caused you and Sudarshan such discomfort. At any opportunity I shall try to set the record straight — but nobody believes me when I am serious!

Sudarshan continues to be unhappy that circumstances (and perhaps Marshak too, though unwittingly) conspired to deny the credit due to him. The advent of the $V-A$ theory is, according to him, a major landmark in the theory of weak interactions, even as a century earlier Faraday had unified a number of puzzling aspects of electromagnetism. Even more significant, the $V-A$ theory was the stepping stone which eventually led to the Glashow–Weinberg–Salam theory of electro-weak unification (see Chapter 10). Well, there it is, success (of discovery) followed by disappointment (of non-recognition). If there is a moral, it may be that we should practise a certain detachment so that the pangs of disappointment are not that acute.

6

Fun with Mirrors

This chapter tells you a bit more about *P*, *C* and *T* invariances. To start with, these operations are discussed using mirrors as was once done by Wigner. Thereafter, we go into the strange behaviour of neutral kaons, leading to the discovery of *CP* failure. Thus, this chapter is mainly about the consequences of the three important discrete symmetries, namely, *P*, *C*, and *T*.

6.1 Introduction

Mirror, mirror on the wall, Who is the fairest of them all?

I am sure you are familiar with the rhyme from the story of *Snow White and the Seven Dwarfs*. Mirrors don't speak, but this one did — it was a magic mirror. I am now going to introduce you to a few exotic mirrors invented many years ago by Nobel awardee Eugene Wigner. These mirrors don't speak but they do strange things. Wigner is a great name in Physics and it was he who brought home to physicists (along with Weyl) the importance of symmetry. Some years after parity fell, he wrote:

> Nobody was happy with this result (Wu experiment). It is a fact, of course, that most of us are partial towards our right hand as cobalt is towards its left. We feel, however, that radioactive cobalt is not entitled to be partial because it should have forgotten its past and because at the time it emitted the decay particles, it was under no influence that would have favoured one side of the plane of the ring current over the other. This plane was a symmetry plane ...

That tells you how confident the big scientists were about parity and how cheated they felt when it fell!

Although I made a sincere attempt (in the previous chapter), the physics underlying the parity failure experiment is not easy to explain. So, soon after the Wu experiment, Wigner wrote an article explaining what *C*, *P* and *T* were all about and did so by introducing magic mirrors of his own. This chapter is to some extent about those magic mirrors of Wigner.

6.2 What mirrors? Why mirrors?

Recall from Chapter 4 that *P* transforms **r** to −**r**, *C* transforms particle to anti-particle, and *T* reverses particle motion. Now when **r** is reflected in a mirror, we get a vector pointing the 'other way'; that is why the *P* operation is often explained by invoking a mirror. And then waving hands, one uses the mirrors also to explain parity failure, etc. The listener/reader pretends he understands all this, while the lecturer/author thinks he has done a great job! One can no doubt 'do it with mirrors', but one has to be very careful. The person explaining must make it very clear that the mirrors one is talking about are very special mirrors, and the listener must likewise be careful not to confuse physics mirrors with the usual looking glass.

You might ask, "If mirrors complicate, why bring them in at all?" Perhaps one should avoid them, but for now I would like to point out how Wigner handled all this carefully and most effectively.

6.3 How many mirrors?

Our present interest centres around three discrete symmetries, *P*, *C*, and *T*. With each, we can associate a fictitious or magic mirror, which gives us three mirrors to start with. We next make combinations corresponding to *CP*, *CT* and *PT*, which gives us three more mirrors. Finally, there is the grandest of them all, the *CPT* mirror. Remember, with a 'combination' mirror, the order of the operation does not matter; thus a *CP* mirror is the same as a *PC* mirror.

What does each mirror do? Given a sequence of real events or better, physical processes, each mirror produces a sequence of suitably transformed processes. The mirror thus merely 'manufactures' what may be termed an 'image sequence'. Can such an image sequence actually happen? Is it a possible (physical) reality? Is it realised in Nature? These are the questions physicists are interested in. If the 'reflected sequence' is actually observed via experiments, then one says that the concerned transformation — *P*, *C*, *PC*, or whatever — is conserved.

To start with, it was believed that all these seven magic mirrors were perfect. The Wu experiment gave the first shock, showing that the *P* mirror did not quite behave perfectly when events related to W.I. were 'reflected'. People were upset (we just saw what Wigner's reaction was), and soon they found that the *C* mirror also had a flaw, i.e., *C* was not conserved in W.I. This has much to do with the fact that neutrinos have handedness or chirality, as discussed in Chapter 5.

Now, charge conjugation converts a particle into its anti-particle. Applied to a neutrino, it produces an anti-neutrino. Fine. But now let us look at this a little more closely, taking proper account of the handedness of the neutrino. Thanks to P violation in W.I., the neutrino is single-handed; in particular, we have seen that Goldhaber proved it is left-handed. Suppose one applies charge conjugation to the neutrino; it produces a left-handed anti-neutrino. But Nature does not admit of a left-handed anti-neutrino; all anti-neutrinos *must* be right-handed. Thus, the C operation applied to W.I. predicts something that is not observed in nature which means that C is not conserved in W.I.

Fortunately, this C failure in W.I. is a kind of compensating flaw, and that made the CP mirror perfect. Everyone heaved a sigh of relief; but the relief was short-lived, for soon came the news that the CP mirror was also flawed! Pundits shook their heads and said, "My God, this is just too much!" But they took comfort in the fact that the CPT mirror was still there, protected by the Pauli–Luders theorem. Let us examine this closely.

6.4 The three basic mirrors

Let us start with the P mirror. Consider Fig. 6.1; you must study this figure carefully because it illustrates how one can get confused if one thinks of the parity mirror to be the same as the looking glass. Let us start with (a). Shown here is a looking glass or what Wigner refers to as the layman's mirror (LM). Everything is reflected the way we expect from optics. However, physics-wise, things are topsy turvy. Why? Because in this mirror, the magnetic field **B** is dutifully reflected to –**B**, whereas we expect

$$\mathbf{r} \text{ — under } P \longrightarrow -\mathbf{r}$$

$$\mathbf{B} \text{ — under } P \longrightarrow \mathbf{B}.$$

In other words, **B** is an axial vector and should *not* change under P; thus the shaving mirror is *not* the proper representation of P. Let us try to understand the problem. In the real world, one applies the left-hand rule to determine the direction of the magnetic field — if the forefinger is curved in the direction of flow of the electrons, the thumb points towards the magnetic north. The physical mirror faithfully reflects everything, including the thumb — the left hand becomes the right hand. In addition, the mirror reflects the arrows corresponding to the magnetic field, and the field points down, which is wrong. In short, the left-hand rule for the magnetic field becomes the right-hand rule in the mirror world. However, from the physics point of view, the left-hand rule must not change; it must hold true in both the worlds. Thus, the ordinary looking glass would not do to represent the parity operation. The physicist's mirror, *a la* Wigner, shown in Fig. 6.1(c) does not suffer this deficiency and takes care of the problem. Here, not only does **r** become –**r**, but under the P operation, **p** → –**p**. Thanks to this, we must now be a bit more careful. Suppose you observe a table fan in an ordinary

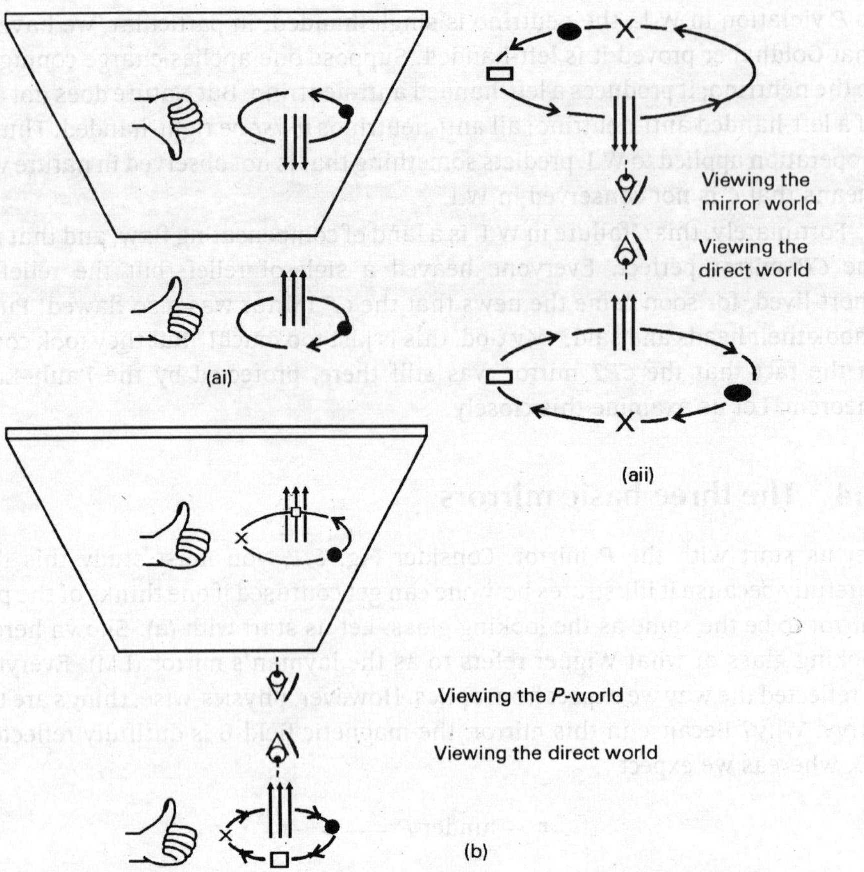

Fig. 6.1 Comparison of the shaving mirror or the layman's mirror (a) and Wigner's *P* mirror (b). (a) reflects according to the laws of optics, and the left hand becomes the right hand in the mirror, and the magnetic field changes direction as shown. But there is something obviously wrong; in the real world, the field lines point according to the left-hand rule but in the mirror world, they appear to follow the right-hand rule which of course cannot be accepted. In (b), the guiding principle is that while **r** is changed to −**r**, **p** is also changed to −**p** by the *P* mirror; this the ordinary mirror does not do and herein lies the difference. As a result, the direction of **B** is unchanged and the left-hand rule works as before. This is symbolically shown by indicating that 'the left hand remains a left hand' when reflected in the *P* mirror. Perhaps you feel that this mirror business is very confusing compared to the equations which are quite straightforward. Indeed! When parity fell, people used to invoke mirrors all the time, and it left me confused. Even today people resort to mirrors, and I thought you should be warned of the dangers of taking these things literally; hence all this explanation.

mirror. If the blades are moving in a clockwise direction in the direct world, they would appear to move in the anti-clockwise direction in the mirror. In the *P* world, the **p** to −**p** transformation annuls this, and so the direction of the magnetic field (which is due to the circular motion of the electrons) ends up being the same. The moral is: the motion of the charge is fundamental; **B** is only a consequence. Thus, one must take care to carry over to the *P* world the basic facts of the direct world, with due regard to the consequences of the transformation. Such caution is required everywhere and at every step.

On, now, to Wigner's descriptions, and like him we shall discuss the actions of the *P, C,* and *T* mirrors by considering the proton–positron (p, e^+) scattering. The proton being massive compared to the positron, does not budge from where it was, whereas the positron is deflected. Scattering not only changes the direction of the positron, but can also flip its spin.

Let **p** and $\boldsymbol{\sigma}$ denote respectively the momentum and the spin of the positron. The golden rules that will guide us are:

$$P\mathbf{p} = -\mathbf{p} \qquad C\mathbf{p} = \mathbf{p} \qquad T\mathbf{p} = -\mathbf{p}$$

$$P\boldsymbol{\sigma} = \boldsymbol{\sigma} \qquad C\boldsymbol{\sigma} = -\boldsymbol{\sigma} \qquad T\boldsymbol{\sigma} = -\boldsymbol{\sigma} \qquad [6.1]$$

Observe that $CPT\,\mathbf{p} = \mathbf{p}$ and $CPT\,\boldsymbol{\sigma} = \boldsymbol{\sigma}$, just as we expect.

Now study Figs. 6.2–6.4. They offer a picture of the scattering when viewed in the three basic (Wigner) mirrors. We are, of course, assuming for the present that the mirrors are *not* flawed, i.e., they are perfect.

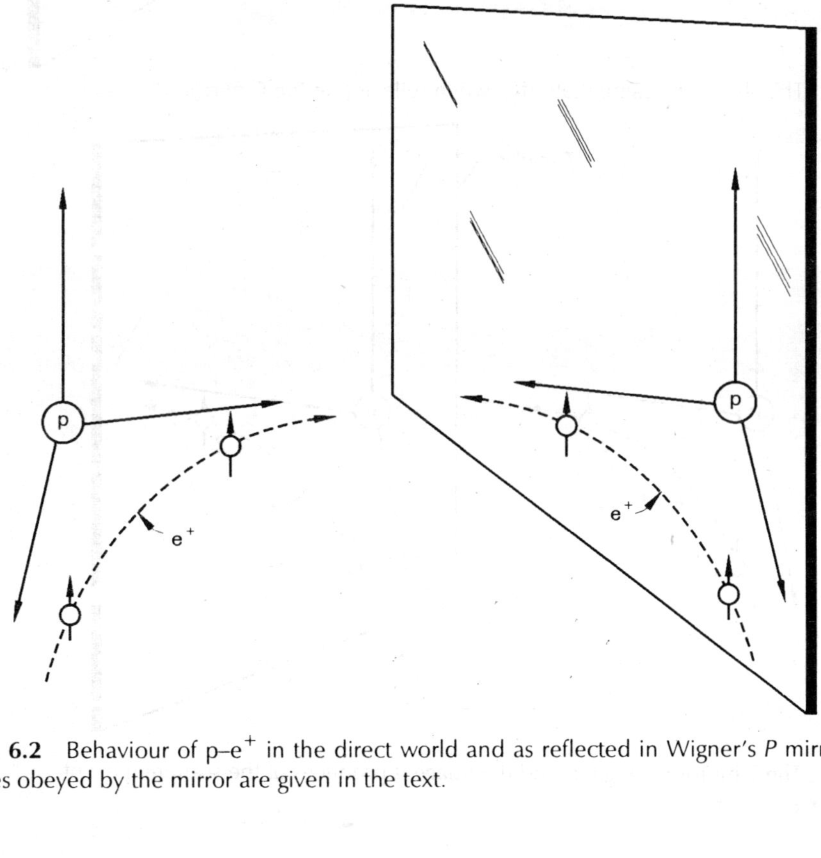

Fig. 6.2 Behaviour of p–e^+ in the direct world and as reflected in Wigner's *P* mirror. The rules obeyed by the mirror are given in the text.

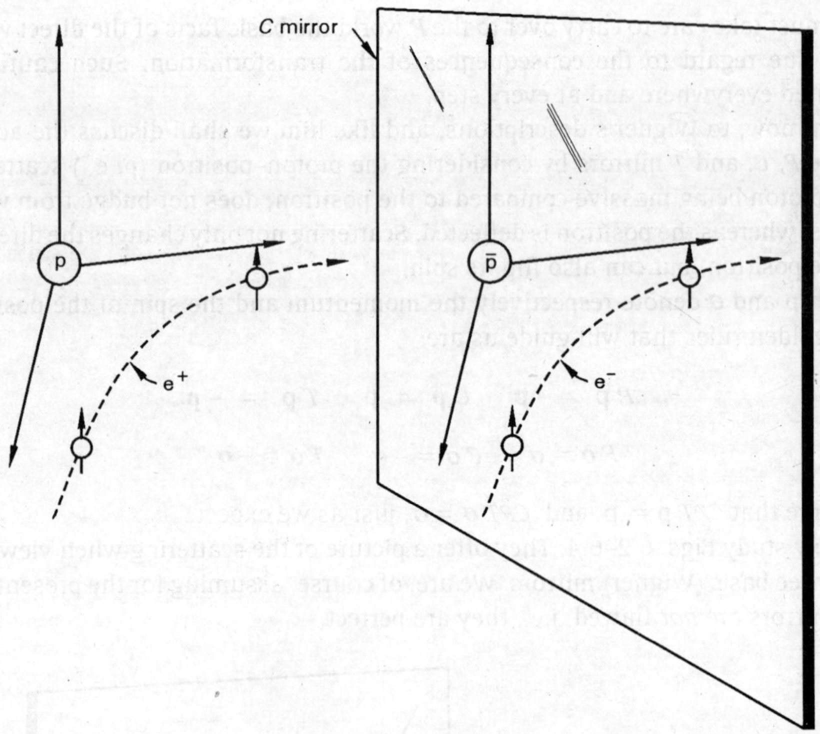

Fig. 6.3 This shows the same scattering when reflected in the C mirror.

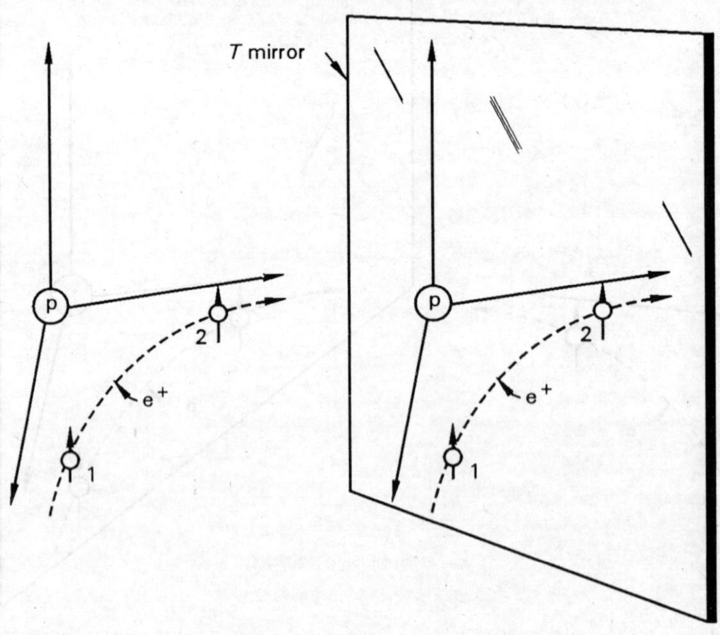

Fig. 6.4 The *T* mirror of Wigner, and the image it produces of the scattering event.

6.5 The K-meson story

This story is exciting but can also be confusing because the same particles appear in various disguises! However, there will be no problems if you keep your sights straight; you will be amazed at how cleverly Nature plays delicate and subtle games!

It all starts with our friend the K meson. When you encountered it earlier, it had two names, tau and theta but once parity fell, people knew these two particles were one and the same, i.e., the K. In the $\tau - \theta$ puzzle we were interested in K^+, but here we focus on the neutral counterpart, i.e., K^0 and its anti-particle, $\overline{K}^0$. These two particles are different from each other with regard to the strangeness quantum number. For the first, $S = +1$, while for the second $S = -1$. Notice that the anti-particle of K^0 is not the same as K^0: it is different, i.e., $\overline{K}^0$. But the anti-particle of π^0 is π^0 itself. This might seem odd but finds a ready explanation in the quark model (to be discussed later).

Our story will take you through the sequence

$$\begin{pmatrix} K^0 \\ \overline{K}^0 \end{pmatrix} \rightarrow \begin{pmatrix} K_1 \\ K_2 \end{pmatrix} \rightarrow \begin{pmatrix} K_S \\ K_L \end{pmatrix}.$$

In other words, we will start with the pair of particles K^0 and $\overline{K}^0$, then discover that they have become K_1 and K_2, and finally that they change into K_S and K_L respectively. Watch carefully!

Let us do it all slowly, starting with the question, "How does one distinguish between K^0 and $\overline{K}^0$?" Not easily since practically all the attributes are the same, but as just noted, there is a difference, and that is the strangeness.

Next we study the decay of the K's. Experiments show that both K^0 and $\overline{K}^0$ decay into π^+ and π^-. So we have

$$K^0 \longrightarrow \pi^+ + \pi^-,$$

$$\overline{K}^0 \longrightarrow \pi^+ + \pi^-.$$

Let us now look into the *C* mirror and see how these two reactions proceed, i.e., let us apply the *C* transformation to both sides of the above reaction equations. We then get

$$C|K^0> = |\overline{K}^0>,$$

$$C|\overline{K}^0> = |K^0>, \qquad [6.3]$$

$$C|\pi^+\pi^- > = |\pi^+\pi^- >.$$

Study the above equations carefully, remembering that K^0 and $\overline{K}^0$ are *different* particles. You will then notice that when the final state, i.e., $\pi^+ \pi^-$, is reflected in the *C* mirror there is no change but when the initial states, i.e., K^0 and $\overline{K}^0$ are, there is a change. You may wonder how this is possible. Are we having another crisis like the τ–θ? Not really. The experts assure us as follows, "*C* will preserve the initial state only if *S* is a good quantum number. In fact, *S* is a good quantum number in S.I. but here the decay proceeds via W.I.; and in W.I., *S* is not a good quantum number." Fine. In that case, can we expect the following?

$$K^0 \to \pi^+ + \pi^- \to \overline{K}^0 \ ? \qquad [6.4]$$

What this says is that K^0 first decomposes into a charged pion pair which then recombine to form K^0. In the Feynman language we would represent this process as in Fig. 6.5, from which we see that pions appear only in the virtual state. Equation [6.4] is interesting because it tells us that K^0 can play a Jekyll and Hyde game — sometimes it can appear as K^0 and at other times as $\overline{K}^0$! This is not all, for more is to come.

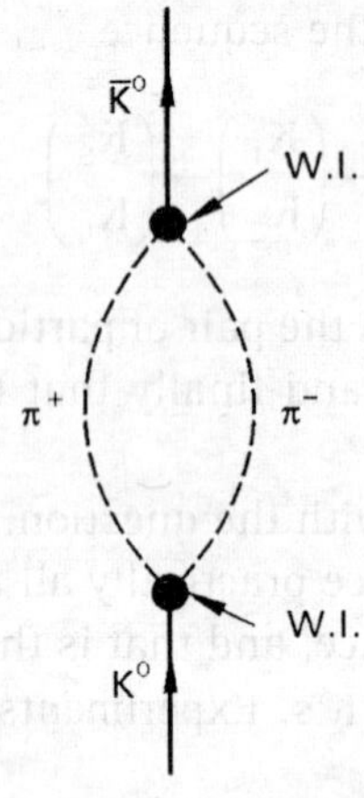

Fig. 6.5 Feynman diagram to illustrate the $K^0 \to \pi^+ \pi^- \to \overline{K}^0$ transition.

Denote the amplitude for the decay of the K's as *A*. We suppose now, and there are good reasons for doing so as we shall soon see, that K^0 and $\overline{K}^0$ are really 'mixtures' of two other particles K_1 and K_2, the mixtures being given by

$$K^0 = \frac{1}{\sqrt{2}}(K_1 + K_2),$$

$$\overline{K}^0 = \frac{1}{\sqrt{2}}(K_1 - K_2). \qquad [6.5]$$

I must confess I have been a bit sloppy and should really have written

$$K^0 = \frac{1}{\sqrt{2}}\{\,|K_1> + |K_2>\}\ ,$$

etc., but in the interest of simplicity, I trust you will allow me some liberties. Getting back to Eqn. [6.5], by simple algebraic manipulation we then obtain

$$K_1 = \frac{1}{\sqrt{2}}\left(K^0 + \bar{K}^0\right),$$

$$K_2 = \frac{1}{\sqrt{2}}\left(K^0 - \bar{K}^0\right). \qquad [6.6]$$

So we have these two pairs, (K^0 and $\bar{K}^0$) and (K_1 and K_2), each being a mixture of the other. The question naturally arises as to which of these is primary and which is secondary. As it will become clear shortly, the pair (K_1, K_2) is primary and the other one is secondary, i.e., the two particles K^0 and $\bar{K}^0$ are really mixtures.

Now comes another interesting fact. Since both K^0 and $\bar{K}^0$ decay into two pions with the same amplitude A, we can infer that K_1 will decay in a similar manner with an amplitude ($A/\sqrt{2}$) — see Eqn. [6.5]. By the same token, K_2 *cannot* decay into two pions because the amplitude in this case is zero—see Eqn. [6.6]. This difference has an impact on the lifetimes of the two particles K_1 and K_2. K_1 decays quickly (via the two–pion mode) but K_2 which is denied this channel has to look for other avenues. So, one would expect K_2 to live longer, and indeed it does. Whereas the lifetime of K_1 is $\sim 0.86 \times 10^{-10}$ sec, that of K_2 is $\sim 0.5 \times 10^{-7}$ sec. All this leads to some interesting experimental consequences, but to understand these, a digression is necessary.

6.6 A refresher on quantum tunnelling

What I am going to say about quantum tunnelling is probably familiar to you from ***QR-I***. Anyway, let us go over it again quickly. Figure 6.6(a) shows a quantum well with infinite walls on either side. A particle in this well will have a discrete spectrum of energy levels (recall Fig. 3.4); we focus on the lowest one, the ground state.

We next consider two wells side by side, as in Fig. 6.6(b). There are two ground states which we denote as $|L>$ and $|R>$ respectively. If a particle is introduced into one of these two wells, it would remain there forever.

Next, we modify the infinite barrier between the two wells and create a situation as in Fig. 6.6(c). We can think of this new situation as having been created by a suitable interaction H_I. We now use the machinery of perturbation theory and go to work. What happens then is the following: firstly, the baseline shifts by an

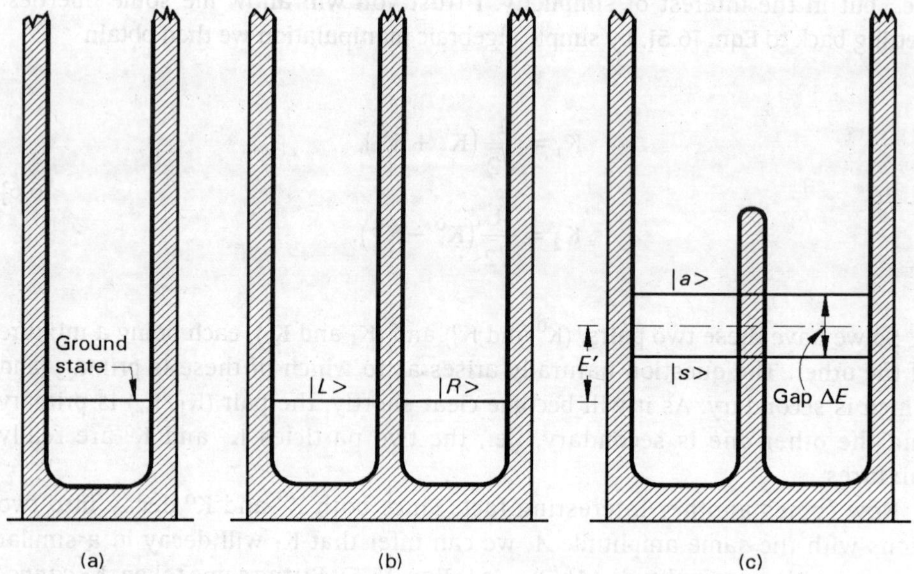

Fig. 6.6 A recap of quantum tunnelling. (a) shows a well that is infinitely deep. (b) shows two such wells in close proximity. However, because of the infinite barrier, the two levels $|L>$ and $|R>$ are totally unaffected. When the barrier is reduced as in (c), the states $|L>$ and $|R>$ mix to produce new states $|s>$ and $|a>$ as shown.

amount E'. Next, two levels appear, separated by an energy gap ΔE. And finally, the new states $|s>$ and $|a>$ are related to the old by:

$$|s> = \frac{1}{\sqrt{2}}\{|L> + |R>\},$$
$$|a> = \frac{1}{\sqrt{2}}\{|L> - |R>\}. \qquad [6.7]$$

Working backwards, we get

$$|L> = \frac{1}{\sqrt{2}}\{|s> + |a>\},$$
$$|R> = \frac{1}{\sqrt{2}}\{|s> - |a>\}. \qquad [6.8]$$

Suppose a particle is put in the left well at time $t = 0$. What happens to it later? Quantum mechanics has the answer.

Let $|\psi(t)>$ denote the state of the particle at time t. Of course

$$|\psi(0)> = |L> = \frac{1}{\sqrt{2}}\{\,|s> + |a>\}\,.$$

As time passes, the particle is partly in the state $|L>$ and partly in $|R>$. You will recall that quantum mechanics permits this kind of odd behaviour by means of the superposition principle. In short, the probability of finding the particle either in the left or the right well at various times would oscillate as sketched in Fig. 6.7(a). If the particle is radioactive and can decay, the probability will get modified as in Fig. 6.7(b).

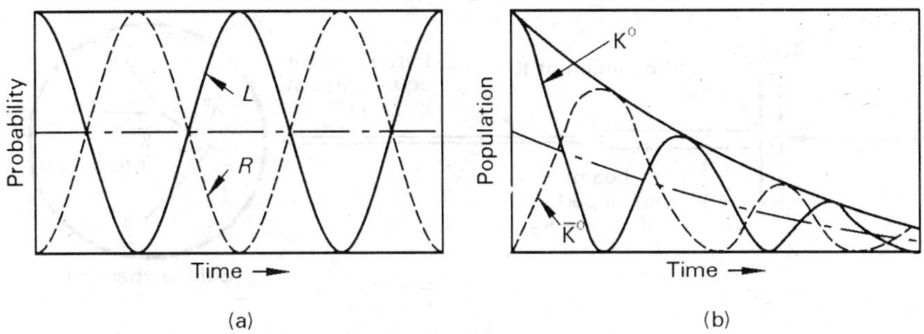

Fig. 6.7 Suppose a particle is dropped at time $t = 0$ into the left well in the complex shown in Fig. 6.6(c). Due to quantum effects, the probability of finding the particle in the two wells L and R will oscillate in time as shown in (a). If the particle itself decays, the probability curve will dampen. This situation is relevant to neutral kaons, and is sketched in (b). Further explanations can be found in the text.

By now you are possibly imagining all sorts of things that could happen to the K mesons. Well, if you have been thinking like that, then I should say that you have got the hang of what research is all about!

6.7 Back to kaons

We now suppose that to start with, the particles K^0 and $\overline{K^0}$ are stationary states of the Hamiltonian $H_{S.I.}$. We next suppose that the perturbing Hamiltonian $H_{W.I.}$ mixes these to produce K_1 and K_2. Based on what was discussed in the previous section, we could then make the following analogies:

$$K^0 \longrightarrow |L> \qquad K_1 \longrightarrow |a>$$

$$\overline{K^0} \longrightarrow |R> \qquad K_2 \longrightarrow |s>. \qquad [6.9]$$

Just as a particle dropped into one well rattles back and forth between the two wells, the K^0 particles can likewise change back and forth between the $|K^0>$ and the $|\overline{K^0}>$ states. If at time $t = 0$ we start with a pure K^0 beam, then as the beam travels, it would lose its purity, becoming a mixture of K^0 and $\overline{K^0}$; the relative contents would in fact oscillate; this is referred to as *hypercharge oscillation*.

Is all this true? Yes, because it has all been tested experimentally. The simplest experiment to perform is the following: start with a pure K^0 beam. In reality, this is a mixture of K_1 and K_2; but the former is short-lived (it decays into pions). Therefore downstream, it would all be K_2 and no K_1 at all. But K_2 is made up of K^0 and $\bar{K}^0$ — see Eqn. [6.6]. Thus, though we start with a pure K^0 beam, down the line $\bar{K}^0$ mesons join to give it company! Experiments show that this *does* happen (see Fig. 6.8).

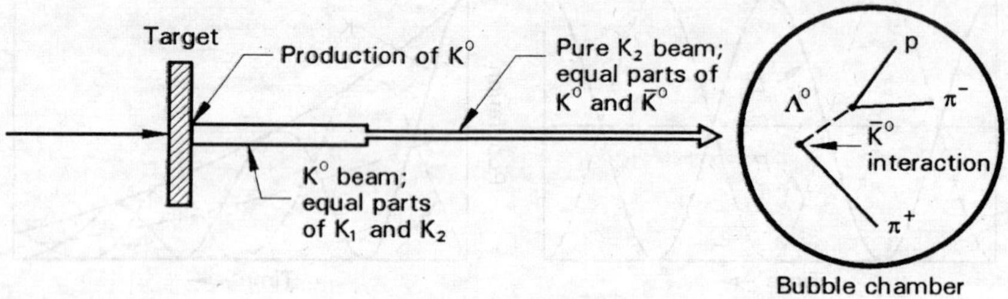

Fig. 6.8 This shows how the process illustrated in Fig. 6.5 is experimentally tested. The presence of $\bar{K}^0$ is detected using a bubble chamber. $\bar{K}^0$ can produce a Λ^0 particle which leads to a V-like prong. (Λ^0 is the V-particle of cosmic ray physics, and is produced by the interactions of $\bar{K}^0$ with nuclei.)

So $\bar{K}^0$ appears later, even if one starts with a pure K^0 beam. But what about the oscillations in the populations of K^0 and $\bar{K}^0$? Those too have been observed; it is all like what the textbooks on QM say about ammonia maser, benzene resonance, etc.; only, this time it is about exotic mesons. As Feynman observes:

> It is not based on elegant hocus-pocus ... We have taken the superposition principle to its ultimate logical conclusion ... It works!

Is that all there is to it? Not quite!

6.8 The fall of *CP* invariance

Let us go back to the $K^0 \leftrightarrow (K_1 + K_2)$ dichotomy. We start with a K^0 beam. This really is a mixture of K_1 and K_2. So, as the K^0 beam travels, there are really two beams going together, one of K_1 and the other of K_2. However, on the way, all the K_1's disappear due to decay into pions and thereafter the K_2 beam travels alone.

Now the rule book says that K_2 cannot decay into pions. I made this point earlier in passing but would now like to dwell on it a bit. Observe firstly that

$$P|K^0\rangle = |K^0\rangle$$

$$C|K^0\rangle = |\bar{K}^0\rangle.$$

$$\text{Hence } CP|K^0\rangle = |\overline{K}^0\rangle. \qquad [6.10]$$

Recall now Eqn. [6.6]. This gives

$$CP|K_1\rangle = \eta_{CP}|K_1\rangle = |K_1\rangle, \qquad [6.11]$$

which implies that the eigenvalue of the *CP* operator, viz., η_{CP} equals unity. Similarly,

$$CP|K_2\rangle = -|K_2\rangle, \qquad [6.12]$$

showing that $\eta_{CP} = -1$ in this case. We further have

$$CP|\pi\pi\rangle = |\pi\pi\rangle, \qquad [6.13]$$

implying $\eta_{CP} = 1$ here.

Now let us reflect on all this. The states $|K_1\rangle$ and $|\pi\pi\rangle$ are both associated with $\eta_{CP} = +1$. Hence the decay $K_1 \longrightarrow 2\pi$ is allowed. By contrast, if *CP* is conserved, we cannot have K_2 decaying into two pions because $\eta_{CP} = -1$ for K_2 and $+1$ for 2π.

In June 1964, Cronin and Fitch found that K_2 *did* decay into two pions though only occasionally (about two in one thousand cases), which meant that *CP* invariance did not hold. *CP* invariance was clearly violated in kaon decay, though rather feebly.

6.9 Enter K_S and K_L

Earlier we had K_1 and K_2 as the pure particles out of which came the mixtures K^0 and $\overline{K}^0$ — that was in the good old days when *CP* invariance still reigned. But Cronin and Fitch showed (effectively) that K_1 and K_2 also could mix, i.e., they were not as pure as one had thought. In a sense, K_1 and K_2 then came out of our belief in *CP* invariance. Once *CP* invariance was knocked out, there was only one invariance left intact — the *CPT* invariance. One necessarily had to lean on that. When that was done and the entire problem was analysed, it emerged that two new particles K_L and K_S replaced K_1 and K_2. These are defined by:

$$K_S = A\{(1+\varepsilon)K^0 - (1-\varepsilon)\overline{K}^0\},$$

$$K_L = A\{(1+\varepsilon)K^0 + (1-\varepsilon)\overline{K}^0\}, \qquad [6.14]$$

where

$$A = \frac{1}{2\left(1+|\varepsilon|^2\right)^{1/2}}. \qquad [6.15]$$

Earlier we had

$$\begin{pmatrix} K^0 \\ \overline{K}^0 \end{pmatrix} \rightarrow \begin{pmatrix} K_1 \\ K_2 \end{pmatrix};$$

now there is a further extension

$$\begin{pmatrix} K_1 \\ K_2 \end{pmatrix} \rightarrow \begin{pmatrix} K_L \\ K_S \end{pmatrix}.$$

The complex quantity ε represents the effect of a tiny *CP* failure. Experiments have shown that

$$\varepsilon \sim 1.6(1 + i) \times 10^{-3}.$$

6.10 An optical analogy

This whole business of K mesons can be understood through a beautiful analogy to the behaviour of polarised light. To start with, we recall that light can be either linearly or circularly polarised (see Fig. 6.9). Further, the two types are related via the relations

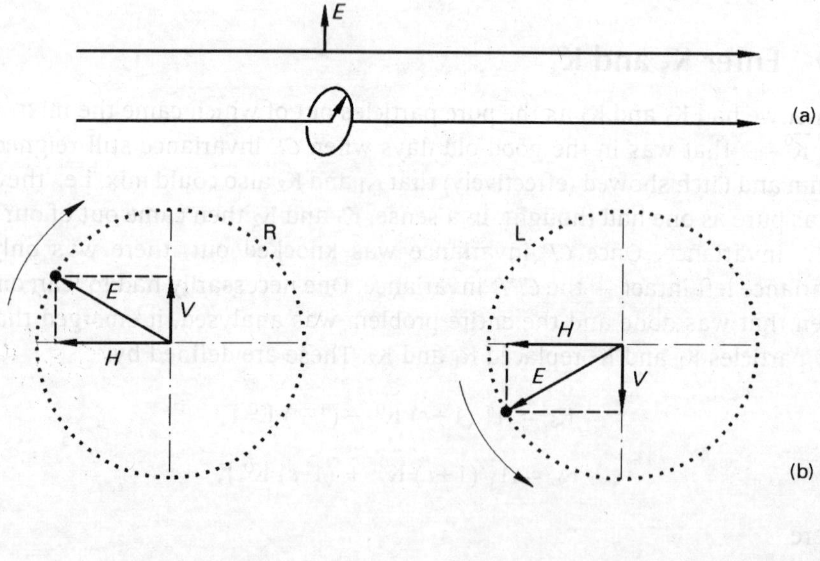

Fig. 6.9 (a) shows linearly and circularly polarised light while (b) indicates the relationship between circular and linear polarisations. In other words, circular polarisation can be regarded as the superposition of two orthogonal linear polarisations with a suitable phase relationship between them.

$$|R> = \frac{1}{\sqrt{2}}\{|H> - |V>\},$$

$$|L> = \frac{1}{\sqrt{2}}\{|H> - |V>\}. \qquad [6.16]$$

Next we make the analogies

$$|H> \rightarrow K_1, \qquad |V> \rightarrow K_2,$$

$$|R> \rightarrow K^0, \qquad |L> \rightarrow \bar{K}^0. \qquad [6.17]$$

Now circularly polarised light can be converted to a state of linear polarisation by transmission through a suitable filter as illustrated in Fig. 6.10.

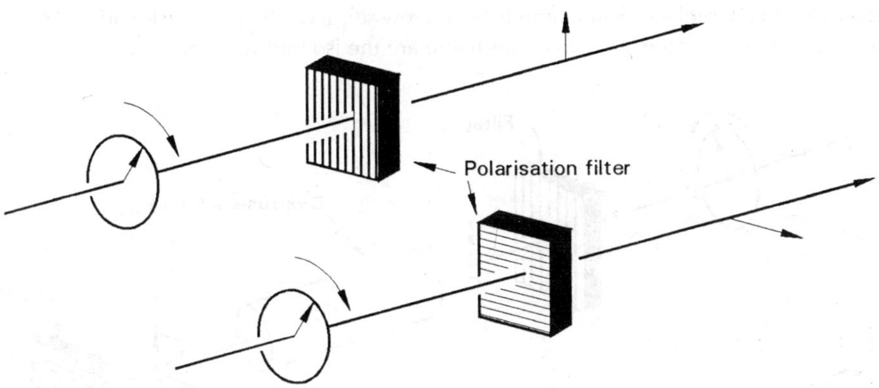

Fig. 6.10 As we saw in the previous figure, circular polarisation is equivalent to the superposition of two linearly-polarised components perpendicular to each other. By placing a polarisation filter in the path of the beam, one of the two components can be eliminated, as this figure shows.

Next we consider the rotation of the plane of polarisation as light traverses what is called an optically-active medium. Such media are made up of molecules which have a screw sense as illustrated in Fig. 6.11. Just as we have right- and left-handed screws, we can have molecules of opposite *chirality* (spiral sense). Understandably, they turn the polarisation vector in opposite directions. Molecules which rotate the polarisation of light to the right are called *dextro-rotary* while those which do it the other way are termed *levo-rotary*.

Against this background, consider the progress of the light beam in Fig. 6.12. It starts off in a circularly-polarised state, goes through various other states and finally ends up in a linearly-polarised state. No doubt it goes through many changes, but every one of these is readily understood. Also sketched in Fig. 6.12 is the fate of a K^0 beam and one can see parallels. To start with, we have just the K^0's and that is like having a right-circularly polarised light beam (recall

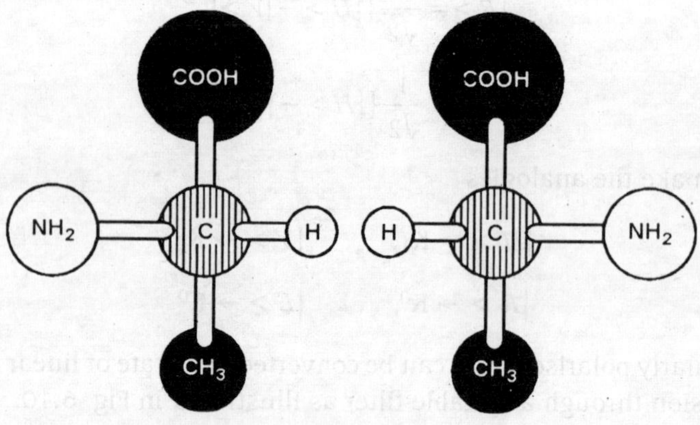

Fig. 6.11 Many molecules in nature have a screw sense or chirality. Such pairs of molecules are called optical isomers. Shown in this figure are the isomers of alanine.

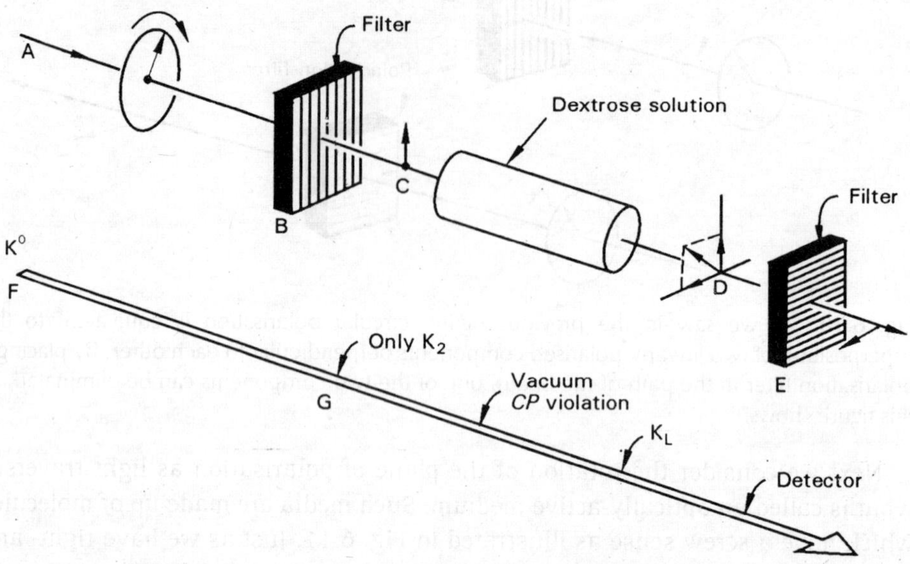

Fig. 6.12 This figure offers an analogy between the behaviour of polarised light and the behaviour of K^0 mesons, with *CP* violation being operative. Explanations in the text.

Eqn. [6.5]), which tells us that K^0 is really a mixture of K_1 and K_2. Downstream, there is a selective disappearance of K_1 (via two-pion decay), and this is like passing the circularly polarised light beam through a polarisation filter. What remains then of the original K meson beam is just K_2, and that is like linearly-polarised light. Thereafter, while travelling further, the K_2 beam acquires a small contamination of K_1 — this admixture is due to *CP* failure. The corresponding step in optics is the rotation of the plane of polarisation by an optically active medium. Thus this medium mimics *CP* failure. Out of this failure are born K_S and K_L of which K_L survives. This K_L is really our old friend K_1 in a new avatar. It undergoes two-pion decay which is what Cronin and Fitch detected, and for that they were rewarded with the Nobel Prize.

Notice that *CP* failure adds only a small amount of K_1. Hence the population of K_L is also small — but it *is* there, and that is what really matters. And it is there only because of *CP* failure.

Question: "*CPT* is conserved, but there is *CP* failure. Does it mean that there must be *T* failure as well?" Yes, if we believe in *CPT* invariance, which we do. Unfortunately, *T* failure is quite difficult to establish experimentally. Many experiments have been performed but there is no unambiguous experimental evidence as yet for *T* failure.

6.11 Chapter summary

- This chapter is almost entirely about neutral K mesons, and the story of *CP* failure.
- To start with, we revisited *C*, *P* and *T* and their various combinations.
- Following Wigner, we saw how we could use magic mirrors to describe what happens. We counted seven such mirrors, including one for *CPT*.
- Each mirror defines a transformed sequence of events corresponding to those in the 'direct' or real world.
- The transformed sequence is a reality if the corresponding transformation is conserved, i.e., an experiment can be done to actually observe the transformed sequence.
- If the transformed sequence is ruled out by Nature, then we say that the corresponding transformation fails.
- Experiments have shown that *P* fails in W.I.
- Working with neutral kaon beams, Cronin and Fitch showed that *CP* also fails in W.I.
- It is believed that *CPT* cannot fail — that is the Luders-Pauli theorem.
- Since *CP* fails, and since *CPT* is conserved, it follows that *T* invariance must fail.
- However, the failure of *T* invariance is difficult to establish experimentally.
- The kaons play a Dr. Jekyll and Mr. Hyde game. If *CP* is conserved, then this game can be neatly understood via the QM superposition principle.
- This is almost true, except that *CP* failure spoils things just a wee bit.

The kaon story is briefly as follows. Thanks to the K^0 to $\overline{K}^0$ conversion and the superposition principle, a pure K^0 beam can get transformed into a mixture of K_1 and K_2. K_1 decays into pions but if *CP* is conserved, K_2 is not supposed to. However, in about two cases out of a thousand, K_2 does decay into pions, which means that there is a weak *CP* failure.

There remain the questions: "Why is *CP* violated? And, why is this violation so feeble?" Concerning these, Cronin said in his Nobel acceptance address:

> We must continue to seek the origin of *CP* violation by all the means at our disposal ... We are hopeful that at some epoch, perhaps distant, this cryptic message from Nature will be deciphered.

And so the search goes on!

Appendix to Chapter 6

Some care is needed in going over from the textbook tunnelling problem discussed in Section 6.6 to the case of the K mesons. Firstly, we note that of the two states $|s\rangle$ and $|a\rangle$ born out of $|L\rangle$ and $|R\rangle$, the state $|s\rangle$ is lower in energy. It is in fact the ground state of the perturbed system, whereas $|a\rangle$ corresponds to the excited state of the system. Now excited states have finite lifetimes. Remembering this, we carefully make the following identifications:

$$|L\rangle \longrightarrow |K^0\rangle \qquad |R\rangle \longrightarrow |\bar{K}^0\rangle,$$

$$|s\rangle \longrightarrow |K_2\rangle \qquad |a\rangle \longrightarrow |K_1\rangle.$$

Notice that it is K_2, the longer-lived particle, that is identified with the symmetric state and not K_1. In the presence of *CP* violation, K_2 transforms to K_L as discussed in the chapter.

7

Big Money for Small Particles

This chapter offers a glimpse of the fascinating world of charged-particle accelerators or 'machines', as they are often called, used in high energy physics experiments. Also briefly described are the detectors used in such experiments. The contents of this chapter will help one appreciate better some of the exciting discoveries that are described later.

7.1 The basic requirements

You are probably wondering how physicists probe deep into the nucleus/ nucleon — recall Figs. 1.1. and 1.2. In this chapter, your curiosity will be satisfied.

Suppose we want to measure the length of a football field. We would use a long tape graduated in centimetres — centimetre accuracy would be fine in this case. Next, let us say that we are interested in the diameter of a nail. Here we would use a screw gauge with a least count of a fraction of a millimetre. If we want to measure the size of a bacteria, we would have to use a microscope with sufficient magnification. What I am driving at is that, as we start looking at smaller and smaller objects, we must improve the resolution.

In particle physics, matter is usually probed by shooting projectiles at an appropriate target. A particle of momentum **p** has a de Broglie wavelength λ ($=\hbar/\mathbf{p}$) associated with it. The smaller the region we wish to probe, the smaller must be the value of λ; in a sense, the value of λ determines the resolution. Now, small λ in turn implies that **p** must correspondingly be large and hence also the energy of the projectile. So, in order to probe deep into the nucleus/nucleon, one must first have a machine that can boost the energy (and therefore also the momentum) of the probe particle. Machines which do this are called (particle) accelerators.

7.2 Electrostatic acceleration

Particle accelerators made their first appearance in the early thirties; since then, accelerator technology has made phenomenal progress, reaching almost mind-boggling proportions. Particle accelerators can accelerate only charged particles, and they do this by employing electric and magnetic fields in ingenious combinations. No means are available for accelerating neutral particles like the neutron. However, it is possible to use appropriate reactions to produce beams of neutral particles; this is how neutrino-induced reactions are studied. Later, we shall consider a few such reactions.

In the very first machines, the energy attained was only about 1 MeV or so. In about sixty years, this figure has multiplied a million-fold, a truly fantastic increase. Readers who have accompanied me through this series will recall a brief reference to accelerators in ***WATT***.

Before we start talking about the different types of accelerators, there is one point you should appreciate. Consider an electron of energy 1 TeV. As you will soon see, it requires a very big machine consuming an *enormous* amount of energy to produce electrons of this energy. If we express 1 TeV in CGS units of energy, i.e., in ergs, the electron energy is ~1.6 ergs. Consider now a mosquito moving at a speed of 10 cm/sec. Let us say the mass of the mosquito is 0.1 gm. The kinetic energy (1/2) mv^2 is then ~5 ergs. In other words, the kinetic energy of a 1 TeV electron is less than that of a mosquito! So why are particle accelerators such great guzzlers of energy? I leave you to figure that out!

The basic principle underlying a (charged) particle accelerator is quite simple, as can be seen from Fig. 7.1 (see also, if possible, Fig. 2.1 in ***WATT***). Suppose we place an electron on the bottom condenser plate — see Fig. 7.1(a). The electron will be attracted towards the top plate and in reaching it, would acquire an energy of 1 eV. So that is all that is required to accelerate a charged particle, an electric field. Many accelerators use electric and magnetic fields. In all cases, the energy acceleration is *always* due to the electric field; the magnetic field plays other important roles but it can never accelerate a charged particle in the sense of increasing its speed.

The principle of the acceleration of a charged particle is thus quite simple — all one needs is an electric field. But in practice there are many difficulties, and the simple condenser-type system cannot be used beyond a few MeV at best — see Fig. 7.1(b). What does one do then, especially if one wants to go to the TeV region? The answer is, "Give resonant kicks".

7.3 The linac

The trick just mentioned is quite simple conceptually. Suppose we have a swing and a person on one side regularly gives it a gentle push at the right time (as sketched in Fig. 7.2). As you know, the amplitude of oscillation and thereby the energy of the swing would steadily increase. In a resonance-type accelerator, this is what one essentially does — accelerate the particle by giving it controlled doses of energy increases at the right time. In each dose the energy increase is small, but then little drops of water make a mighty ocean, don't they?

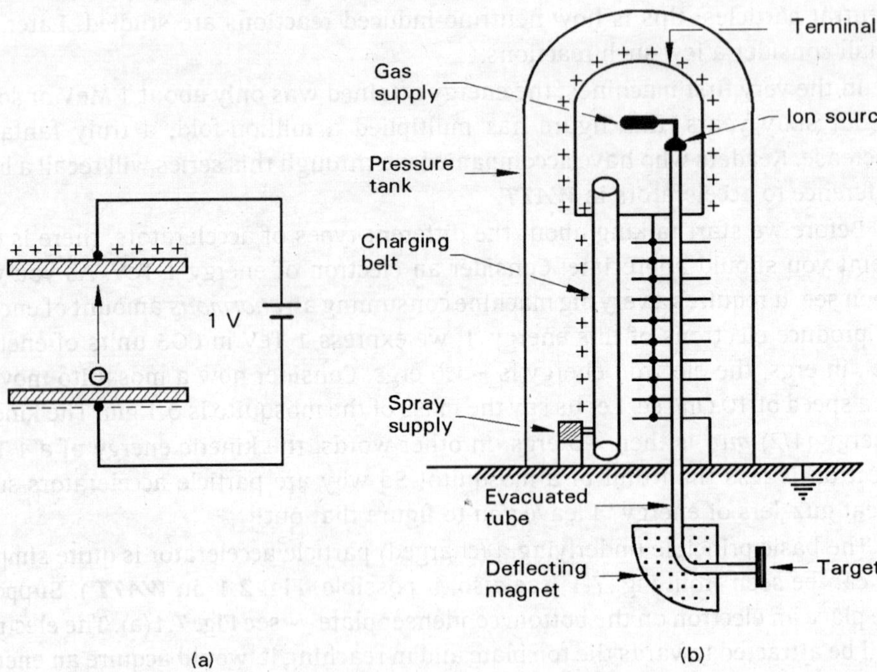

Fig. 7.1 (a) shows an ideal parallel plate condenser arrangement, with a potential difference of 1 V. An electron placed on the bottom plate would be accelerated upwards by the electric field, and arrive at the top plate with an energy of 1 eV. (b) shows the schematic of what is called a van de Graaf accelerator. Here a potential difference is created between the terminal and the ground. The terminal is charged using a conveyor belt that carries a charge from the bottom; this charge is sprayed onto the belt. A potential of upto 5–6 MV can be built this way. To prevent electrical breakdowns, the terminal is placed in a pressurised tank filled with insulating gas (usually SF_6). The potential is used to accelerate positive ions (usually p or α) down the evacuated tube to the target. A machine of this type exists in BARC. A more modern version of the van de Graaf is the Pelletron. There are three Pelletrons in India, in Mumbai, Delhi and Bhubaneshwar.

There are two major types of resonance accelerators, the linear and the circular. The schematic diagram of the linear accelerator is illustrated in Fig. 7.3. Consider Fig. 7.3 (a). Suppose an electron is placed at A; it will then be accelerated towards B. At B it enters the drift tube 1. The inside of the drift tube is a field-free region, where the electron does not receive any further acceleration. Therefore the electron moves through the tube with a constant velocity, say v, which it acquired while moving towards B.

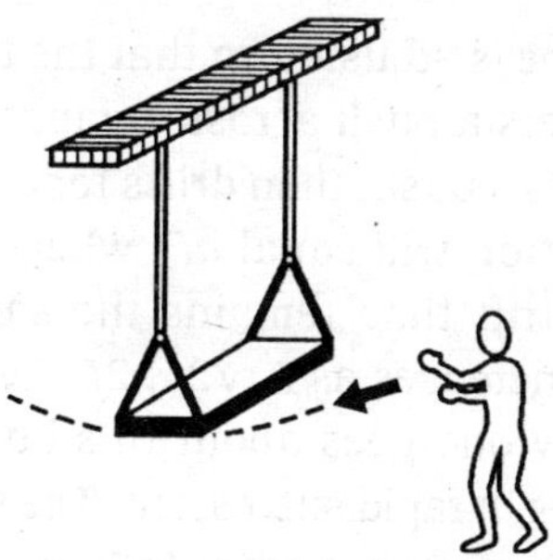

Fig. 7.2 Increasing the amplitude of a swing by giving it resonant pushes.

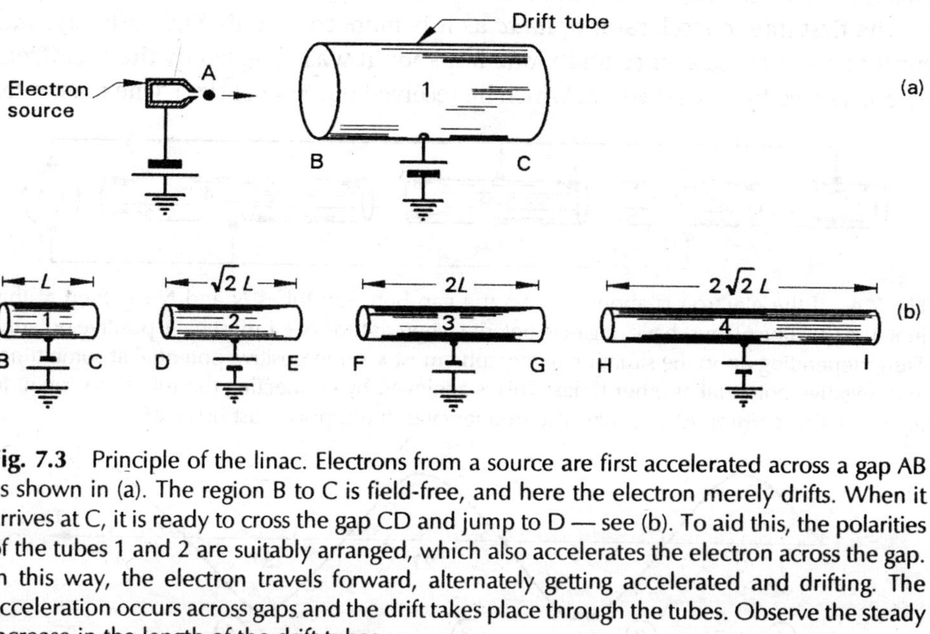

Fig. 7.3 Principle of the linac. Electrons from a source are first accelerated across a gap AB as shown in (a). The region B to C is field-free, and here the electron merely drifts. When it arrives at C, it is ready to cross the gap CD and jump to D — see (b). To aid this, the polarities of the tubes 1 and 2 are suitably arranged, which also accelerates the electron across the gap. In this way, the electron travels forward, alternately getting accelerated and drifting. The acceleration occurs across gaps and the drift takes place through the tubes. Observe the steady increase in the length of the drift tubes.

Now consider the electron just as it arrives at C (with a velocity v). The time T for traversing through the drift tube 1 is given by (L/v). Suppose at the instant the electron arrives at C the polarities of tubes 1 and 2 are arranged as in Fig. 7.3 (b). This implies that the polarity of tube 1 must be reversed compared to what it was in (a). The electron will then be accelerated in going from C to D. Since the potential jump is the same as before, the electron energy will now be doubled and the velocity would increase to $\sqrt{2}v$. Suppose the length of the tube 2 is $\sqrt{2}L$. Then the transit time through tube 2 equals $\sqrt{2}L/\sqrt{2}v = (L/v) = T$. The electron now arrives at E and simultaneously we arrange conditions so that the potential of tube 2 is now negative while tube 3 is at a positive potential. As the electron goes from E to F, its energy again increases; this is followed by a drift through tube 3.

Once again the length of this tube is adjusted so that the transit time is T, and so on. In a nutshell, the electron gets a push across a gap, then drifts for a time T, arrives at the next gap, again gets a push, then drifts for a time T, and so on. After n pushes, the energy of the electron will equal nE, where E is the energy gained across one gap. Although the drift time remains the same after each push or acceleration, the drift distance increases as L, $\sqrt{2}L$, $2L$, $2\sqrt{2}L$, ..., etc.

You might be wondering how one goes about this business of reversing the polarities of the various drift tubes in rapid succession. The scheme used is sketched in Fig. 7.4. The idea is to use alternating current (AC) and link the alternate drift tubes as shown. The voltages then switch back and forth as explained in Fig. 7.5, facilitating a sequence of resonant pushes after equal time intervals T.

The first linear accelerator or *linac* as it is more commonly known today, was built in 1931 by Lawrence and Slone but soon it was eclipsed by the cyclotron, also invented by Lawrence (for which he received the Nobel Prize). One reason for

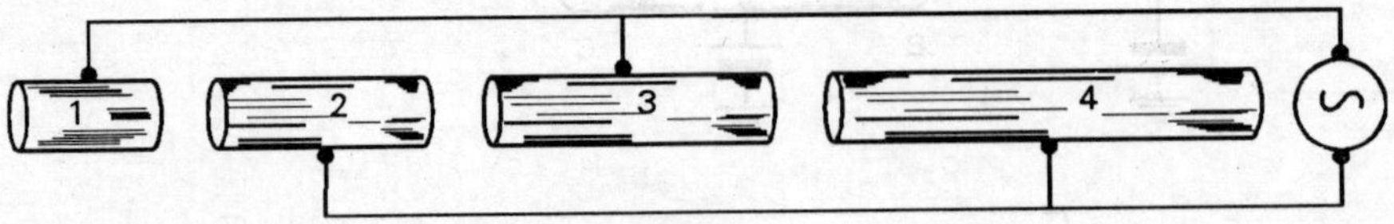

Fig. 7.4 If the electron is about to cross the gap between tubes N and $N+1$, then at that juncture the tube N must have negative polarity whereas tube $N+1$ must have positive polarity. Thus, depending upon the situation, every tube must assume positive potential at some times and negative potential at other times. This is achieved by connecting the tubes as shown, to an alternating current (AC) supply. The frequency of the supply must be $(1/2T)$.

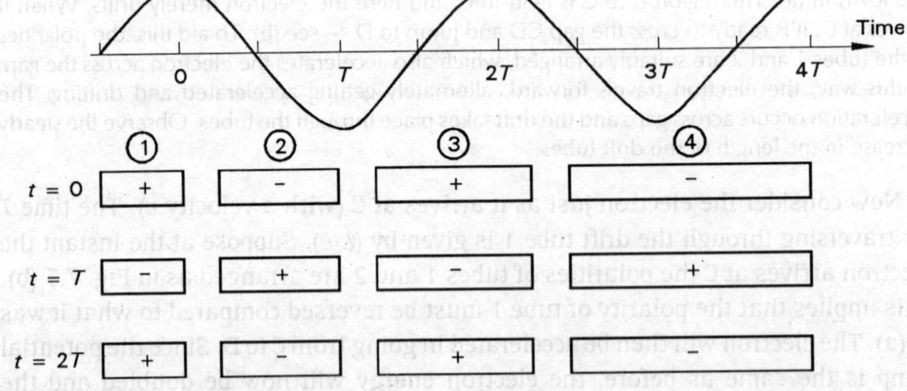

Fig. 7.5 This figure amplifies Figs. 7.2 and 7.4 by showing how the AC affects the potential of the drift tubes at various instants of time. At the top is shown the time variation of the potential; observe that the potential switches polarity after every T seconds. Below the voltage waveform are sketched the polarities of the various drift tubes at different instants of time. Note that T refers to the transit time of the electron through the drift tube, and not the period of the AC; the latter is equal to $2T$.

this eclipse was that efficient and powerful radio frequency (RF) sources needed for building large linacs, were not available at that time. After World War II the scene changed completely, thanks to the development of the radar, which in turn made available strong RF sources and other RF systems. The linac then came into its own, and at Stanford University, linac technology has been pushed to astounding limits. In fact, the Stanford Linear Accelerator Centre (SLAC) boasts of the world's longest linac — it is over 3 km long and produces electrons of over 30 GeV. It is popularly known as the two-mile accelerator.

What about building still longer accelerators, leading to still higher energies? Well, there are many problems (starting with money of course) and one of these is that the drift tubes become longer and longer with increase of energy. Further, at some stage there would be problems with the curvature of the Earth. However, to do experiments at higher energies, one does not necessarily have to make the linac still longer. There is a ingenious trick known as *colliding beams*, but that I shall discuss later.

7.4 The synchrotron

We now turn to cyclic machines, i.e., accelerators, in which the electron has circular motion. The earliest machine of this kind is the cyclotron which has already been mentioned. In this book we are interested in energies much higher than can be achieved with cyclotrons. Such energies are attained using *synchrotrons*, which are also cyclic accelerators. The principle of the synchrotron was independently discovered by Vekslar in Russia and MacMillan in America.

A schematic diagram of the synchrotron is shown in Fig. 7.6. The electron moves in a *fixed* circular path along which it is guided by what are called *bending magnets*. Although to understand the principle of the machine we shall consider the behaviour of one electron, in a real machine there will be bunches of electrons, each bunch containing as many as 10^{10} or more electrons. When dealing with a group of electrons like this, one has to worry about *focusing*, as otherwise the electrons would spread themselves all over the place. Thus, as in optics where one always deals with a bundle of rays, one has to take care in the focusing of the particle beam, which is done with special focusing magnets.

Before we get down to the principle of the synchrotron, we should remind ourselves of the following basic fact. Suppose an electron moving with a velocity v is subjected to a magnetic field H applied perpendicular to the direction of motion as sketched in Fig. 7.7. The electron would then move in a circular path of radius r, say. The magnetic force acting on the electron is equal to the centripetal force, the force equation being

$$\frac{evH}{c} = \frac{mv^2}{r}. \qquad [7.1]$$

Hence,

$$\frac{eH}{mc} = \frac{v}{r} = \omega, \text{ say}. \qquad [7.2]$$

The quantity ω has the dimension of frequency and is called the *cyclotron frequency*. Observe from the last equation that if r is kept fixed but v is increased (i.e., the energy of the electron is increased), then H also has to be correspondingly increased.

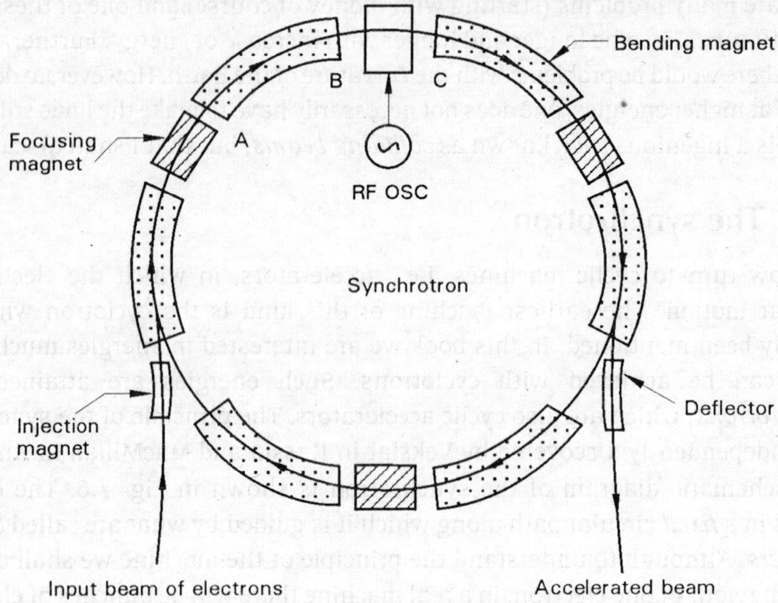

Fig. 7.6 Schematic diagram of the synchrotron. The particles enter as shown and thereafter keep moving on a near-circular track. Every time the particle goes from B to C, its energy increases by ΔE. Despite this energy increase, the particle moves on the same circular track because the field provided by the bending magnets is also increased correspondingly. When the particles have been accelerated to a sufficiently high energy, they are extracted using a deflector. They may then be led suitably to the experimental station. Both electron and proton synchrotrons are known; naturally, these machines can also handle e^+ and $\bar{p}$ respectively.

Let us get back to Fig. 7.6 and the principle of the synchrotron. We start with an electron at A where it has been injected with an energy E_0. Due to the action of the magnetic field, the electron is steered along the circular path and let us assume it has arrived at B. Here we arrange for a quick energy boost of ΔE, so that the electron arrives at C with energy $E_0 + \Delta E$. The magnets now take over and steer the electron back to B via A, all around the circle. Once at B, we again arrange for an energy increase ΔE; as a result, when the electron now arrives at C, its energy would be $E_0 + 2\Delta E$. The circular journey CAB starts all over again; once again there is an energy increase in going from B to C, and the electron energy now goes up to $E_0 + 3\Delta E$. In this way, after each circular trip, the electron energy keeps on increasing in steps of ΔE.

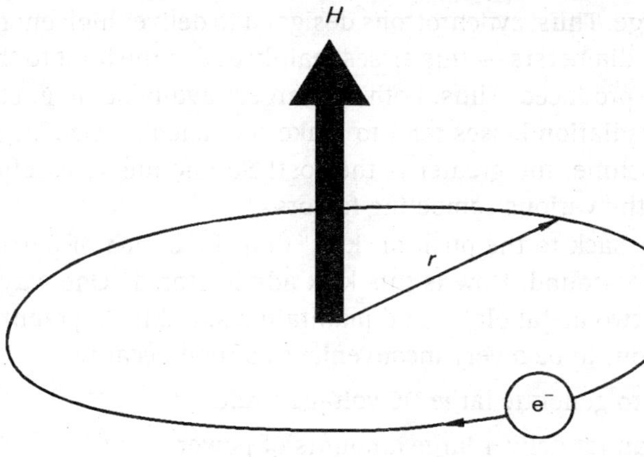

Fig. 7.7 Circular motion of an electron under the influence of a magnetic field H.

A few subtle points must be noted at this stage. First is that as the electron energy increases, so does its velocity v. Thus, from Eqn. [7.2] it becomes clear that the magnetic field must be correspondingly enhanced in order to keep the electron moving along the same circle. In other words, after every circular trip, the magnetic field too must be given a suitable jump. Secondly, since the velocity increases after each energy boost, the time taken for the circular trip gets reduced.

When the desired energy is reached, the bunches of particles are ready for extraction. The beam is suitably extracted and steered away to some other spot for experimentation. Synchrotrons have been built to accelerate electrons as well as protons. Several points need to be appreciated here. Firstly, when a charged particle moves in a circular orbit, it experiences acceleration, and an accelerated charged particle radiates energy. The radiation emitted by charged particles in a synchrotron is called synchrotron radiation. I have made a brief reference to this in the volume companion ***Relativity***. For a given energy, the synchrotron emission is stronger for electrons than for protons. Now synchrotron emission is sometimes considered beneficial and there are in fact many synchrotron radiation facilities in the world (one of these is nearly operational at the Centre for Advanced Technology, Indore). But in high energy physics, energy loss due to synchrotron radiation is a nuisance and for electrons it becomes very serious as one goes to higher and higher electron energies. In the case of protons, at the currently attainable energies, the loss is not yet that serious. One way to minimise losses due to synchrotron radiation is to make the circular orbit as large as possible (can you figure this one out?).

The next point relates to the requirement of the magnetic field. Since the circular orbit in which the particle moves is fixed, the magnetic field needed to keep the particle in orbit progressively increases as the particle energy increases. But then, there is a limit to the highest field one can achieve; this is the limitation of available technology. However, if the radius of the circle is large, the field required is not large. Thus, synchrotrons designed to deliver high energy particles tend to have large diameters — this arises mainly due to the limit to the magnetic fields that can be produced. Thus, both the largest available magnetic field and the synchrotron radiation losses tend to make the machine size large; but then the larger the machine, the greater is the cost! So one has to carefully strike a balance between the various competing factors.

Let us now get back to the push or 'kick' that the bunch of particles receive every time they go around. How is this kick administered? One way of doing it would be to have two metal plates and maintain a suitable DC potential between them. This turns out to be a very inconvenient method because

(a) it is not easy to generate large DC voltages, and

(b) DC systems cannot deliver large amounts of power.

One therefore uses a suitable high power RF source.

Table 7.1 gives some of the important characteristics of the large proton synchrotron at the Fermi National Accelerator Laboratory (FNAL) in America.

Table 7.1 Characteristics of the proton synchrotron at FNAL

Item	Value
Maximum proton energy	500 GeV
No. of protons per second	$\sim 6 \times 10^{12}$
Repetition rate (pulses/sec)	1/12
Ring diameter	2000 m
Steel weight	9000 tons
Copper weight	850 tons
Power input to magnets	36 MW
Cost	~ $ 250 million

Remark

The above figures pertain to Mark I of the machine, completed in 1972. Since then, the machine has been considerably upgraded. Currently, the beam energy is 1 TeV. Further, facilities for (p, $\bar{p}$) collisions have been provided, corresponding to a collision energy of 2 TeV.

7.5 Fixed target beams v/s colliding beams

What does one do with a beam of accelerated charged particles? Simple, allow the particles to smash against a suitable stationary target, and then watch what happens. This might seem like a casual approach but in practice, this is pretty much what is often done. Of course, one starts by defining as clearly as possible what one wants to study. Next, choosing the right particle to accelerate — often the choice is restricted by the machine that is available. Then one decides the energy to which the particle is to be accelerated — this again is decided by the machine. Lastly, one designs a detector system suited to see what one wants or hopes to see. When the charged particles in the accelerated beam collide with the particles in the stationary target, all kinds of reactions take place depending upon the conditions. One then selects, observes and studies the process of interest.

When an accelerated charged particle collides with a particle in a stationary target, it is like a moving car hitting a parked car. The energy released in the collision is referred to as *collision energy*. For several decades particle physics experiments involved such a scheme, i.e., stationary targets. If one wanted more collision energy, then one had no option but to increase the energy of acceleration. As time went on, it became increasingly difficult to accelerate the charged particle but the unfortunate part was that even after increasing the particle energy, the collision energy hardly increased (see Fig. 7.8). In fact, the collision energy E'

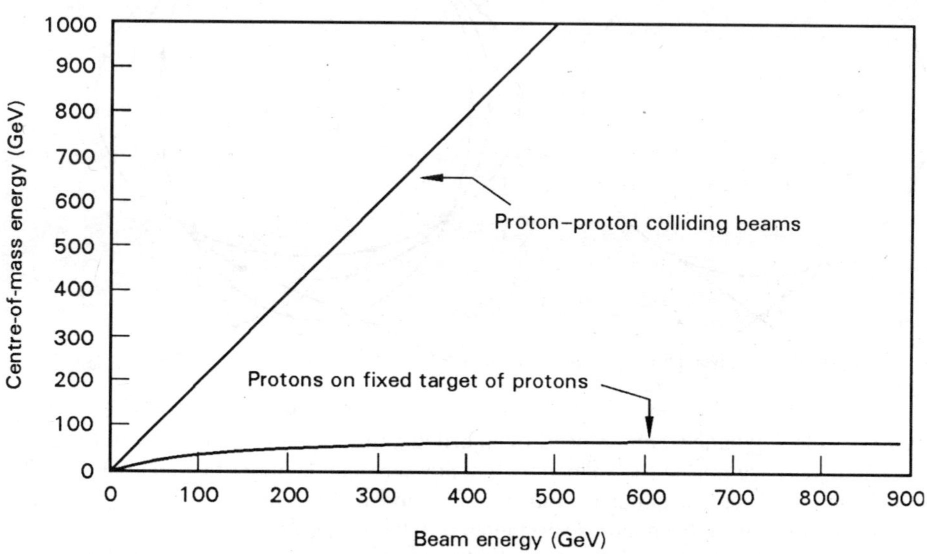

Fig. 7.8 This shows the variation of the centre-of-mass energy or the collision energy as it is often called, when a proton beam collides with (i) stationary protons, and (ii) protons (or anti-protons) coming from the opposite direction and with the same energy. Notice the tremendous gain in the case of colliding beams.

available when a particle of mass m and energy E collides with a stationary particle of the same mass, is proportional to $\sqrt{E}$.

Someone then came up with a bright idea, perhaps by watching car accidents! This bright genius mused, "Supposing two cars speeding in opposite directions were to collide head on; then one has a terrific smash. Why not do the same thing with charged particles?" And thus was born the idea of colliding beam machines. In this, one essentially arranges for a charged particle accelerated to energy E, to collide with its anti-particle, also accelerated to the same energy (using essentially the same machine). Further, the two particles are arranged to come from opposite directions. Does it improve matters? Yes it does, as can be seen from Fig. 7.8 — in fact there is a immense improvement since the collision energy is now proportional to $2E$. Clearly, colliding beams are worth it! Surely there must be a price to pay? Yes there is; try and figure it out!

7.6 Colliders

How does one go about this collision business? Figure 7.9 offers a schematic representation:

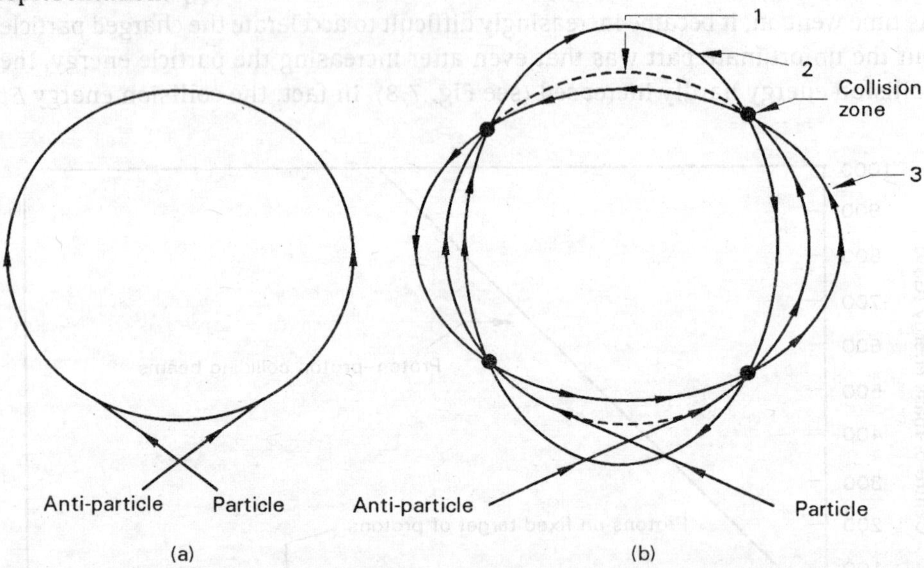

Fig. 7.9 Colliding beams are arranged by injecting particles and anti-particles of the same energy into a storage ring, in such a manner that they circulate in opposite directions. Usually, the colliding particles are either e^- and e^+, or p and $\bar{p}$. Of course the particles and the anti-particles must have been already accelerated to the desired energy — this is often done using a synchrotron. (a) shows an idealised scheme for the oppositely circulating particles. (b) spells out some more details. Here orbit 1 shows the ideal circular path while orbits 2 and 3 show the non-circular paths followed in practice by the particle and the anti-particle respectively. These two orbits intersect at select points, where the detectors are arranged to study the result of the collisions. The deviation from circularity is highly exaggerated for clarity.

This is more or less what people do all over the world. Let us say the charged particle used is a proton. One first accelerates the protons using a synchrotron. The bunches of accelerated protons are then fed into a *storage ring* where they keep going round and round. This is much like what happens in a synchrotron, except that in this case the particles do not receive energy boosts — in other words, the energy of the particle remains constant. Of course, ultra-high vacuum ($\sim 10^{-10}$ atmosphere at least!) is needed so that the bunches keep circulating for hours. Meanwhile, anti-proton bunches are prepared as discussed in Fig. 7.10 and these too are injected into the storage ring. Since the anti-protons have negative charge, they respond differently to the magnetic field, and end up circulating in the opposite direction. The design of the storage ring is such that the tracks of the protons and the anti-protons cross each other at specified places. This is where 'accidents' are expected, and scientists place complex observational equipment nearby to catch the debris of the accidents. They would be most disappointed if there were no accidents (one presumes that their expectations would be different when they are speeding in a car!).

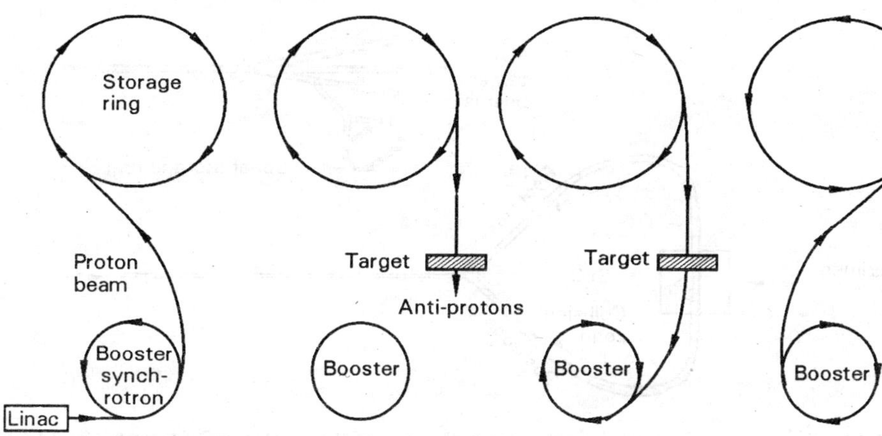

Fig. 7.10 This figure illustrates how p and $\bar{p}$ beams may be prepared and injected into a storage ring so that collisions occur. Protons are first boosted to the desired energy using a booster. The protons are then injected into the storage ring, where they circulate in a clockwise direction as shown in (a). A part of this circulating proton beam is extracted and allowed to impinge on a suitable target as shown in (b). This results in the production of anti-protons. The extraction is carried out till the required number of anti-protons have been produced. The remaining protons continue their circulation in the ring. The anti-protons are then injected into the booster and their energy is boosted as shown in (c). Observe that the anti-protons circulate in the opposite direction in the booster, as compared to the protons. The energy-boosted anti-protons are then injected into the storage ring from the opposite direction as shown in (d). The ring now has protons and anti-protons of the right energy circulating in opposite directions, and now all set for high-energy collisions!

Today, most of the giant machines are of the collider type. Of these, two deserve special mention. First is the Stanford Linear Collider (SLC) which is sketched in Fig. 7.11. This is an offshoot of the Stanford thrust in linacs which has been going on for nearly fifty years. The unique feature of the SLC is that there is no storage ring here. The electrons and the positrons cross each other just once; hits, if any, must occur then. Does this not reduce the collision probability compared to the storage ring? Yes it does, in a sense, but one can compensate by making the 'highway' narrow, so that the electrons and positrons have little option of escaping collisions. Why all this bother? Why not the good old storage ring? There is an answer of course and I suggest that you try and find it out — a hint: synchrotron radiation losses.

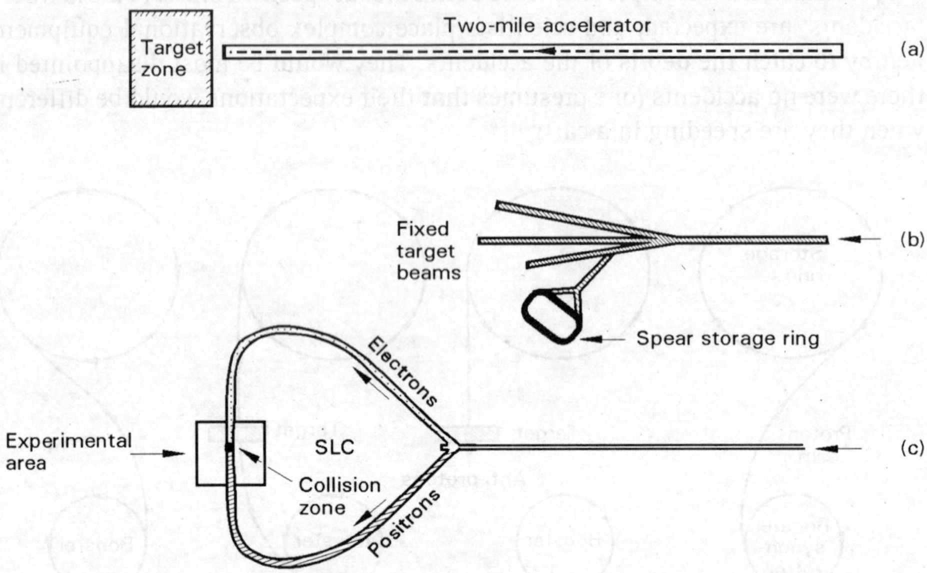

Fig. 7.11 This figure gives a brief schematic diagram of the evolution of the Stanford linac facility. First there was the two-mile accelerator, whose beam impinged on a fixed target as in (a). Later this arrangement was improved so that many experiments could be prepared, and the beam steered to the desired experiment. The advantage was that while one experiment was going on at one target station, other teams could get their experiments ready at the other stations. Also included was a small storage ring called SPEAR for producing $e^- e^+$ collisions — see (b). (c) shows the linear collider SLC.

Getting back to cyclic colliders, the world's biggest one is at CERN, Geneva (more about CERN soon). The storage ring is in an underground tunnel 28 km in circumference. The tunnel itself is a remarkable achievement, and a whole book can be written on it. It suffices to say that this tunnel technology came in handy while building the Chunnel, the tunnel under the English Channel which connects England and France.

CERN has many machines but the one just referred to is its latest and best. It is called LEP (large electron–positron collider). In it, electrons and positrons are fed in at an energy of 50–60 GeV which means a collision energy of over 100 GeV, a whacking big amount for e^- and e^+. Already, scientists are dreaming of installing in the same tunnel, a proton–anti-proton collider called the LHC (large hadron collider), which would deliver much higher collision energies.

7.7 Big science in America

This is perhaps an appropriate place to get back to what I earlier called big science (see Chapter 4). As one probed deeper, the particle energies rose sharply and in turn this meant more money, more people to design and build the machines, and also increased electric power consumption. The LEP, for example, would consume as much power as a fairly large city in India! It was no longer possible to handle this within a University campus, and it called for new administrative structures. Even in Stanford, a separate centre, i.e., SLAC (Stanford Linear Accelerator Centre) had to be created. In America, the first generation of post-war machines came up in the Brookhaven National Laboratory (BNL) near New York, and the Argonne National Laboratory (ANL) near Chicago. But the machines kept getting bigger, and even multi-purpose labs like BNL and ANL were no longer adequate. Thus in 1968, a new lab named the Fermi National Accelerator Laboratory (FNAL) was created just to house America's biggest machine, the tevatron. As the name itself indicates, the tevatron is meant to accelerate particles upto TeV energies.

I should not let you get the impression that it was only the size of the machine that mattered and nothing else. Clever physicists have always been trying to coax more juice out of older-generation machines; many have succeeded and even picked up the Nobel Prize — so it was not always brawn that won. Still, energy *is* a sledgehammer and people *do* tend to flock to the most powerful machine available.

Big machines don't spring up just like that. Often technological breakthroughs have to be engineered first and many people work quietly in the background for this purpose. And then there are the highly visible (and vocal) physicists who go on 'sales tours', trying to convince politicians, bureaucrats, the press and the public that high energy machines are a great thing for society to have.

7.8 Enter CERN

For several decades after the Second World War, America dominated high energy physics — no surprise in that. Devastated as they were, the European countries were expected to be non-starters in this big game but Europe proved to be a dark horse! What the European countries did was simple. They said, "Individually, we can't afford these monster machines. But why don't we join hands and build them? It will give our designers work to do; it would boost our industries; it would improve our technology; and in the process it would also keep our young physicists happy." One of course always knew that unity is strength, but in real life unity is rare. Where scientists with their king-sized egos are concerned, it was simply not expected to work at all, but wonder of wonders, it did, and in an unimaginably superb way. This collaboration of nineteen or so countries produced the CERN, the European organization for high energy physics research, and CERN went steadily from strength to strength, the latest being the LEP.

Encouraged by the fantastic success of CERN, the European countries have attempted similar collaboration in other areas like microelectronics, computers and space technology but somehow, success on the scale of CERN's has eluded them. One wonders why.

7.9 The counter-challenge and its demise

With CERN steadily notching success after success, America could no longer be complacent about its number one position in high energy physics. And so in the early eighties, American physicists came up with a proposal for a super-duper dream machine, the *Superconducting Super Collider* (SSC). If CERN's LEP had a circumference of 28 km, the SSC would beat it hollow by going up to about 90 km. If CERN had (e^-, e^+) collisions, the SSC would do better with $(p, \bar{p})$ collisions. Further, the SSC was to leave the LEP way behind in collision energy, attaining 40 TeV, instead of the mere 200 GeV or so that the LEP would offer. American physicists assiduously cultivated their politicians, lecturing them on how the SSC would boost America's prestige. The politicians bought that line, and finally the US Government authorised what was termed the biggest civilian project in human history — the funding was to be in billions of dollars.

With great gusto, American physicists and engineers went to work. LEP might have gone into operation earlier and even dazzle for a while but when the SSC came up, it would make the LEP look like a toy, or so the Americans thought. But alas, things didn't quite happen that way. The main problem was that the American economy was not doing well at that time, and people soon began to wonder whether America could afford such an expensive machine. In the scientific community also, many opposed the SSC, saying that too much research money was being spent on too few physicists. Politicians now got into the act and started serious rethinking about allocating funds for the SSC while high energy physicists waited anxiously, biting their nails. Finally in September 1994 came the word, "Close down the SSC project!" This command was like a bombshell to the American high energy community. So much spadework had already been done and now the politicians wanted to kill their dream project. In fact the project was so far advanced, that it was going to cost the tax payer $ 500 million just to shut down the project. But the US Congress said, "Never mind; it is cheaper to bury the project than to complete it!"

The Europeans made polite noises. And why not, now that the Americans were queueing up at CERN and talking about international collaboration?!

7.10 Detectors

Building machines is only half the job; using them properly and effectively is the other half — in fact, it is the harder part.

How does one actually use the available high energy beams? Roughly as follows. One first plans on what one *wants* to observe — it could be the W^+, it could be what are called neutral currents, or anything else of interest. Whatever it is, one must first be very clear about it. Naturally, this decision about what one wants to observe would be guided by:

(a) the burning issues of the day, and

(b) the capability of the machine at one's disposal.

Having thus decided the goal to be achieved, one next plans the various details of the experiment. Firstly, one must arrange for the reaction process to take place, be it scattering or whatever. One must obviously plan this in such a way that the undesired events are considerably reduced. High energy reactions always produce a lot of debris, and one must design the detector system suitably so that it catches only the anticipated event. The detectors must be supported by complex, high-speed electronics systems, which are in turn backed by high-speed computers, TV cameras and what not. These days, each experiment costs several millions of dollars. They involve large teams of scientists, often distributed between different research centres. Thus one hears words like Beijing–Hamburg–Paris collaboration, Moscow–Rome–Liverpool collaboration and so on. In the recent and much-advertised discovery of the top quark, the TIFR was a participant institution. Exciting results are promptly published in either the *Physics Letters* (published in Europe) or the *Physical Review Letters* (published in America). The paper may be only four pages in length but often, one full page is required just for the names of all the authors! Suppose one hits the jackpot in an experiment; with a hundred or so scientists milling around, who takes the bow? Who goes to Stockholm to collect the coveted Nobel? Think about it!

Sorry for that digression; let us again get back to physics. In the fifties and the sixties, the detector that was invariably used was the bubble chamber invented by Donald Glazer who later won the Nobel Prize. The bubble chamber is a close cousin of the cloud chamber, the workhorse of cosmic ray physicists. It works on a somewhat similar principle too. The chamber contains a low boiling-point liquid whose pressure is suddenly lowered, bringing the liquid to a super-heated condition. The liquid is then in an unstable state; its temperature is above boiling point but it is not boiling yet (why?). Suppose that at this stage a charged particle passes through the liquid. The particle produces ionisation all along its path, and these ions act as nuclei, exactly as in the case of the cloud chamber. Local boiling occurs wherever there are ions, resulting in the production of tiny bubbles. Thus the trajectory of the charged particle appears as a track of tiny bubbles, and the bubble trail is duly photographed. The chamber is sensitised by expansion which is arranged to occur just when the pulse of accelerated particles is about to enter the chamber. When the particles in the incoming beam enter the chamber, they react with the nuclei of the liquid, and these reactions are studied. Figure 7.12 shows a schematic representation of the sequence of these events.

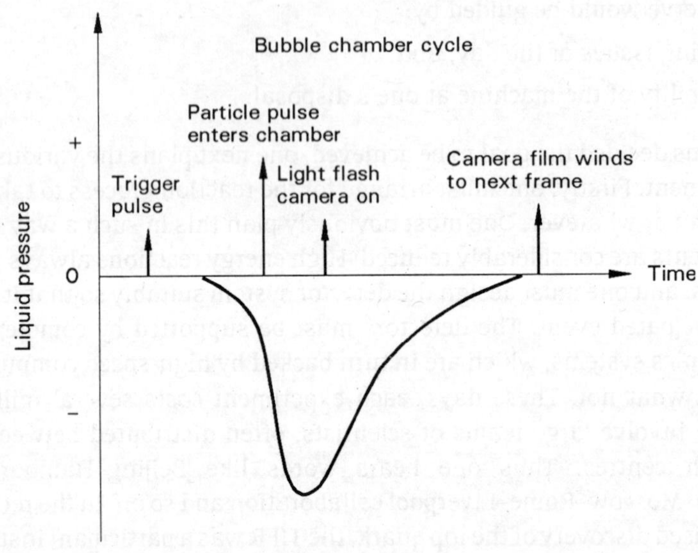

Fig. 7.12 This figure gives a schematic representation of the sequence of events that are pertinent to the bubble chamber operation.

As you can see, the bubble chamber is rather like the good old cloud chamber. Compared to the latter, it has one major advantage and also one major disadvantage; however, the former overwhelms the latter. The advantage is that the density of the liquid is very much higher than that of water vapour; high energy events are therefore more easily triggered. Luckily for high energy physics, the bubble chamber was invented in 1952, just when the big machines were making their appearance. It is no exaggeration to say that for nearly a quarter of a century thereafter, the bubble chamber reigned supreme, revealing many extraordinary processes.

What about the disadvantage? Unlike the cloud chamber, it cannot be triggered by select preconditions. Thus, whereas the cloud chamber could be sensitised only when the conditions were ripe for the particular event sought, the bubble chamber was like an eager host, welcoming one and all. Using the bubble chamber was therefore rather like a lottery, and if the event sought after was rare, one had to hunt for it in thousands if not in millions of bubble chamber pictures. Despite this drawback, the bubble chamber acquitted itself most credibly.

And then came the spark chamber. It started on a modest note but kept improving steadily until one day it simply edged out the bubble chamber, turning the latter into a museum piece. The simplest version of the spark chamber is shown in Fig. 7.13. As in the case of the bubble chamber, the tracks were initially photographed. This was no doubt cumbersome but unavoidable. Meanwhile, electronics improved by leaps and bounds and people also managed to obtain electronic signals from the entire track of the charged particles. All this information was fed into a computer which obligingly carried out all the required analysis, reconstructed the tracks and displayed them on the screen. The spark chamber could be triggered precisely when one wanted, and this was a great plus. But what about going back to gas? Wasn't that a great penalty? True, one did lose there compared to the liquid, but the ability to make the detector respond only when one wanted (as in the case of the cloud chamber) was a tremendous advantage; it meant that one no longer had to wade through thousands of (bubble chamber) pictures, rather like hunting for a needle in a haystack.

The spark chamber shown in Fig. 7.13 illustrates only the basic principle. Today's spark chambers are not only larger but also much more evolved and capable of superior performance. Figure 7.14 illustrates schematically, some of the advancements. Like the machines in which they operate, the detectors too cost a lot of money, although not as much as the machines.

7.11 Concluding remarks

Machine technology is not static. Suppose an accelerator is built; this would be version 1. Detectors are then built to go with the machine, experiments performed and some information extracted. Meanwhile, the machine builders are not idle. They are already dreaming up improvements to version 1 and typically after about a five-year wait, they get a chance to improve their 'baby'; this leads to version 2. Simultaneously, detectors also are improved and this cycle of marginal improvement goes on for some years. A campaign is then started for a new machine. However, when the new machine is sanctioned, the old one is not quite abandoned. It is used as a booster or an injector, as appropriate.

So where do we stand today? At present a plateau appears to have been reached as far as the very big machines are concerned. The current leader is the LEP at CERN. CERN is also committed to the upgradation of the LEP to the LHC.

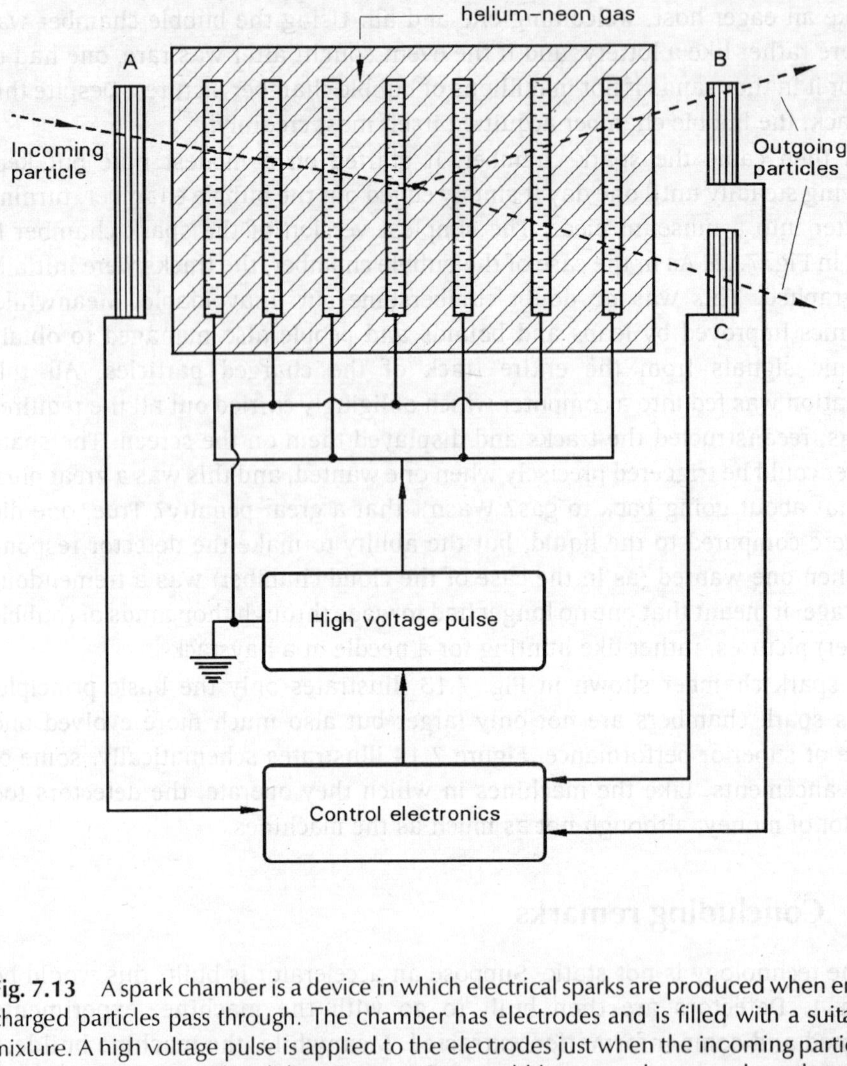

Fig. 7.13 A spark chamber is a device in which electrical sparks are produced when energetic charged particles pass through. The chamber has electrodes and is filled with a suitable gas mixture. A high voltage pulse is applied to the electrodes just when the incoming particle goes through, whereupon a breakdown occurs. One could be more choosy and say that not only must a charged particle enter the chamber, but two charged particles must also leave it. The watch-dog counters A, B and C enable this control. These counters send their signals to the control system and if the latter decides that the conditions are OK, then a signal is sent to the power supply to produce a high voltage pulse. The camera also is alerted to shoot a picture of the track, made visible by the spark. What is shown here is just a rough schematic. In practice, spark chambers can be both huge and complex.

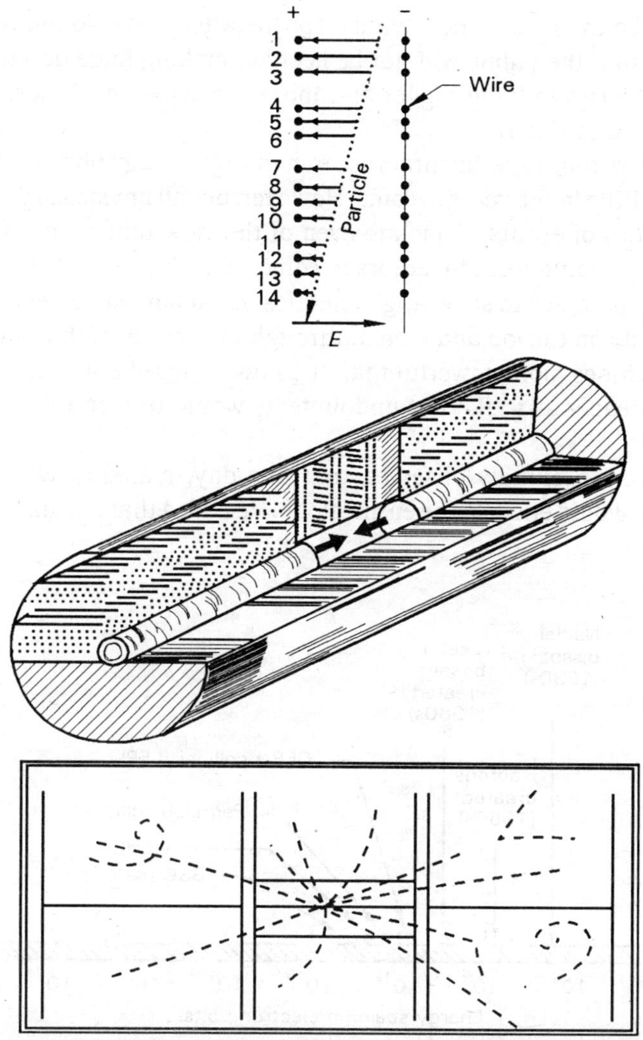

Fig. 7.14 This figure gives a rough scheme of a modern spark chamber, and illustrates its complexity. (a) shows a series of wires (they are all perpendicular to the plane of the paper). An electric field E is maintained between the wires. As a charged particle makes its way, it produces ionisation all along its path. The electrons produced by the ionisation then travel towards the positive wires and when they arrive there, they give a signal. Each wire gives its own signal. One records not only the serial number of the wire but also the time of the signal. Knowing the time the particle entered the chamber — there are ways to do this — one can literally calculate the track of the charged particle, as it went through the chamber. (b) shows the cutaway view of the detector used by Carlo Rubbia to detect the W^- and the W^+ intermediate bosons. The big bold arrows show the p and the $\overline{p}$ colliding. The surrounding region is full of wire grids stacked in an intricate matrix. These wires are connected to suitable electronic circuits. And when the event occurs, all particles are tracked and trajectories computed using a high-speed computer which then produces a display of the event, shown schematically in (c).

So LHC would be the next leader. What happens after that? No one knows. It is almost certain that the public will not be keen on making huge donations of tax money for physicists to build bigger and more expensive machines, which has already happened in America.

It thus appears that experimental research in high energy physics would have to slow down a little in the years to come. However, not all physicists are depressed by this likely turn of events. Some are even of the view that such a setback is a good thing, for it would force inventors to come up with new and clever schemes for accelerating particles to super-high energies but at a much lower cost. Indeed, many are already on the job and a breakthrough may occur. If that happens, one can expect machines more powerful than the now-demised SSC, very much more compact and cheaper as well. That undoubtedly would trigger a new era in high energy physics.

Long ago, Fermi jokingly observed that one day, mankind would build an accelerator that would encircle the entire globe. He added that one half of mankind

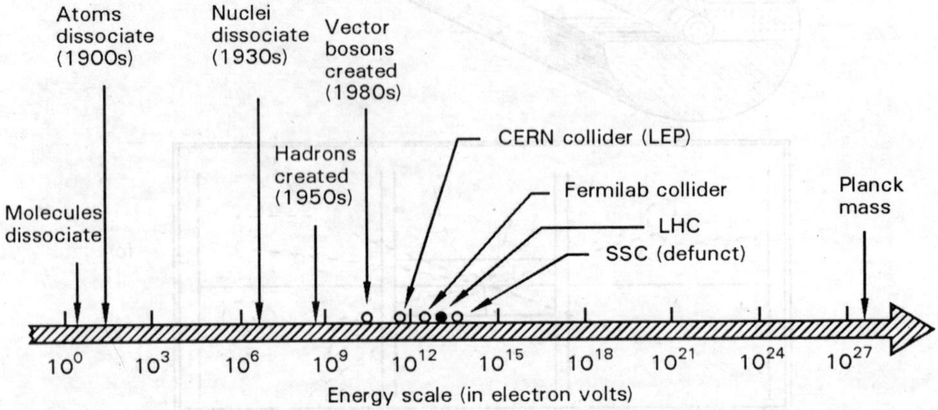

Fig. 7.15 This figure must be studied along with Fig. 3.3. At first one was able to break molecules apart into their constituent atoms. Then the electrons in the atom could be isolated. By 1930, there were particle beams of sufficient energy to smash nuclei. In the fifties, one entered a new phase — smash as well as create. Thus, a whole lot of hadrons were created, which led to particle spectroscopy. Still higher energies took one to W^-, W^+ and Z^0. Today, the big collider is CERN's LEP. Soon to follow would be the LHC, also at CERN. From AD 1900 to 2000, we have moved from just 1 eV to 10^{12} eV! This has undoubtedly revealed a lot. But when one talks of the early Universe (as we shall in Part II), one must deal with energies $\sim 10^{27}$ eV. Even if we are to leapfrog from the present 10^{12} eV to say 10^{14} or even 10^{15} eV, we would still be a long way off from 10^{27} eV. There would thus be a huge gap between what man can physically reach and what he must reach to explore small distances comparable to the size of the infant Universe. This gap seems like an infinite desert that can never be crossed.

would be engaged in building and maintaining the machine while the other half would be busy using it! That, of course, is unlikely to happen, but even if one were to build a gigantic machine like that, be assured that the energy it would deliver would be small in comparison to the scale of energies needed for exploring very small distances, i.e., of the order of the Planck length. Physics research at such incredibly small distances would throw much light on the history of the infant Universe (see Fig. 7.15). It is estimated that an accelerator that can probe with a spatial resolution equal to the Planck length would be as big as the Universe itself! So clearly there is a ceiling on the biggest accelerator that one can design and build. Does that mean that there is also a boundary in physics which man can never cross? Think about it.

8

The Eight-fold Way

This chapter describes how in the sixties, Gell-Mann brought order into chaos by invoking SU(3) symmetry and classifying elementary particles into families, based on concepts like isospin, strangeness and hypercharge.

8.1 Introduction

As explained earlier, in the thirties, the world was simple; one believed that the entire Universe could be explained in terms of just the electron, the proton and the neutron, with possibly the neutrino and the positron included to take care of radioactivity. However, as the years rolled by, it turned out that the world wasn't all that simple. More and more elementary particles came to be discovered, and by the sixties the number of these so-called elementary particles was so high that one began to wonder whether these particles merited being called elementary (see also Fig. 8.1). Things were in a mess but there was a breakthrough and out of this chaos came not only order but also beautiful physics. This chapter is about the beginnings of that success story.

8.2 Particles and resonances

For a moment let us forget elementary particles and consider instead the hydrogen atom. It is a composite system made up of a proton and an electron, and has a spectrum of energy levels which we have discussed many times in this series, e.g., in ***Saha***. The atom has a ground state and a whole lot of excited states above the ground state which, however, have a finite lifetime — an atom in an excited state emits a photon after a while and returns to the ground state. The excited state thus decays and has an energy width $\Delta E \sim (\hbar/\tau)$, where τ is the lifetime. See also Figs. 3.4 and 3.5.

In the early fifties, Fermi was shooting π^+ and π^- from the Chicago cyclotron at (stationary) protons to see what happens. That story is told in greater detail in ***Bhabha***, and here we confine ourselves to the essential physics. Fermi's experimental

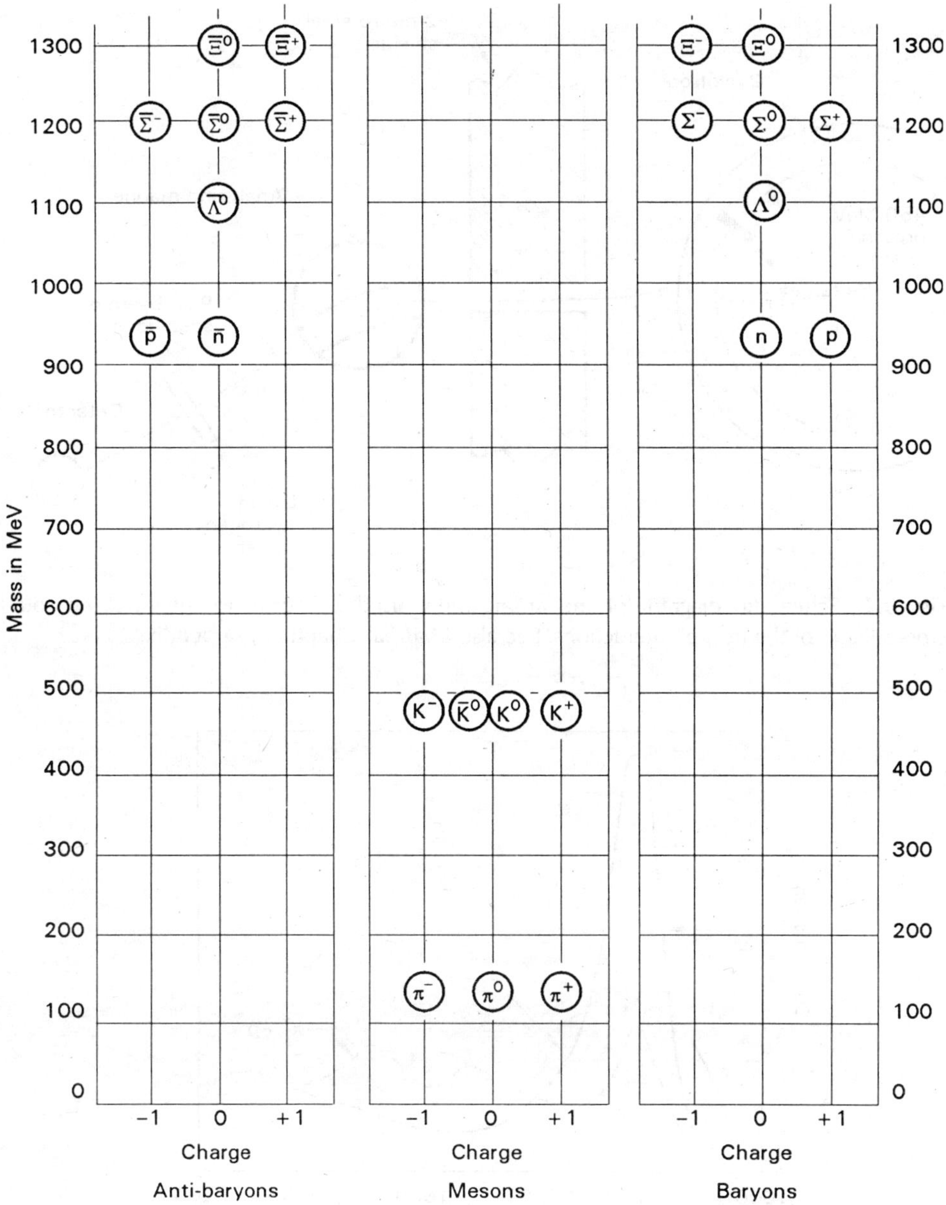

Fig. 8.1 Chart of the anti-baryons, mesons and baryons known in the mid-fifties. By the mid-sixties, the number of known particles had swelled to over a hundred. The corresponding chart would be quite over-crowded, which is why the situation at an earlier point is presented.

arrangement is sketched in Fig. 8.2, and the results he obtained are shown in Fig. 8.3. What is striking is the presence of resonances; prominent among them is Δ^{++},

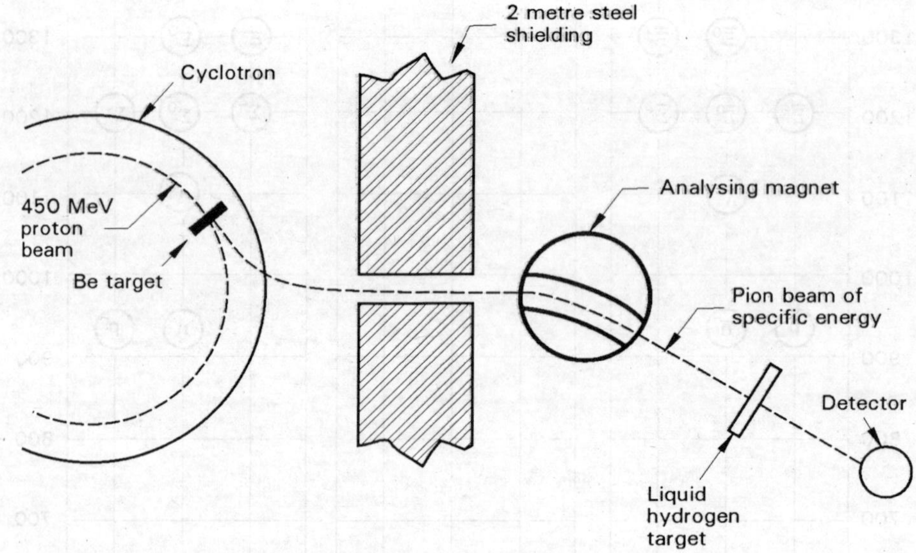

Fig. 8.2 Schematic diagram of the arrangement used by Fermi to measure the total cross-section of the ($\pi^{\pm}$, p) interactions. See also ***Bhabha***, Chapter 4, Appendix.

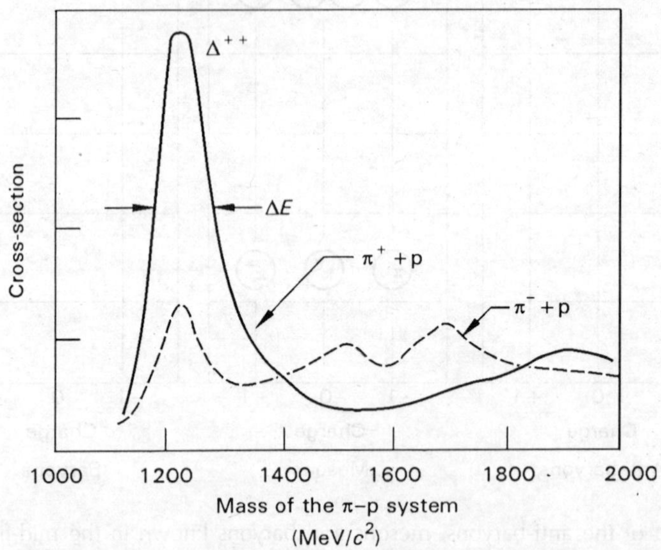

Fig. 8.3 Resonances observed by Fermi in the (π^+, p) and the (π^-, p) reactions. The prominent resonance at 1238 MeV is now identified as the (short-lived) Δ^{++} particle, designated as Δ^{++}(1238). The width ΔE determines the lifetime of the particle via the Heisenberg uncertainty relation.

which occurs at an energy of 1238 MeV. Resonance is no doubt a valid description of the observed peak, but Δ^{++} (1238) can also be viewed as a (short-lived) particle. In today's parlance, Fermi's experiment could be paraphrased as

$$\pi^+ + p \longrightarrow \Delta^{++} \longrightarrow \pi^+ + p$$

and visualised as illustrated in Fig. 8.4(a). In brief, Δ^{++} is a particle which is produced in the (π^+, p) reaction. It lives for a short time $\tau \sim (\Delta E/\hbar)$, where ΔE is the resonance width, and then decays into a π^+ and a p. If Δ^{++} is a particle, surely it must be possible to produce it in other ways? Indeed, and Δ^{++} also appears, for example, in high energy (p, p) reactions,

$$p + p \longrightarrow n + \Delta^{++}$$

$$\longrightarrow n + p + p$$

(see also Fig. 8.4(b)). Thus a large number of short-lived elementary particles can also be looked upon as resonances, implying that there is really nothing elementary about these particles. Incidentally, the idea of resonance is quite universal in physics, and some aspects pertinent to the present case are discussed in Box 8.1.

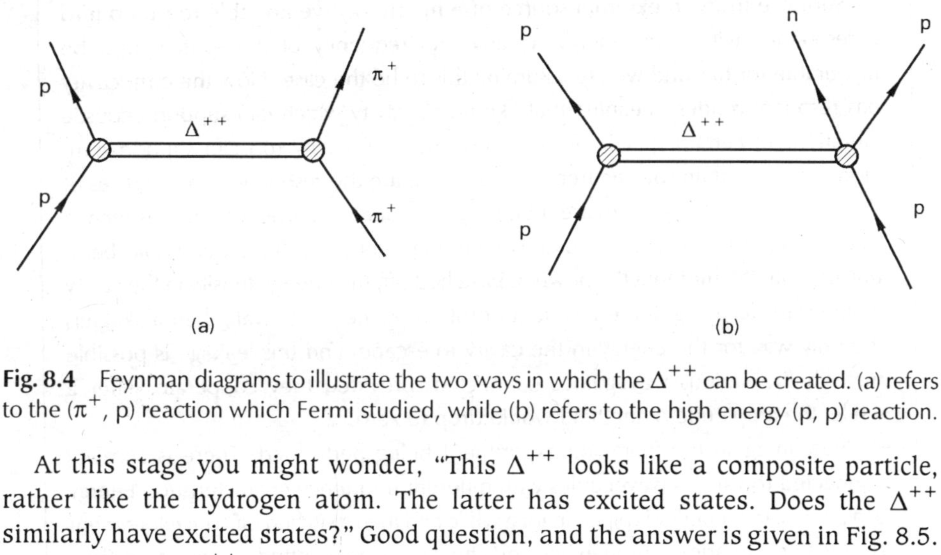

Fig. 8.4 Feynman diagrams to illustrate the two ways in which the Δ^{++} can be created. (a) refers to the (π^+, p) reaction which Fermi studied, while (b) refers to the high energy (p, p) reaction.

At this stage you might wonder, "This Δ^{++} looks like a composite particle, rather like the hydrogen atom. The latter has excited states. Does the Δ^{++} similarly have excited states?" Good question, and the answer is given in Fig. 8.5.

Particles like Δ^{++} can not only be formed in different ways but can also decay in different ways, as Fig. 8.6 highlights. What emerges from all this is that the particles that once were passing off as 'elementary', are not really elementary in the true sense of the word. What are they then? Are they composed of something else which is elementary? Is there any way of comprehending this big, complex, sub-atomic zoo? Yes indeed, and we shall explore these questions step by step.

Box 8.1

When one speaks of the Δ^{++} (1238) resonance, one essentially thinks of it as a particle. But thanks to the wave–particle duality, one can also understand the behaviour and the properties of the Δ^{++} and indeed all other resonance particles by focusing on the wave aspect. Here, the analogy of the electromagnetic resonant cavities proves very useful. Such cavities are used in linacs, for example, and are used to generate electromagnetic waves of a particular frequency. Suppose a cavity of this type is switched on and e.m. waves are generated; if the power is now switched off, the wave field in the cavity would slowly diminish in amplitude, eventually becoming zero; this happens due to leakage from the cavity as well as absorption of energy by the walls of the cavity.

Let us now try to make analogies between resonance particles and the resonant modes of electromagnetic cavities. To start with, we imagine that long pipes are connected to the cavity, through which waves can be guided into the cavity or made to exit from the cavity; the technical name for such tubes is *waveguide*. Now these waveguides are rather choosy; if the wavelength is larger that a certain critical value, the waveguide refuses to transmit the wave. In other words, there is a threshold frequency above which only the guide transmits.

Suppose from an external source of e.m. energy we are able to pump into a resonant cavity, e.m. energy; clearly the frequency of the source must be favourable for this and we are assuming this to be the case. Now the e.m. cavity has resonant modes, meaning that like an air cavity which can support acoustic vibrations of certain specific frequencies, the e.m. cavity too can support e.m. vibrations of certain specific frequencies; these are the resonance frequencies of the cavity. Let us say we are pumping in energy at one of the resonance frequencies. As long as the source is pumping in energy, the cavity would be in full form but the moment the power is switched off, the energy density in the cavity would start decaying. If there is no absorption by the cavity wall, then leakage is the only way for the energy in the cavity to escape, and this leakage is possible through the various waveguides connected, provided they cooperate. After a while, the amplitude of the field would drop to zero.

Now imagine the following experiment being performed. There is a cavity connected to various waveguides with different frequency characteristics. Energy at a particular frequency is piped in for a while and then switched off. The experiment is repeated at various frequencies and the results are plotted. One would then obtain a graph as in Fig. (ii), the explanation for which is available in Fig. (i). What Fig. (ii) shows is that there are resonances at certain frequencies — these are resonance frequencies of the cavity. Further, some resonances are strong while others are weak, and all resonances show a width which is determined by how

Contd...

Box 8.1 *...contd.*

fast the energy stored in the cavity leaks out, once the external power is switched off. The leakage rate, in turn, is decided by how many waveguide channels would obligingly carry away the energy from the cavity, i.e., *how many channels are open at that frequency*. Based on all this, the meaning of Fig. (ii) should be obvious.

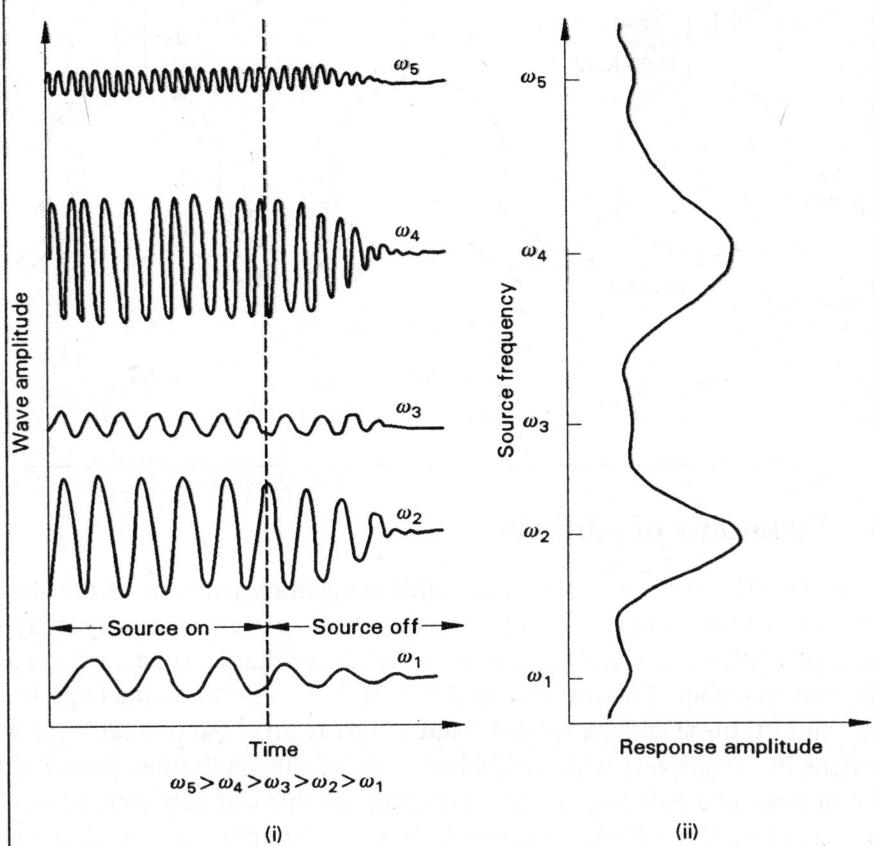

(i) (ii)

Let us now apply all these concepts to resonance particles. Suppose one shoots a pion at another pion. Then, at the energy of 750 MeV, the resonance particle ρ (rho) with spin parity $J^P = 1^-$ is formed. This particle is unstable, and after a brief period, decays into two pions. The decay occurs via only one channel. If the energy of the colliding particles is increased, then at 1600 MeV, there is another resonance — this can be viewed as an excited state of the ρ. In this case, there are two decay

Contd...

Box 8.1

...contd.

channels open, meaning that the decay is either via the formation of just two pions or via the formation of four pions. Possibly, at still higher energies, other channels would open up. These concepts are illustrated in Figs. (iii) to (v) with suitable cavities; I trust you can easily understand this figure.

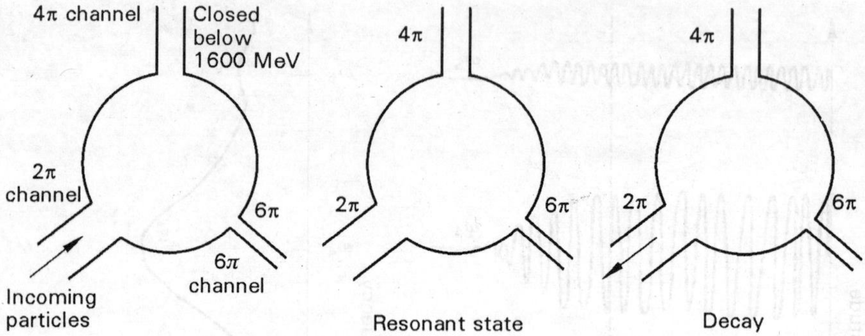

(iii) (iv) (v)

8.3 Taxonomy of particles

The proliferation of the number of so-called elementary particles in the sixties was reminiscent of a situation that had prevailed earlier in chemistry. By studying thousands of chemical reactions, it had been deduced that all solids, liquids and gases were made up of atoms, and further that there were 92 distinct species of atoms. In fact, these various species could be neatly arranged in a table (as was first done by Mendeleev) which exhibited some of the similarities between the different types of atoms (e.g., Li, Na, K, Cs, Rb). All this was very nice, no doubt, — one could explain millions of compounds as combinations of just 92 different species of atoms — but would Nature be really that uneconomical as to have 92 different types of building blocks? Remember, in those days people believed that atoms were the indivisible building blocks of matter. Everything changed after the discovery of the electron, proton and the neutron, and people now realised that Nature was both economical and smart, and that all the matter in the Universe could be explained in terms of these three particles, or so it seemed.

In the early sixties, particle physicists faced a situation similar to that faced by chemists earlier. There was this big zoo of the so-called elementary particles or resonances, whatever one chose to call them. Could these be neatly organised into families like Mendeleev did earlier with atoms? Further, could they be explained in terms of a few basic entities, even as the 92 different atoms could be understood in terms of just the electron, the proton and the neutron? The answer was a clear YES, and it emerged out of a careful, imaginative analysis of the large volume of data accumulated through many experiments.

It sometimes happens in physics that great breakthroughs are preceded by a period in which a lot of data is patiently collected, some of which does not make

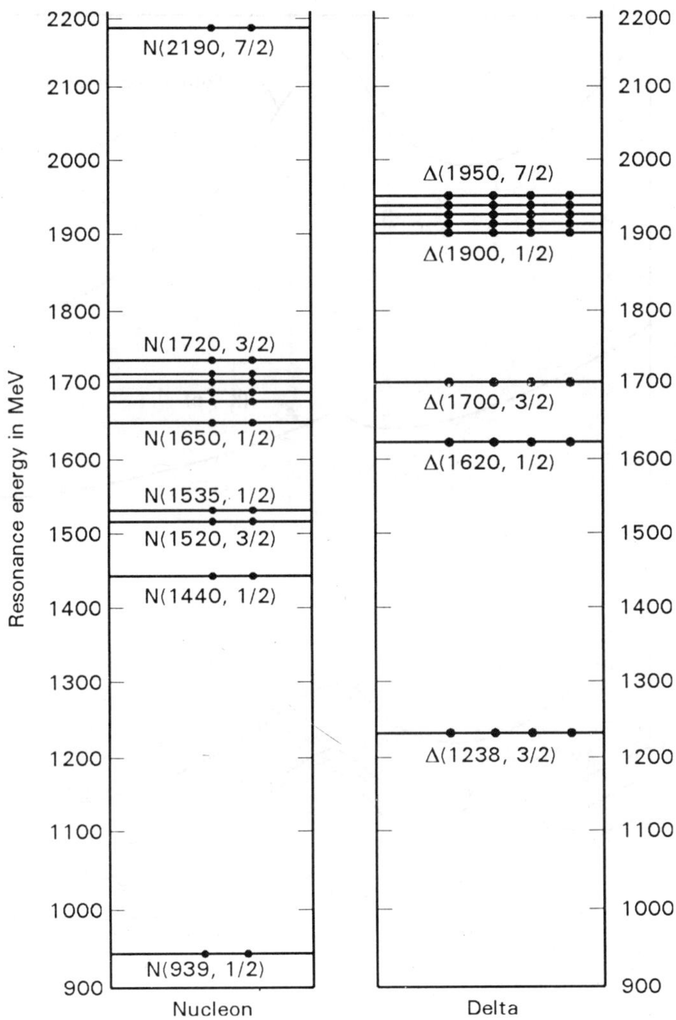

Fig. 8.5 The various states of excitation of the nucleon (N) and the delta (Δ) particle. The lowest state is the ground state which, in the case of the nucleon, corresponds to the proton and the neutron. Note that the term nucleon represents two particles while the term delta represents four particles (Δ^-, Δ^0, Δ^+ and Δ^{++}).

much sense at the time of collection. Myopics often superciliously dismiss data accumulation as 'stamp collection', but data accumulated in a frontier area is seldom wasted. Centuries ago Tycho Brahe sacrificed much sleep to patiently track the planets, which subsequently helped Kepler to see patterns, and still later enabled Newton to formulate a new theory altogether (see ***Chandra***). The story was repeated in the case of spectral lines of the hydrogen atom. In ***Saha***, I have recalled that story in some detail — how Lyman, Balmer, etc., grouped the

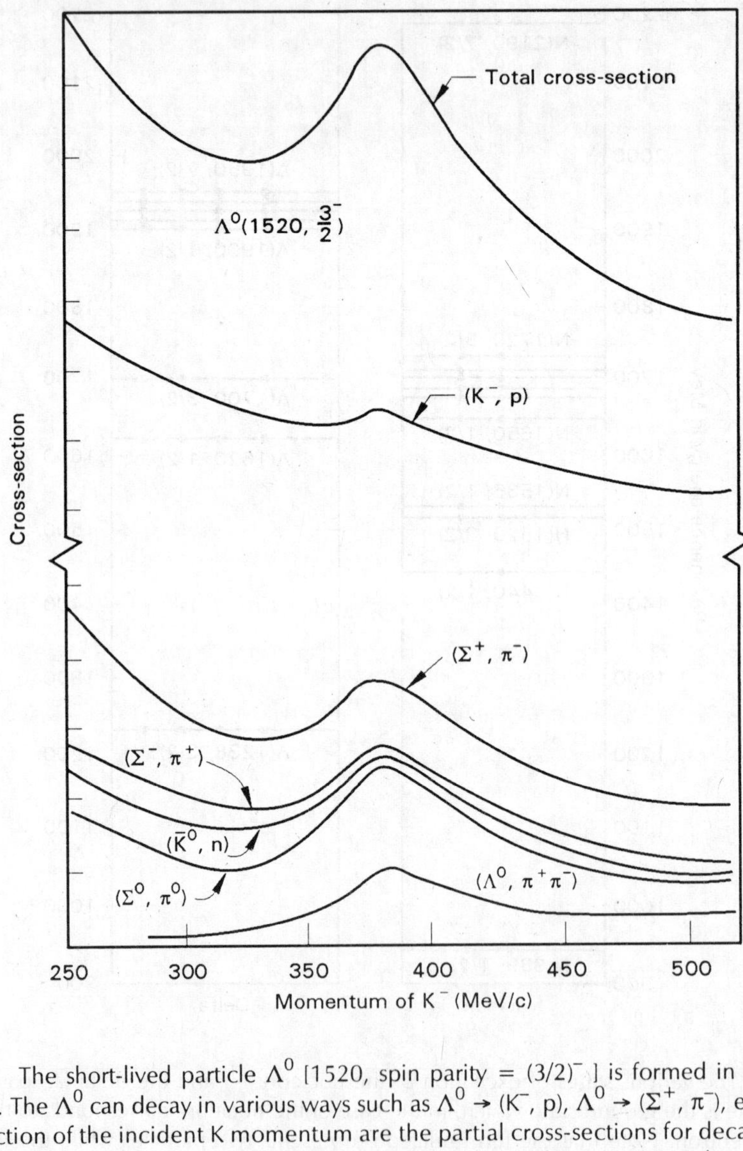

Fig. 8.6 The short-lived particle Λ^0 [1520, spin parity = $(3/2)^-$] is formed in the (K^-, p) reaction. The Λ^0 can decay in various ways such as $\Lambda^0 \rightarrow (K^-, p)$, $\Lambda^0 \rightarrow (\Sigma^+, \pi^-)$, etc. Plotted as a function of the incident K momentum are the partial cross-sections for decay via these various channels, as also the total cross-section which is a sum of the various partial cross-sections.

spectral lines into distinct families and even gave empirical formulae, and how later Bohr explained all these various empirical series in one shot with his atomic model. Subsequently of course, the Bohr model was put on firmer footing, first with QM and then with QED. Where elementary particles were concerned, history was now about to be repeated, and the first question that arose was, "How about patterns in the particle zoo?"

Have you heard the word taxonomy? It means classification. Whenever there are objects seemingly belonging to a large number of species, the first step in bringing order is to introduce classification — one does this in the library, in the plant kingdom, and in the animal kingdom. In all cases, the basic principle is the same, i.e., one first assigns some attributes to the species and then does the sorting. What attributes can we employ to classify elementary particles? Table 8.1 gives the answer. However, some further explanatory remarks are called for and these follow. No doubt you are already familiar with some of these but they are worth repeating in the present context.

Table 8.1 A brief recall of some of the attributes/quantum numbers, used for classifying elementary particles.

Attribute	Symbol
Charge	Q
Spin	J
Isospin	T
Baryon number	B
Strangeness	S
Hypercharge	Y

Remarks:

1. The list is not complete.
2. Most of the terms have been explained before.
3. This table is tailored for a discussion of mesons and baryons.

Charge conservation

Conservation of electric charge is something we all are familiar with. For reasons not yet understood, the charge Q of all known elementary particles (except the quarks) is an integral multiple of e, where e is the magnitude of the electron charge. Thus $Q = Ne$, where N is an integer. In all reactions, the total charge is always conserved. Suppose, for example, there is the reaction

$$\mathrm{a + b \longrightarrow c + d + e.}$$

Charge conservation implies

$$Q_a + Q_b \longrightarrow Q_c + Q_d + Q_e,$$

i.e.,

$$N_a + N_b \longrightarrow N_c + N_d + N_e.$$

Baryon number conservation

Many years ago, Stuckelberg pointed out that in nuclear reactions, the number of nucleons remains the same, just as in chemical reactions the number of atoms is conserved. Later, a number B called the *baryon number* was assigned to each particle such that B is 1 for baryons, −1 for anti-baryons, and zero for all other particles. Like charge, the baryon number is also conserved. For example, in the β decay of the neutron we have:

$$n \longrightarrow p + e^- + \bar{\nu}.$$

In terms of B this reads

$$1 \longrightarrow 1 + 0 + 0,$$

showing that B is conserved.

Isospin

Isospin is the short form of the word isotopic spin (a word once popular, but now replaced by the shorter form). For a minute let us go back to the ordinary spin (angular momentum) of the electron. Denote it by J. As explained in *QR-I* and also elsewhere, one can never measure the absolute orientation of J in space; instead, one can only measure its projection along any specified axis. By convention we take this axis as the z-axis. In particle physics one does not denote the three components as x, y, and z but as 1, 2, and 3. This means that one can measure only the J_3 component of J. In units of $\hbar$, the allowed values of J_3 for the electron are +(1/2) and −(1/2). These two states are often denoted by drawing an arrow pointing either up or down.

With this background, let us consider isospin. Heisenberg noted that the neutron and the proton could be regarded as two states of the same particle, i.e., the nucleon (N). To describe these two states, one first introduces an operator T which is an analog of J. The neutron and the proton are then states obtained by projecting T along the 3-axis. In slightly more sophisticated language, the proton (p) and the neutron (n) are nothing but the eigenstates of the operator T_3 — see also Fig. 8.7. The concept of isospin is very useful in grouping particles similar in mass and other properties into a single family. For example, like the proton and the neutron, the trio π^+, π^- and π^0 belong to a family with isospin = 1. Like electric charge, baryon number and spin, isospin is an additive quantum number.

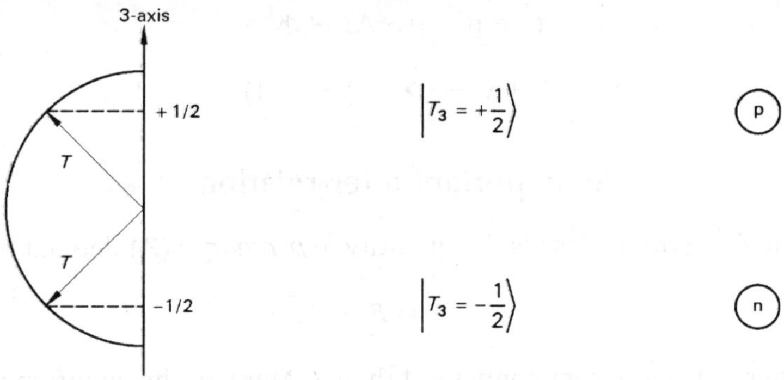

Fig. 8.7 This figure seeks to illustate how the proton (p) and the neutron (n) can be regarded as two states of the same particle, i.e., the nucleon (N). The states are distinguished via the isospin operator T, which is like the angular momentum operator $\boldsymbol{J}$. As in the case of angular momentum, one can observe only the projection of T along the 3-axis. For the nucleon, $T = (1/2)$, which means that T_3 can take on only the values +(1/2) or −(1/2). These states correspond respectively to p and n.

There is one other point that deserves mention. As you are aware, in the absence of any external field, an electron in the spin state +(1/2) has the same energy as an electron in the spin state −(1/2). Correspondingly, we expect that the neutron and the proton should have the same rest mass, but they do not. Why? The electron example provides the answer. If the electron is placed in a magnetic field, then the two spin states have different energies, i.e., the magnetic field removes the spin degeneracy. The isospin degeneracy just mentioned is strictly with respect to strong interactions alone; it ignores other interactions. If the effects of electromagnetic interactions are considered, then the isospin degeneracy is removed; in other words, the small mass difference between the proton and the neutron is largely due to e.m. effects.

Strangeness

Earlier, I told you that first in cosmic rays and later in accelerator experiments, a number of particles were discovered, whose behaviour was rather peculiar. These particles were produced in strong interactions in times as short as 10^{-23} sec but while decaying they took an enormously long time $t \sim 10^{-8}$ sec, characteristic of weak interactions. Such particles were called strange particles. No one knew at that time why they behaved the way they did. The answer to this riddle came eventually out of some astounding physics, and alongside, Gell-Mann and, independently, Nishijima, introduced a new (additive) quantum number called *strangeness*. Strangeness is conserved in strong interactions, as can be seen from the following example:

$$\pi^- + p \longrightarrow \Lambda^0 + K^0$$

$$S: \quad 0 + 0 \longrightarrow (-1) + (+1).$$

An important inter-relation

Closely related to strangeness is the quantity *hypercharge* (Y) defined by

$$Y = B + S. \qquad [8.1]$$

These days, Y is more commonly used than S. Many of the quantum numbers used for characterising elementary particles are not completely independent, an important inter-relation connecting them being:

$$Q = T_3 + (S + B)/2. \qquad [8.2]$$

At this stage, you should study Table 8.2 which takes you a little further than Table 8.1 and gives you a better feel for these quantum numbers. Table 8.3 follows it up with a list of a few particles and their attributes.

Table 8.2 A few families of particles, and some of their attributes

	Particle	Symbol	J	B	S	Y	T
	Photon	γ	1	0	0	0	0
Mesons	Pion	p^+, p^0, p^-	0	0	0	0	1
	Kaon	K^+, K^-	0	0	+1	+1	$^1/_2$
	Kaon	$K^0, \bar{K}^0$	0	0	−1	−1	$^1/_2$
	Omega	ω	1	0	0	0	0
	Rho	ρ	1	0	0	0	0
	Eta	η	0	0	0	0	0
Baryons	Nucleon	p, n	$^1/_2$	1	0	1	$^1/_2$
	Delta	$\Delta^-, \Delta^0, \Delta^+, \Delta^{++}$	$^3/_2$	1	0	1	$^3/_2$
	Lambda	Λ^0	$^1/_2$	1	−1	0	0
	Sigma	$\Sigma^+, \Sigma^0, \Sigma^-$	$^1/_2$	1	−1	0	1
	Cascade	Ξ^-, Ξ^0	$^1/_2$	1	−2	−1	$^1/_2$
	Omega	Ω^-	$^3/_2$	1	−3	−2	0

Remarks:

1. Observe that in each family, there are $(2T + 1)$ particles or members, corresponding to the $(2T + 1)$ distinct values that T_3 can take.
2. Use this table to verify Eqn. [8.2].
3. Particles with $S \neq 0$ are strange particles. Examples are provided both in the meson as well as baryon categories.

Table 8.3 A chart of some of the important quantum numbers

Conserved quantity	Symbol	Observed values	Examples: Proton	Examples: Negative Pion
Electric charge	Q	$0, \pm 1, \pm 2, \pm 3, \ldots$	$Q = +1$	$Q = -1$
Baryon number	B	$0, \pm 1$	$B = +1$	$B = 0$
Hypercharge	Y	$-2, -1, 0, +1$	$Y = +1$	$Y = 0$
Isospin	T	$0, 1/2, 1, 3/2$	$T = 1/2$	$T = 1$
Spin	J	$1/2, 3/2, 5/2, \ldots$ $0, 1, 2, \ldots$	$J = 1/2$	$J = 0$
Parity	P	$-1, +1$	$P = +1$	$P = -1$

Remark:

The quantities shown are conserved in strong interaction but not necessarily (at least some of them) in electromagnetic or weak interaction.

8.4 Regge trajectories

Among the earliest patterns to be detected in the particle zoo was the so-called *Regge trajectory*. The basic idea is quite simple. Suppose one-ton spacecrafts are placed in circular orbits around the Sun at radii corresponding to each of the nine planets. The binding energy of the satellite to the Sun would be highest when it is in the orbit of Mercury and least when it is in the orbit of Pluto. On the other hand, the angular momentum increases as one goes outwards. Thus if a plot of the angular momentum versus binding energy is made, it would be as in Fig. 8.8. You should be able to calculate and plot this curve yourself. It is easy; do try it. A similar plot could be made for the hydrogen atom where we have an electron orbiting a proton. The Italian physicist Regge asked, "Is there such a behaviour in the case of particles bound by strong interactions?" The answer turns out to be yes (see Fig. 8.9). Here each curve links particles which share the same quantum numbers for all attributes but differ only in spin (J) and parity (P). Each Regge trajectory corresponds to what might be called a Regge family. For a while, identifying Regge families was a popular pastime but soon it gave way to a more powerful family classification based on symmetry groups.

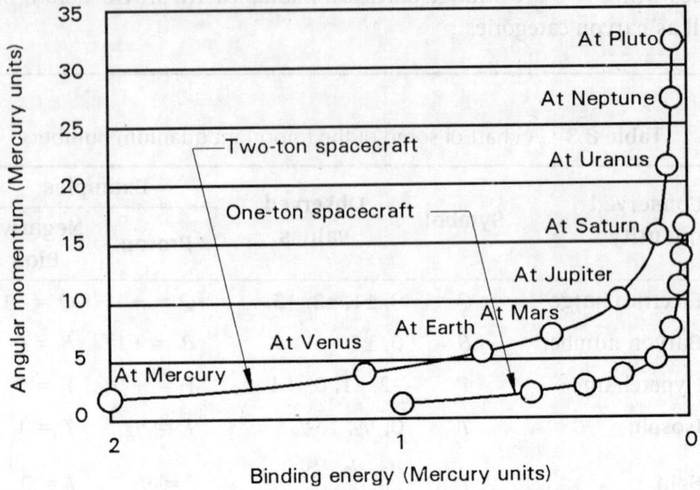

Fig. 8.8 Regge trajectories for a one-ton and a two-ton spacecraft. The orbits correspond to the nine planetary orbits. The binding energy of the one-ton spacecraft in the orbit of Mercury is taken as unity. The angular momentum is plotted as a function of the binding energy.

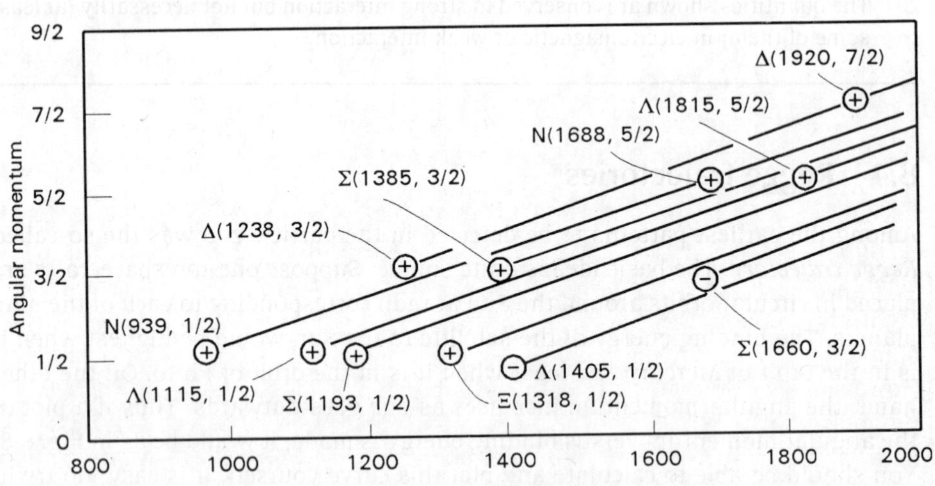

Fig. 8.9 This plot is like the previous one, but is for baryons. Each trajectory corresponds to a family whose members have the same values for all the quantum numbers except spin and parity, i.e., J^P.

8.5 A prelude

In 1949, there were not that many elementary particles but already, Fermi and Yang had begun to wonder about composite particles and the basic building blocks of elementary particles. For example, they postulated that the pion is a bound state of a nucleon–anti-nucleon pair. After the discovery of strange particles, Sakata in Japan pointed out that if in addition to the n and p one also considers the Λ^0 as a basic building block, then the occurrence of strange particles too could be explained (just as the pion was explained by Fermi and Yang). The triplet (p, n, Λ^0) were collectively referred to as sakatons. The Fermi–Yang model and its extension by Sakata were good beginnings, no doubt, but their predictions did not agree with experiments and so they both fell by the wayside. However, they did give some important clues, to understand which, a reference to Fig. 8.10(a) is first necessary.

Suppose two electrons are knocked out of a neutral nickel atom; this gives a Ni^{++} ion with eight 3d electrons in the outermost (occupied) shell. A reference to ***Saha*** might help you to understand what I am saying, in case you have any difficulty. If one calculated the energy of Ni^{++} ignoring the Coulomb interactions amongst the eight 3d electrons, one would get a single energy level as shown on the extreme left in Fig. 8.10(a). Although a single level is shown, the line drawn is a superposition of many coincident lines, implying *degeneracy*. If the effect of Coulomb interactions is taken into account, the single level splits into many levels leading to what is called *term splitting*. Next, let us imagine that the ion is placed in a crystal lattice. The ions of the lattice exert an electrostatic field which leads to further splitting of the levels of Ni^{++}; this is called *crystal field splitting*. If now we allow for what is called *spin–orbit interaction* (a relativistic effect), there is still more splitting. And if you are not satisfied with all this, you could apply a magnetic field which leads to *Zeeman splitting* (see ***Bose***).

The message I am trying to get across is simply this: in the simplest picture, many levels coalesce into a single one, leading to high degeneracy. However, this degeneracy may be partially or even fully lifted by perturbations.

How is all this relevant to us? That should become clear from Fig. 8.10(b). Using a symmetry-based scheme which I shall presently describe, one can group eight mesons — K^0, $\overline{K}^0$, etc. — into a single family as shown on the left. In the simplest approximation, all members of this family would have the same mass. However, on allowing for various perturbations, the mass degeneracy is progressively lifted and the mass spectrum comes close to reality. Notice that in both (a) and (b), the state on the left is the most symmetrical one. As one moves to the right, the symmetry is gradually lowered by perturbations. This lowering of symmetry is also referred to as *symmetry breaking*. Of special importance is what is called *spontaneous symmetry breaking*. This will come up for discussion when we deal with electro-weak unification.

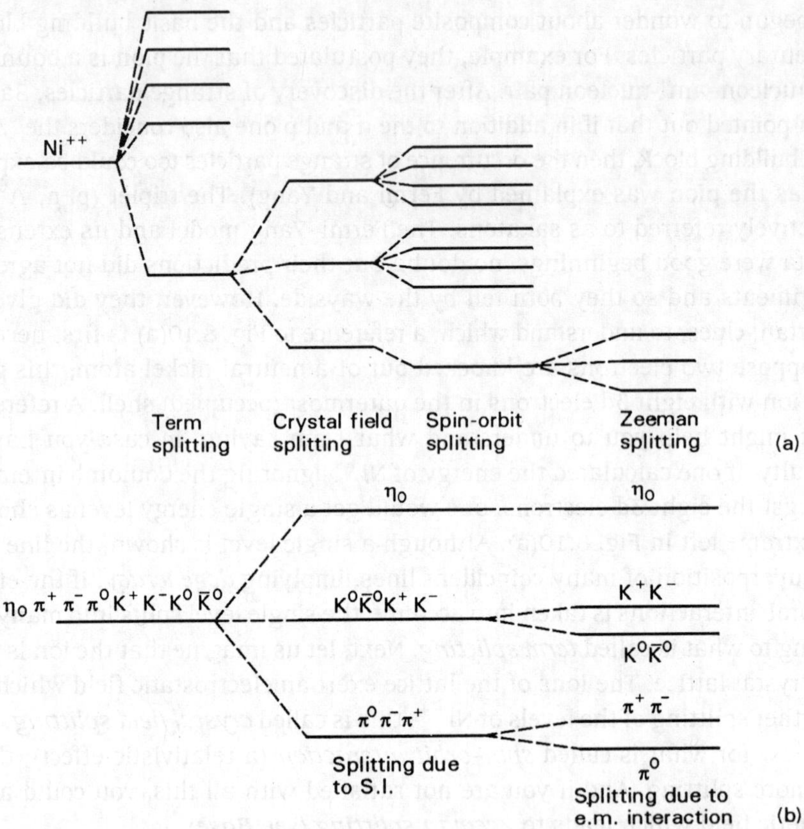

Fig. 8.10 (a) shows the progressive splitting of the degenerate energy level of Ni^{++}. The level at the extreme left corresponds to that obtained when all perturbations are ignored. Inclusion of Coulomb interactions between the eight 3d electrons leads to term splitting, followed by other splittings as discussed in the text. (b) depicts a similar story in particle physics. At the left is a family of eight mesons, forming a Gell-Mann octet. In the simplest, i.e., pure SU(3) scheme, all these mesons have the same mass. Allowing for perturbations due to S.I., some mass difference arises. If in addition e.m. perturbations also are allowed for, then the masses of $\pi^{\pm}$ become different from that of π^0. There is thus a parallel between (a) and (b).

So it all boils down to this, "How can one organise all these diverse particles into groups or families, based on certain common features? Are there any guidelines from, say symmetry? Further, is there any fundamental basis for such a scheme, if it exists?" In the sixties, Gell-Mann gave a firm and clear answer to all such questions. We turn now to that story, dealing in this chapter with some of these questions. The rest will be dealt with in a later chapter.

8.6 The eight-fold way

The story I am about to narrate has two parts actually — the first deals with a scheme, and the second with a (phenomenological) basis for that scheme. Gell-Mann is the star of both. Ne'eman of Israel (see Box 8.2) and George Zweig (who will feature in the next chapter) independently proposed the same scheme. Today, Ne'eman and Zweig are barely remembered, which is not surprising because Gell-Mann squeezed so much out of both the scheme as well as the model.

Box 8.2

Yuval Ne'eman is one of the unusual characters in physics. He is an Israeli but unlike most Israelis who have emigrated from either Europe or America, he was born in the old state of Palestine, before Israel was created in 1948. Ne'eman loved physics but he loved his country as intensely and therefore got involved in various military activities. He could not devote as much time to physics as he wanted and this was gnawing at him till one day he managed to get a posting as a Military Attaché in the Israeli Embassy in London. Moshe Dayan the then Israeli Minister for Defence and therefore also Ne'eman's boss, told the latter that he could dabble in physics while in London, provided it did not interfere with his official work; Ne'eman agreed to the condition.

Having arrived in London, the next job was to find a guru who would guide him in research. Luckily for him, Abdus Salam agreed to be his mentor. Ne'eman knew some mathematics but little of high energy physics and that proved a constant problem. Salam would set him some task but Ne'eman would turn up with some other calculation in his briefcase. Everytime this happened, Salam would shake his head and say, "Oh, you have been wasting your time again on these things. Why don't you do the calculation I gave you instead?" A puzzled Ne'eman would ask what was wrong with what he had done and Salam would explain that Ne'eman's 'discovery' had already been made a few years earlier by someone else.

Far from being discouraged, Ne'eman felt confident and happy that he was smart enough to reinvent the wheel. And then one day, Salam told him about a Russian mathematician named Dykin who had done some important work in group theory. Salam perhaps did not expect Ne'eman to do much but the latter promptly digested Dykin's thesis and came up with the SU(3) scheme for particle classification. Salam pondered over it and said, "Well, this looks like what Sakata and the others have done. The Japanese model is not very popular you know." Ne'eman was not in the least upset; instead, he typed up his article and sent it to the journal *Nuclear Physics* but unfortunately, the paper was returned because the manuscript was typed in single space instead of double spacing as required.

Contd..

Box 8.2

...contd.

So the manuscript had to be typed all over again and the paper resubmitted, and in all this, valuable time was lost.

Meanwhile, Ne'eman called on Salam again and the latter received him with an astonished look. Why? Because in his (Salam's) hands was a preprint from Gell-Mann, describing the same SU(3) model which Salam had earlier pooh-poohed when Ne'eman showed it to him! For once, Salam's student had not reinvented the wheel but was actually ahead.

The story of the scheme really starts with particle taxonomy. As particles turned up in bewildering numbers, people at first wondered what they all meant. Then they realised that the first thing they ought to do was to classify the particles, and this is where the various quantum numbers proved very useful. When this grouping into families was done, the number 8 kept turning up again and again, i.e., many families had eight members. Did this number mean anything and if so what? Given the fact that particle taxonomy relied very much on quantum numbers which were conserved, and also the fact that conservation laws spring from symmetry, it was clear that the 'magic' number 8 was in some way related to symmetry. But what exactly was the relationship?

The subject that provides the mathematical basis for symmetry is called *group theory*. I have already introduced you to this in ***Bhabha***; in this volume we will dwell more upon groups but that comes later. Recognising the importance of group theory, physicists took crash courses on the subject and then tinkered with all sorts of groups and their combinations, but the magic numbers found in particle taxonomy remained elusive. One of those involved in this was Murray Gell-Mann. At that time he was in Paris, the Mecca of mathematics. Like others, Gell-Mann too attended lectures by famous mathematicians but it didn't help — instead he was turned off! As he later said:

> Mathematicians tend to present things in such a — what is the word? — such a *non-constructive* way. They like to prove that there is something, but not actualy *show* you what it is. When they give examples, they are so trivial that you don't learn anything from them.

See also Box 4.1 in ***Bhabha*** for more about the differences in the style of mathematicians and physicists.

Although his initial efforts proved futile, Gell-Mann persisted and back in America one day he hit the jackpot when he tried out the group SU(3) (pronounced 'ess you three'). The technical name for this group is 'special unitary group in three dimensions'. Can we, without going into technical details, understand in a simple way what did the trick? Yes we can.

As you know, the electron has a spin of 1/2 and its two possible spin states can be schematically denoted as |+> and |– >or as | ↑ > and | ↓ >. Suppose we bring two electrons together (as would be necessary while forming a hydrogen molecule out of two hydrogen atoms). The combined system would have a total spin of either 1 (= 1/2 + 1/2) or 0 (= 1/2 – 1/2). Let us denote the various states associated with the united system as |+>|+>, |+>| – >, |– >|+> and |–>|– >. It turns out that these possibilities mix and re-emerge as:

|+>|+> (1/√2) { |+>|–> + |–>|+> } |–>|–>	Triplet, corresponding to total spin +1
(1/√2) { |+>|–> – |–|+ }	Singlet corresponding to spin 0.

Now the original pair {|+>, |–>} form a doublet which respects the group SU(2). And we have just seen that when two doublets combine, they form a singlet and a triplet, i.e.,

$$2 \times 2 = 1 + 3.$$

In the case of SU(3), there is a basic triplet and Gell-Mann investigated what happens when a triplet is combined three times over (later you will understand why he examined a triple combination). Gell-Mann found:

$$3 \times 3 \times 3 = 1 + 8 + 8 + 10. \quad [8.3]$$

When a triplet combines three times over according to the rules of SU(3), one gets a singlet, two octets and a decimet (also called decouplet).

Once Gell-Mann arrived at Eqn. [8.3], he promptly analysed all known particles to see if they fitted into the above pattern. For this purpose he made plots as in Figs. 8.11 and 8.12, and lo and behold, he had several octets. Elated, Gell-Mann called this the 'eight-fold way', an allusion to the teachings of the Buddha who said:

> Now this, O monks, is noble truth that leads to the cessation of pain; this is the noble Eight-fold way; namely right views, right intention, right speech, right action, right living, right effort, right mindfulness, right concentration.

Gell-Mann's eight-fold way was not an overnight success. No doubt it packaged many mesons and baryons into neat octets but there was also a decimet. At the time that Gell-Mann first worked out the theory, the four delta particles (which form a part of the decimet) were known and could be berthed but the other six particles which were supposed to form the decimet were not known then. With six major gaps in the family, no wonder there were not many takers for Gell-Mann's SU(3) idea. But one by one the missing particles turned up to occupy their slots. Finally, only one particle was missing and at a conference in CERN in

1962, Gell-Mann boldly declared that efforts must be intensified to hunt for the missing particle, i.e., Ω^- (omega minus) with strangeness equal to –3.

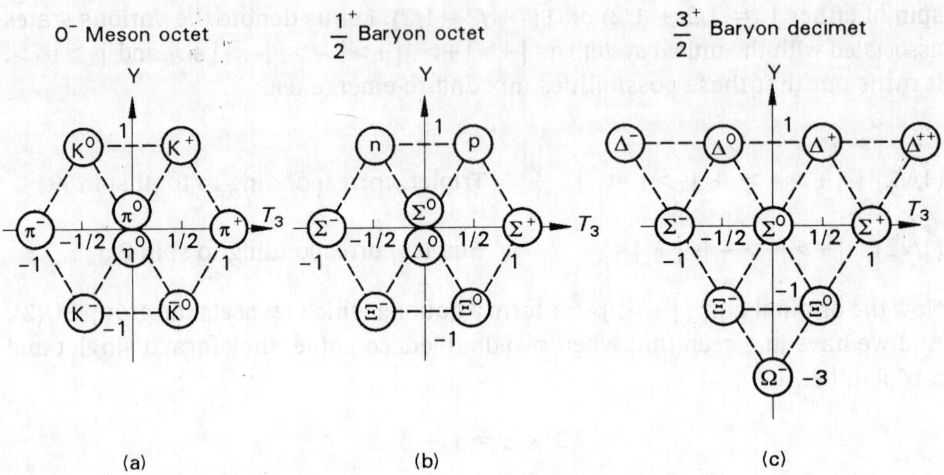

Fig. 8.11 These are some of the arrays that made Gell-Mann famous. Plotted here in the $T_3 - Y$ plane are the families of particles having the same spin–parity combination. Notice that the masses of the particles forming the octet are shown in (a) and (b). The mass difference arises on account of perturbations discussed in the previous figure. When Gell-Mann published his scheme, only four particles were known in the array (c). Later, other particles turned up and occupied their positions. The last to show up was Ω^-, and when it was detected, it was a great triumph for Gell-Mann and SU(3).

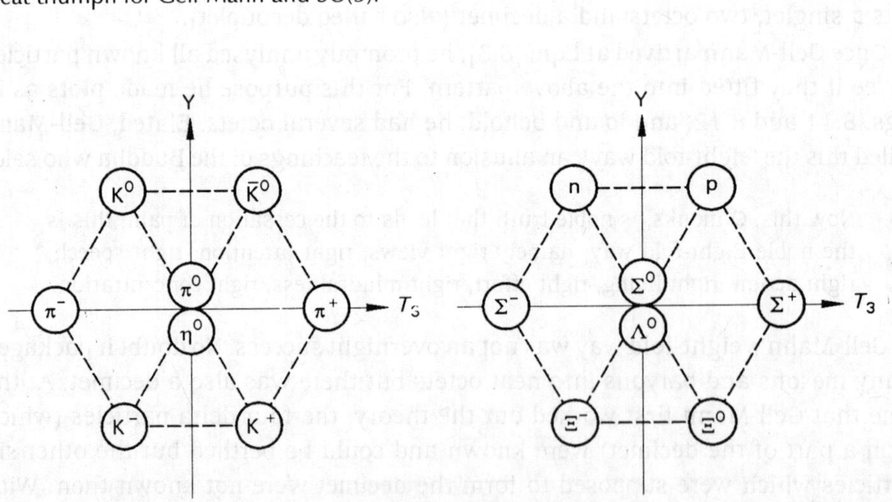

Fig. 8.12 Two more octets are shown here, and they are similar to those in (a) and (b) of the previous figure. However, the particles shown here correspond to the excited states of the particles shown earlier, i.e., these particles have higher mass, and are therefore different.

8.7 The hunt is on

Amongst those attending the CERN conference was Nick Samios of the Brookhaven National Laboratory (BNL). He cornered Gell-Mann and asked how this Ω^- was expected to decay. Gell-Mann scribbled the answer on a paper napkin which Samios carefully folded and put into his pocket. A plan was taking shape in his mind — shoot a K- beam into a liquid hydrogen bubble chamber, and watch the production of the Ω^- in a (K^-, p) interaction. Samios was not the only one to think along those lines. Some British scientists also heard Gell-Mann advocating the search for Ω^-, and they too decided to pitch in. But unfortunately they had problems with their bubble chamber and their experiment simply did not get off the ground.

Samios did not have it easy either. He too had bubble chamber problems and at one stage things looked so bad that there was the danger of a huge explosion. Should he open up the chamber and examine it to see if there was any problem or should he trust his luck and simply press on, hoping that there would not be an explosion? Samios opted for the latter and luck sure was on his side. Nothing terrible happened to the chamber and millions of photographs could be taken. Scanning them to pick the selected event was an equally arduous task, and went on non-stop day after day, week after week. And then one day, in the early hours of the morning, Samios spotted the event that he had slogged all along to detect. There it was, the much-wanted Ω^-, with all the expected signatures (see Fig. 8.13).

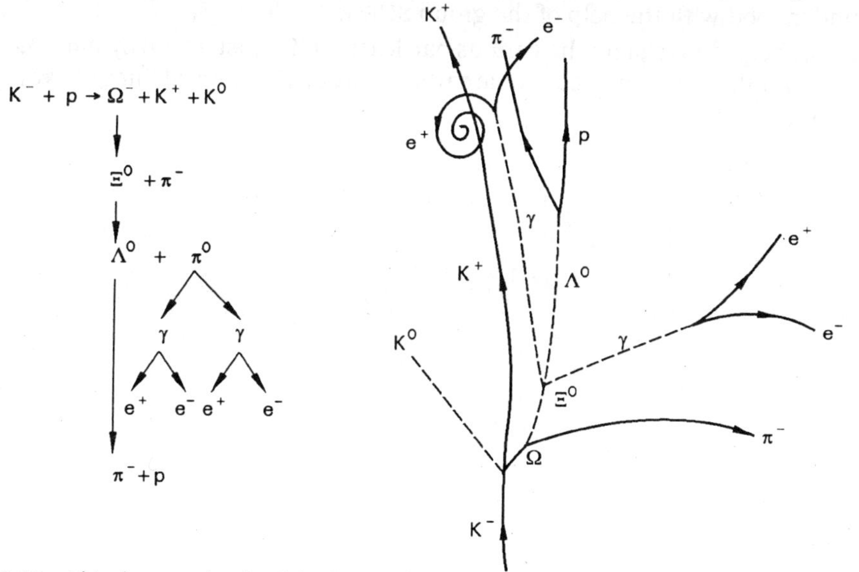

Fig. 8.13 This shows a sketch of the famous bubble chamber photograph that revealed to the world the existence of the Ω^-. In the actual photograph, the dashed lines do not appear. That is because they represent the trajectories of neutral particles, which as you know do not leave tracks. The dashed lines are reconstructed by detailed kinematical analysis. Shown on the left are the various stages of the complex reaction. All these happen in one billionth of a second or even less!

For a moment, Samios was the only man on earth to know that the Ω^- did exist exactly as Gell-Mann had predicted. It had the right mass, right spin, right parity, right strangeness — you name it! And then the news flashed all across America, across the Atlantic, and across the globe — the eight-fold way of Gell-Mann was no hocus-pocus; it was actually true!

Wonderful! The Ω^- exists! SU(3) works!! Great, but then why? It was Gell-Mann again who provided the answer.

8.8 Chapter summary

- Resonances (like the $\Delta^{++}(1238)$) observed in particle reactions are also elementary particles.
- Some of these particles are called strange because for them the strangeness quantum number S is not equal to zero. Historically, there were other reasons for calling them strange.
- Strangeness is conserved in strong interactions.
- When the number of elementary particles swelled to about two hundred, there was an urgent need to group these into families.
- Such a classification was made with the assistance of the various particle attributes.
- Gell-Mann realised that the existence of these various families could be understood with the help of the group SU(3).
- SU(3) helped to explain the various particle families just as the Lyman, Balmer and the other series helped people earlier to group the spectral lines of hydrogen into families.

9

Adding Charm to Quarks

Gell-Mann invented quarks to explain the internal constitution of baryons and mesons. He postulated the existence of the up, down and strange quarks. Are there only three quarks? This chapter deals with that question.

9.1 Introduction

In the last chapter we saw how Gell-Mann invented a taxonomy that helped to bring a great deal of order where chaos reigned earlier. We do not know if Gell-Mann was influenced by the Sakata model but for taxonomy, he did not need anything more than group theory, specifically the group SU(3).

After the first flush of success, the question arose, "Why does SU(3) work?" Gell-Mann said that there was a reason, and went on to reveal what it was. Suppose we consider the neutron (n) and the proton (p) to be two states of the same particle, i.e., the nucleon. Heisenberg showed that these two states can be described via the concept of isospin. Further, the way these states transform under rotations in isospin space is described by the group SU(2). This group SU(2) is no stranger, since it also describes the transformation properties of the two spin states of the electron. Bearing all this in mind, particularly the examples of the n and the p, one can ask, "What is the smallest number of particles (like the n and the p) that can be associated with SU(3) and using which all known elementary particles can be constructed, just as all known nuclei can be represented as suitable combinations of n and p?" The answer turns out to be three. Sakata in fact used three particles, i.e., (p, n, Λ) as the basic building blocks for the hadrons but Gell-Mann said that these were not the basic particles. According to him they were quarks — see also Boxes 9.1 and 9.2.

9.2 The quark model

Gell-Mann had been thinking about quarks for quite some time but his paper on the subject appeared only just when Samios discovered the Ω^- (February 1960).

Box 9.1

Throughout the last century, the term 'quark' was a not-so-often-used poetic term for the cry of a heron or a gull. In 1939, novelist James Joyce introduced the term in a song which occurs in his novel *Finnigan's Wake*. It goes like this:

> Three quarks for Muster Mark!
> Sure he hasn't got much of a bark.
> And sure any he has it's all beside the mark.

Once, when the quark idea was taking shape in his mind, Gell-Mann described the particle in a seminar as quork (rhyming with pork). The particle he was proposing was most peculiar, and he wanted a peculiar name for it! Months later, he was browsing through *Finnigan's Wake*, and he came across the line, "Three quarks for Muster Mark." Gell-Mann then said to himself, "This is it! Three particles make a proton or a neutron. So, let me name these particles quark." Later he said that such christening was just a gag and a reaction against pretentious scientific language.

Box 9.2

The idea that SU(3) implies three basic particles occurred not only to Gell-Mann but independently also to George Zweig, who duplicated the quark model almost exactly at the same time. Zweig called his particle 'ace', but it was the same as the quark of Gell-Mann. When Zweig came up with the ace-model, he was at CERN and according to CERN regulations, CERN staff members must publish their findings in the European journal *Physics Letters*. Zweig sent his paper to the journal but like all path-breaking contributions, it had problems with the referees and the editor. Thus, Zweig's paper remained as an unpublished CERN document. Meanwhile, Gell-Mann published his paper and established priority. Of course, no one took the quark idea seriously in those days, but for Gell-Mann who had already become famous with his SU(3), this did not matter much. For poor Zweig, it was a different story altogether. Returning to America from Europe, he sought a job in an important university on the strength of his unpublished work on the aces, but the chairman of the Physics Department thought this contribution to be nonsense and so Zweig did not get the job he wanted. Later when the quark became established, everyone hailed Gell-Mann but hardly anyone remembered that Zweig too had the same idea at about the same time. Rather like the Sudarshan story, isn't it?

Gell-Mann's scheme is summarised in Table 9.1. One can either think of it as three different quarks or three different states of the same particle (just as the n and the p can be regarded as two different isospin states of the nucleon); in the latter case, one refers to the three different quark states as having different *flavours*. Given the quarks, the cookbook for mesons and hadrons becomes very simple:

1. All mesons are made of one quark and one anti-quark.
2. Baryons are made of three quarks.
3. Anti-baryons are made of three anti-quarks.

Table 9.1 Gell-Mann's quarks, and their attributes

Name	Symbol	Rest mass (MeV)	Charge (in units of *e*)	Spin	Isospin	Strange-ness	Charm	Baryon number
Up	u	310	2/3	1/2	1/2	0	0	1/3
Down	d	310	−(1/3)	1/2	1/2	0	0	1/3
Strange	s	505	−(1/3)	1/2	0	−1	0	1/3

Using these rules, all the known particles in the mid-sixties could be readily explained (see Fig. 9.1). You would think that quarks must have been hailed and given a great reception; nothing of that kind! As Gell-Mann once wryly remarked, quarks went over like a lead balloon. Why? It was all due to the fractional charge. Hundreds of elementary particles were known but not a single one had fractional charge. Indeed, people thought that particles with fractional charge were not possible (see also Box 9.3). Gell-Mann however believed that God had not told anyone that fractional charges were banned and went ahead anyway, particularly as the quark model explained many things unexplained till then. Of course, the question remained as to why no one had ever seen a quark. It was argued that quarks were not seen because they cannot come singly out of the hadrons; they were permanently imprisoned inside — this is the idea of *quark confinement* then dismissed superciliously but now accepted as respected dogma! More about that later; meanwhile we note that in the mid-sixties the reaction to the quark model was, "Thank you Gell-Mann for the SU(3), but you can keep the quark model!"

9.3 Quarks are for real

Evidence for the existence of quarks, though eternally trapped inside hadrons, came from unexpected quarters. For that part of the story, we now move to Stanford.

Box 9.3

This digression is about the fractional charge of the quark. Soon after he proposed the SU(3) scheme, Gell-Mann paid a visit to Columbia University to give a seminar there. It is a custom in American Universities that guest speakers are treated to lunch in the University Faculty Club. Thus, Gell-Mann was taken out to lunch by Serber (his name appears in ***QR-II***). At that time, Gell-Mann was thinking about the quark model but had not developed it in full. He had the gut feeling that three particles would be needed to explain the composition of baryons but was not too sure about mesons. Serber and Gell-Mann were discussing some of these problems over lunch, when Serber asked, "If three pieces can make a proton, why not one piece and one anti-piece make a meson?" Gell-Mann replied that this was not possible because if one admitted this possibility, then these 'pieces' would have to have fractional charges. Serber nodded and slowly said, "Oh, I see why you can't form mesons the same way." The conversation then turned to other matters.

Later that night, Gell-Mann recalled the conversation and told himself, "So what if these 'pieces' have fractional charges? Just because no one has so far seen any particle with fractional charges, does it mean that God has not created any particle with fractional charges?" The next day he made bold to make this assertion during the seminar — that was when he referred to the particle as quork. Later, Gell-Mann published the idea. Today, one accepts quarks with fractional charges without batting an eyelid, even though no one will ever see a free quark (because they are trapped forever in baryons and mesons).

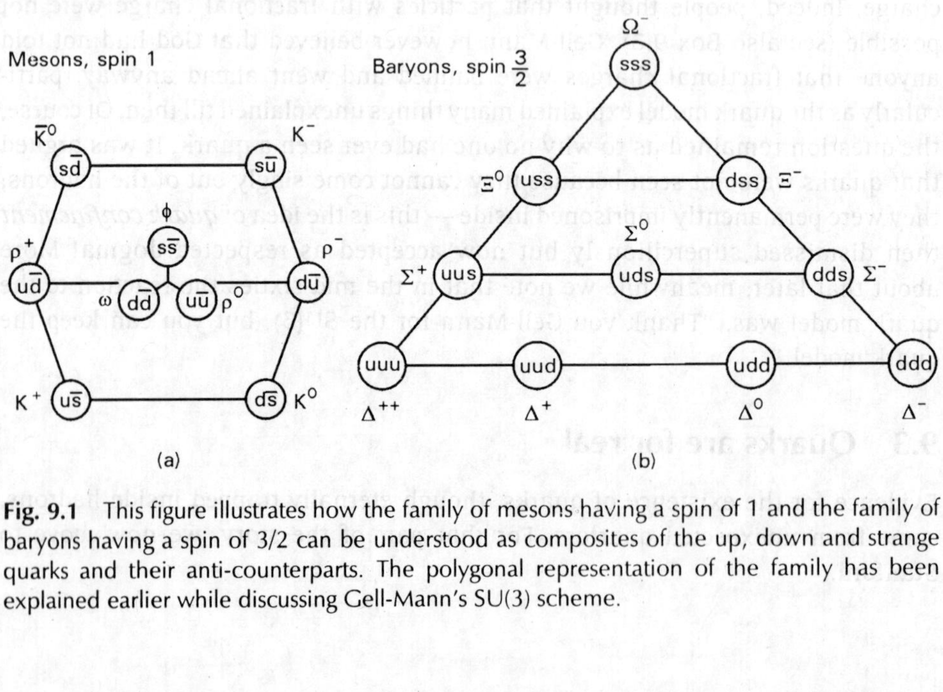

Fig. 9.1 This figure illustrates how the family of mesons having a spin of 1 and the family of baryons having a spin of 3/2 can be understood as composites of the up, down and strange quarks and their anti-counterparts. The polygonal representation of the family has been explained earlier while discussing Gell-Mann's SU(3) scheme.

The Stanford story begins at the end of World War II, when physicists at the University began exploiting war-time advances in microwave technology to build the world's best linac (see Chapter 7); to date, Stanford has maintained its number one position in the linac business. The present machine, known as the two-mile accelerator (that is the length of the linac!) was planned in the late fifties, at which time the project was known as Project M (M for monster!). Stanford scientists were quite clear what they would do with the 'monster' once it was ready — they would first accelerate electrons to very high energies, then shoot them into nuclei and finally into the nucleon itself, to probe the inside. They have been doing this very successfully for several decades now, and the picture they have been obtaining has also sharpened considerably. And one fine day, they discovered that lurking inside the proton was the quark! Before we get to that part of the story, some background is needed.

9.4 Form factor as the signature of the internal structure

Suppose a water wave encounters a pole as illustrated in Fig. 9.2; as you are aware, secondary waves would be generated — this is the phenomenon of diffraction. If there are many poles, each of them would generate its own secondary waves, all of which would then interfere with each other. By studying the interference pattern, one can figure out the way the poles are arranged — the probing of matter by waves follows this general principle. The waves used for probing are either electromagnetic waves or de Broglie waves associated with elementary particles. The schematics of such probing are illustrated in Fig. 9.3. Essentially one studies the angular distribution of the scattered particles, to obtain a signature of the contents of the target particle. Table 9.2 gives some examples of such probings.

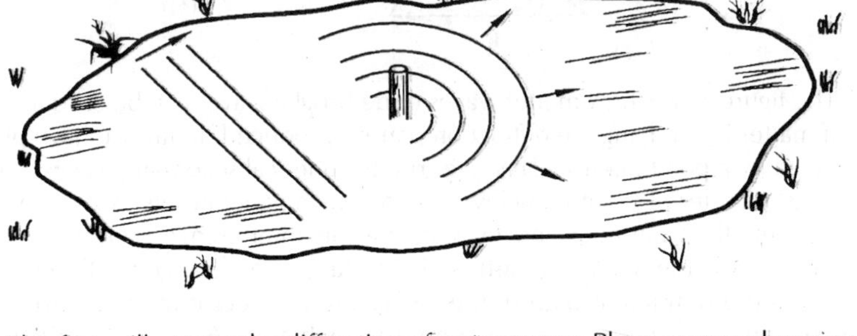

Fig. 9.2 This figure illustrates the diffraction of water waves. Plane waves advancing from the left towards the right encounter the pole from which are generated secondary waves; these arise due to diffraction by the pole. If there are many poles then there would be many sets of secondary waves which would then mutually interfere.

Table 9.2

Projectile	Target	Scale of Probing
X-rays	atom	$\sim 10^{-8}$ cm
Electrons	atom	$\sim 10^{-8}$ cm
α particles	nucleus	$\sim 10^{-13}$ cm
High energy electrons	proton	$\sim 10^{-14}$ cm
Very high energy electrons	proton	$\sim 10^{-16}$ cm

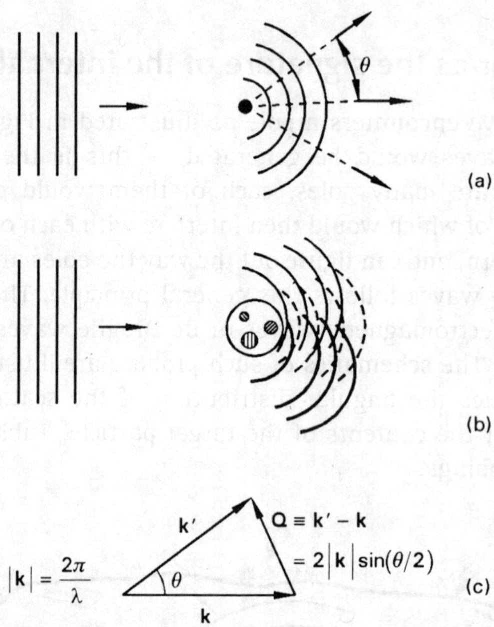

Fig. 9.3 This figure shows how matter waves or de Broglie waves can be used to explore the structure of matter by studying the diffraction pattern produced. In (a), we have plane waves being diffracted by a point particle. This gives rise to spherical secondary waves as illustrated. In (b), one has a cluster of point particles, each of which generates its own set of secondary waves. These sets then interfere and by studying this interference pattern, one can infer something about how the scattering particles in the target are arranged with respect to each other. Studying the diffraction pattern means having a detector far away from the target and counting how many particles are scattered at each angle θ. In (c) is defined the wave vector transfer $\mathbf{Q}$ in terms of the incident and scattered wave vectors $\mathbf{k}$ and $\mathbf{k}'$. The relationship between the magnitude of the wave vector and the de Broglie wavelength λ is also shown.

Let us now try to understand this in slightly greater detail. Whatever be the incident wave and whatever be the target particle, in a simple non-relativistic quantum theory of scattering, one has

$$\left(\frac{d\sigma}{d\Omega}\right) \propto |F(\mathbf{Q})|^2,$$

where

$$F(\mathbf{Q}) = \int e^{i\mathbf{Q}\cdot\mathbf{r}} \rho(\mathbf{r}) d\mathbf{r}$$

is the Fourier transform of the matter distribution $\rho(\mathbf{r})$ within the target particle. $F(\mathbf{Q})$ is often referred to as the *form factor*. Figure 9.4 shows some typical $\rho(\mathbf{r})$'s and the corresponding angular distributions. The connection between a theoretically surmised $\rho(\mathbf{r})$ and experimentally measured $(d\sigma/d\Omega)$ is usually made as follows:

$$\text{EXPT.} \rightarrow \left(\frac{d\sigma}{d\Omega}\right) \rightarrow \text{CONST}\cdot|F(\mathbf{Q})^2_{\text{Expt.}} \rightarrow \text{COMPARE} \leftarrow |F(\mathbf{Q})^2_{\text{Th.}}| \leftarrow [F(\mathbf{Q})]_{\text{Th.}} \leftarrow \rho(\mathbf{r}) \leftarrow \text{THEORY}$$

Historically, the first scattering experiment to give a major clue about internal structure was that of Rutherford. In 1896, J.J. Thomson discovered that the atom was not indestructible but contained tiny particles called electrons. The electron was found to be negatively charged but since the atom as a whole is electrically neutral, it was clear that the atom must also contain a positive charge; but no one knew where this positive charge was. Thomson argued that it was distributed uniformly inside, as illustrated in Fig. 9.5.

Around 1909, Rutherford began to shoot alpha particles from naturally occurring radioactive materials at targets made of different materials, and he studied the angular distribution of the scattered particles. The experimental arrangement used by him is shown in Fig. 9.6. If the Thomson model of the atom were correct, then most of the alpha particles should be scattered in the forward direction. However, Rutherford found that quite a few were scattered through large angles and backwards, many more than expected on the Thomson model. Rutherford then conjectured that the alphas were bouncing off tiny positive particles embedded within the atom, and that these positive particles were all closely huddled together. As you are probably aware, this led to the Rutherford model of the atomic nucleus which later resulted in the Bohr model of the atom.

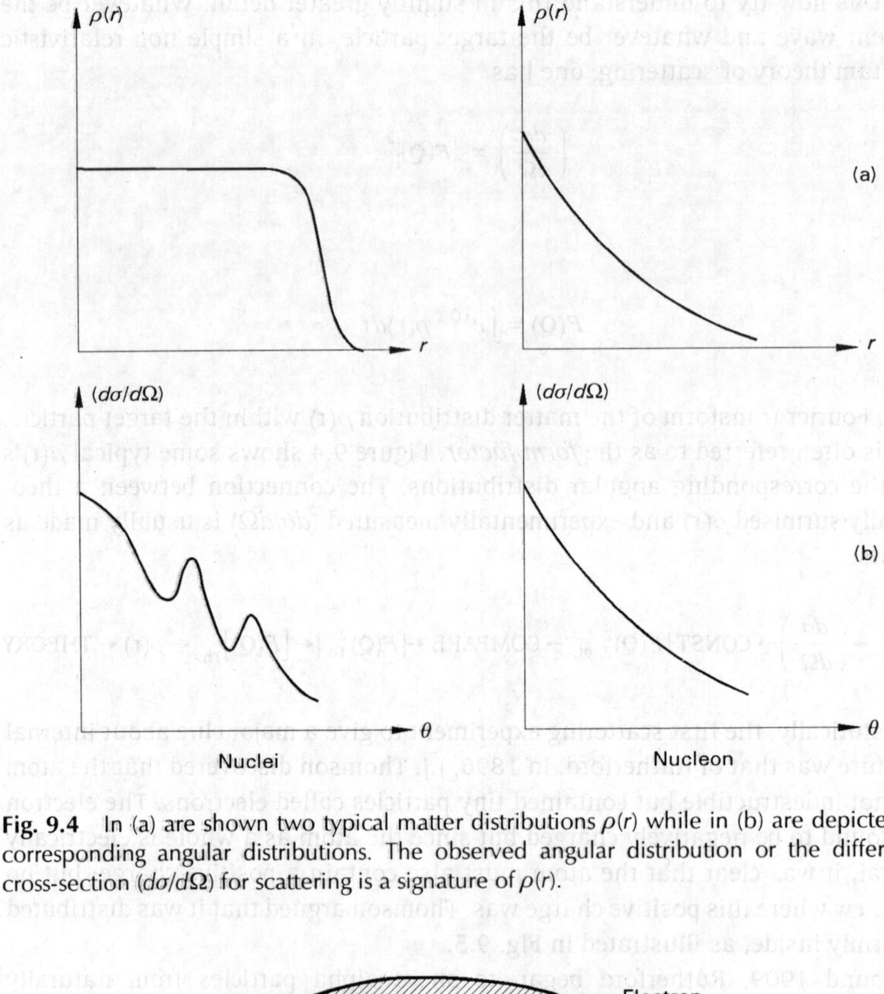

Fig. 9.4 In (a) are shown two typical matter distributions $\rho(r)$ while in (b) are depicted the corresponding angular distributions. The observed angular distribution or the differential cross-section $(d\sigma/d\Omega)$ for scattering is a signature of $\rho(r)$.

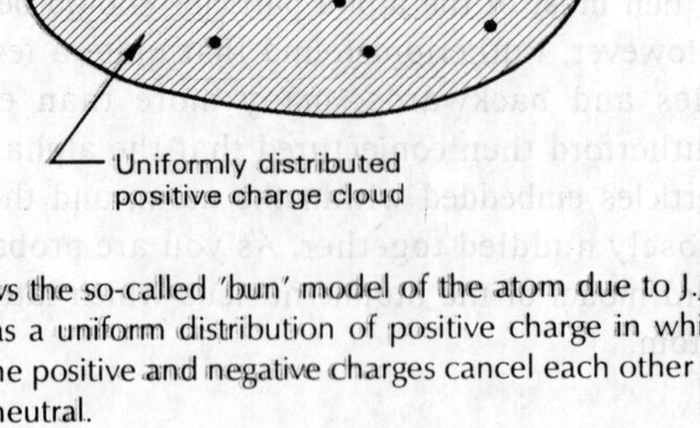

Fig. 9.5 This shows the so-called 'bun' model of the atom due to J.J.Thomson. In this, the atom is visualised as a uniform distribution of positive charge in which are embedded the (point) electrons. The positive and negative charges cancel each other to make the atom, as a whole, electrically neutral.

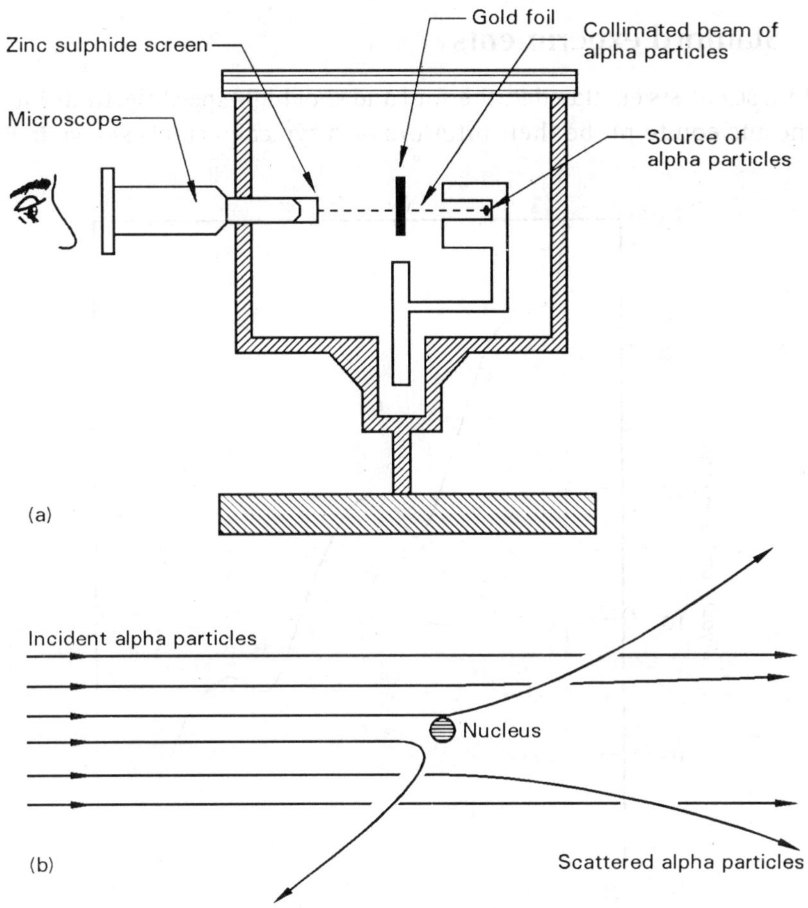

Fig. 9.6 (a) shows a schematic diagram of the apparatus used by Rutherford. Alpha particles from a naturally radioactive source are collimated to form a beam, which is allowed to fall upon a thin gold foil. The α particles are then scattered in various directions, and the scattered particles are observed with a microscope, the front end of which is covered with glass coated thinly with zinc sulphide, which, being a phosphor, produces a flash of light on being impinged by an α particle. This flash is observed through a microscope. The microscope-cum-screen arrangement can be rotated with respect to the foil, which enables the experimenter to vary the angle of scattering. At each angle of scattering, one counts the number of α particles scattered in a fixed amount of time. By plotting the number scattered at each angle θ, one obtains the angular distribution of scattering (the diffraction pattern). (b) shows schematically how the individual α particles are scattered by the nucleus. Now the atom has the size of about an Ångstrom, and if the positive charge were to be uniformly distributed in the atom, as Thomson envisaged, then most of the α particles would be scattered in the forward direction, i.e., at small values of the scattering angle θ. Rutherford found, however, that an 'unreasonably' large number of particles were scattered at backward angles. He then reasoned that the positive charge in the atom must be packed into a very tiny region of diameter 10^{-5} Å, which is now called the atomic nucleus or simply the nucleus.

9.5 Stanford experiments

Stanford scientists emulated Rutherford and shot high-speed electrons into nuclei and the nucleon to probe their interiors — a typical result is shown in Fig. 9.7.

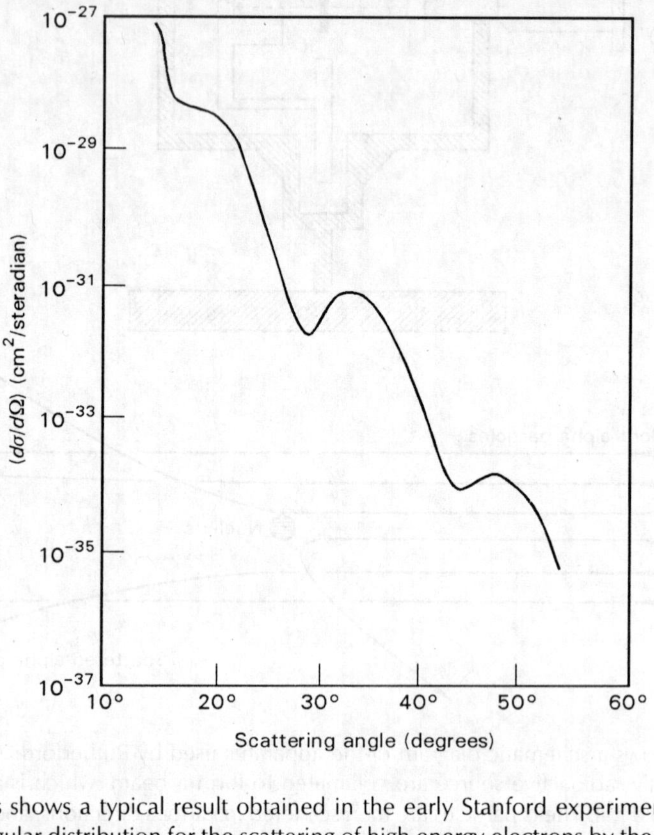

Fig. 9.7 This shows a typical result obtained in the early Stanford experiments. Presented here is the angular distribution for the scattering of high energy electrons by the ^{40}Ca nucleus. The undulations are due to the diffraction of the electron by the positive charge distribution within the nucleus. See also Fig. 9.4(b).

Now consider the e–p scattering. In a casual description, we would merely say, "The electron goes inside the proton, takes a good look, and then comes outside to report what it has seen." In more technical language, the electron interacts with the proton via a virtual photon as in Fig. 9.8. At modest energies, the scattering is purely elastic, i.e.,

$$e + p \longrightarrow e + p\,.$$

The proton recoils due to the impact received but it is still a proton. But if the incident electron energy is very high, all sorts of other stuff comes out — pions, kaons, ..., you name it. Such a process is referred to as *deep inelastic scattering*, and is symbolically represented as

$$e + p \longrightarrow e + X,$$

where X represents the hadronic debris from the exploding proton (see also Fig. 9.8).

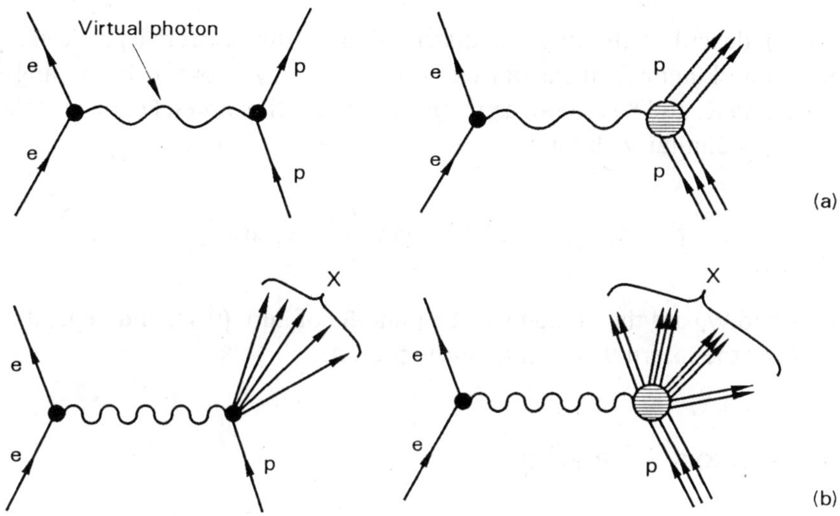

Fig. 9.8 This figure illustrates through Feynman diagrams, the difference between ordinary elastic scattering of electrons by protons, and the deep inelastic scattering of very high energy electrons by protons. Ordinary elastic scattering occurs at moderate electron energies, and this was the phenomenon studied in the early Stanford experiments. The interaction between the electron and the proton occurs via a virtual photon as shown in (a). The momentum of the photon is small and correspondingly the wavelength is large; thus the probing of the proton is done with poor resolution. On the right, the e–p scattering is illustrated in terms of the quark picture of the proton. When the electron energy is high, the momentum of the virtual photon that is exchanged is correspondingly high, and the resolution also improves. At high electron energies, one has an intense reaction as a result of which many particles are produced as shown in (b). The debris is collectively designated as X. The associated quark picture is shown on the right.

The deep inelastic scattering process can be understood roughly as follows. As the energy of the electron increases, the wavelength of the virtual photon decreases, and the proton gets probed on smaller and smaller scales of length. To put it differently, at high energies, the electron does not bounce off the proton as a whole, but ricochets off particles inside, if there are any—this is where it is similar to the Rutherford experiment.

James Bjorken gave the first indication of what one is likely to see in such a deep inelastic scattering experiment. He argued that whether it is elastic scattering or deep inelastic scattering, there are two basic currents which cause the transitions between the quantum states of the system, the lepton current related to the electron and the hadron current related to the proton. For elastic scattering, the theory gives the result

$$\{\quad\}_{ep\to ep} = (.....)\left\{2K_1\sin^2\frac{\theta}{2} + K_2\cos^2\frac{\theta}{2}\right\}, \qquad [9.1]$$

where {....} describes the angular distribution of the scattered particles, (....) contains some kinematical factors we need not worry about, θ is the angle of scattering, and K_1 and K_2 are two constants related to the target. For deep inelastic scattering one similarly obtains

$$\{\quad\}_{ep\to ex} = (....)\left\{2W_1\sin^2\frac{\theta}{2} + W_2\cos^2\frac{\theta}{2}\right\},$$

where W_1 and W_2 are the analogues of K_1 and K_2 of Eqn. [9.1], and depend upon the details of the internal structure of the proton.

9.6 Discovery of scaling

The Stanford experiments showed that a quantity related to W_2 was independent of q^2 (the relativistic analogue of Q^2) — see Fig. 9.9. This insensitivity to q^2 is referred to as *scale invariance* or simply as *scaling*. Now q has the dimension L^{-1}; W_2 being independent of q^2 means that W_2 is the same on all scales of length.

A scale is needed in order to compare quantities that vary in magnitude but are not vastly different from each other. For example, in the world of atoms, the Ångstrom is the scale of length; in terms of the unit Ångstrom, we can quote the diameters of the various atoms, lengths of molecular bonds, etc. In the world of nuclei, the Fermi is the scale of length, and so on. We can always choose a characteristic length in terms of which all lengths can be sized and in terms of which one can talk of big and small. Suppose there is no such thing as big or small; it implies that the object one is dealing with is a point object. Thus, the discovery of scaling in deep inelastic scattering meant that the constituents of hadrons, whatever they might be, were point particles.

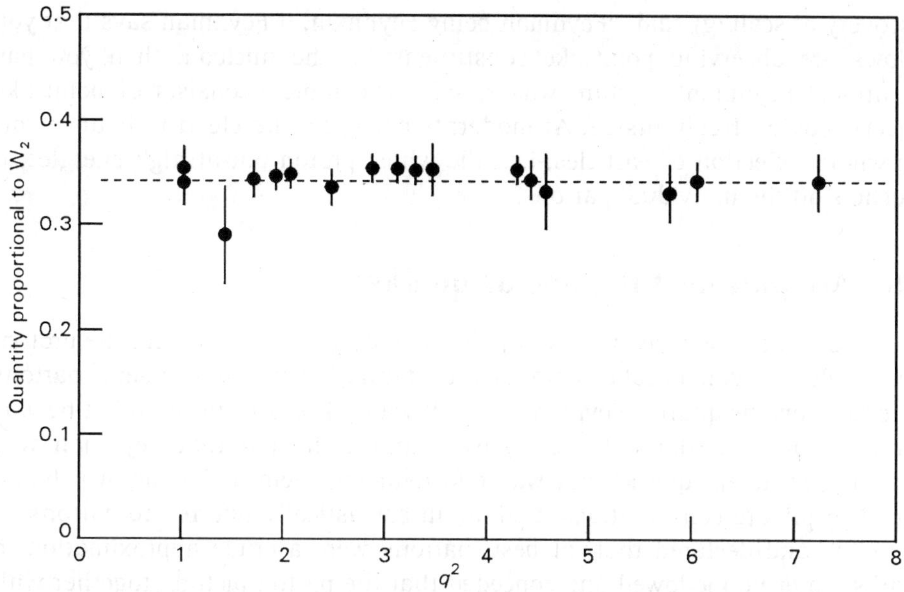

Fig. 9.9 This figure illustrates the scaling property discovered in the Stanford high energy deep inelastic scattering experiments. Plotted here is a quantity proportional to W_2 as a function of q^2. The dots show the results of experiments; the vertical bars are indicative of the range of experimental errors. The dashed line is a guide to the eye, and shows that the quantity concerned does not vary; this is scaling, which Feynman explained by assuming that the proton contains point particles, i.e., the partons.

9.7 Enter Feynman and the partons

It was at this stage that Richard Feynman paid a brief visit to Stanford. Shortly before this, Feynman had become interested in hadronic physics. He had heard about the quark model which was not surprising considering that Gell-Mann's office was just next door to his (in Caltech or the California Institute of Technology). However, Feynman preferred to think about hadrons independently. He assumed that they contained small particles which were possibly point-like, and which were possibly the quanta of some unknown field. Feynman called these particles *partons*.

When Feynman came to Stanford, they naturally rolled out the red carpet and eagerly showed him all their scaling results which had been explained by Bjorken on the basis of a current–current model. Feynman gave them a patient hearing and did not say anything at that time. Later in his hotel room, he suddenly realised that if the protons contained partons, it would automatically lead to scaling; it did not take him long to demonstrate that with a calculation. Conversely, if experiments showed scaling behaviour, it meant that the proton actually contained partons! Herbert Kendall (who later received the Nobel Prize for his discovery of scaling) said, "Feynman being Feynman, if Feynman says that you fellows are observing point-like constituents in the nucleon, then you pay attention." Feynman's picture was simple. The proton consists of point-like objects moving freely inside. At moderate energies, the electron bounces off the whole collection of particles, i.e., the whole proton but at high energies, it ricochets off the individual partons.

9.8 Are partons the same as quarks?

Once people became convinced that point-like objects existed within the proton, objects which Feynman called partons, they naturally began to wonder if partons were the same as quarks. Feynman himself was asked this question but he was not too sure. Gell-Mann, however, was furious; he felt that Feynman was unwilling to accept quarks and was reintroducing them under his own brand name! In private conversations, Gell-Mann sarcastically referred to partons as 'put-ons', and declared that, at best, partons were a crude approximation to quarks. Later he mellowed and conceded that the parton picture, together with scaling, helped to convince people at a crucial time that protons and other hadrons were indeed made of smaller constituent particles which freely moved inside.

Meanwhile, further experiments showed that scaling was breaking down, which meant that the simple point-parton model needed refinement; and that was supplied by the quark model. This brings us to the topic of quark–quark interaction. Later we shall go into details but for the present we note the following:

1. The quark and the anti-quark can interact amongst themselves.
2. Such interaction is the foundation of the strong force, for which Yukawa had given a naive model earlier.
3. Just as the electromagnetic force is mediated by photons, the quark–quark force is mediated by gluons; like the photon, the gluons are massless.
4. Whereas there is only one type of photon, there are eight types of gluons.
5. Unlike the photon, the gluons can interact with each other. This is related to the fact that whereas the photon does not have (electric) charge, the gluons possess (colour) charge.

The step-by-step modification of the simple parton model to the quark model is schematically shown in Fig. 9.10. Here, on the right hand side, are sketched the behaviour of a quantity F_2 which is related to the W_2 we had earlier. This F_2 is plotted as a function of x which is the fraction of the parent proton momentum carried by the constituent particle(s); we need not go further into the technicalities of F_2. What you see on the right is the behaviour of F_2 under different conditions.

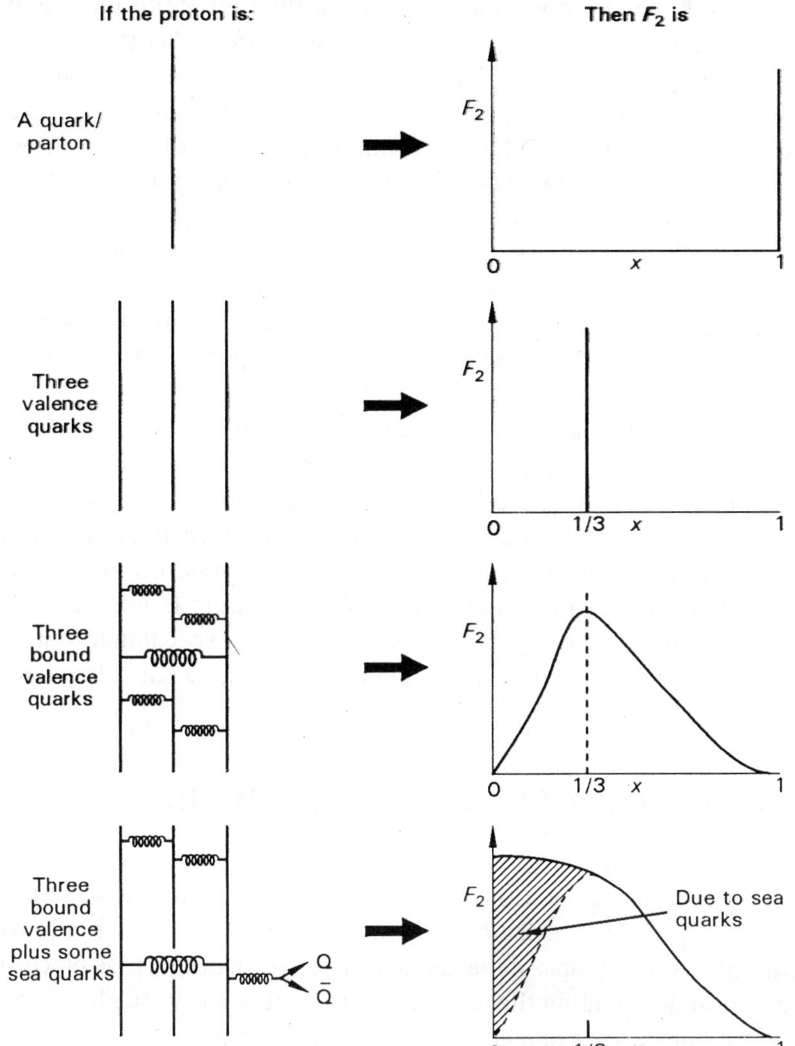

Fig. 9.10 This figure illustrates how one can progressively go from the intuitive parton picture of Feynman to the more complex and realistic quark picture of Gell-Mann. The explanations for this figure can be found in the text.

To start with, we suppose that the proton is just a point particle; let us call it a parton. Then F_2 is the delta function at $x = 1$. According to Gell-Mann, the proton is made up of three quarks. Let us assume that these quarks do not interact; we further suppose that these quarks are point particles; this would mean that the proton can be regarded as made up of three partons. In this case, F_2 becomes a delta function at $x = 1/3$. Next, we suppose that these point particles interact with each other via gluons. We are now coming closer to the Gell-Mann picture and we might as well stop calling the constituent particles partons, because they are not quite the partons that Feynman originally visualised in order to explain scaling. F_2 is now smeared into a continuous distribution which peaks at $x = 1/3$. One more piece of reality must be added next and that is that at high q^2, interaction between quarks produces a lot of debris of quark–anti-quark pairs, etc. The quarks which form the basic Gell-Mann triplet are called *valence* quarks while those in the debris are called *sea* quarks. So you see how far the true picture (which is what Gell-Mann worried about) is, from the model picture assumed by Feynman (strictly for explaining scaling). If you like you might perhaps loosely say in the language of QED that the quark is a 'dressed' parton, dressed by interactions, mediated by gluons.

Before leaving the topic of scaling as well as its breakdown, one important remark needs to be made. At the present time, leptons and quarks are believed to be the ultimate building blocks of matter. While no one is suggesting that leptons have further internal structure, already the idea is being aired that quarks have within them point particles called prequarks. Do prequarks have pre-prequarks within them? No one knows; but if Nature is telescoped in this fashion, i.e., as $(\text{pre})^i$-quarks ($i = 0, 1, 2, \ldots$), then it is speculated that at each level of i, scaling will first appear and then break down, just as it has now done at level $i = 0$.

9.9 A digression on the second type of neutrino

High energy physics may be likened to a triple helix with three intertwined strands, dealing respectively with electromagnetic, weak and strong interactions. Top particle physicists, particularly theorists, constantly do strand-hopping, concentrating on the hot topic of the moment. The parton drama took place around 1969, and for understanding the next part of the quark story, one has to go back several years, hop strands and get back to the weak interaction.

The year is 1957. Parity has been overthrown and the Universal Fermi Interaction has been discovered (see Chapter 5). Schwinger, the hero of QED (see *QR-II*), began to think seriously about weak interactions, and he wondered whether the weak and the electromagnetic interactions might be facets of the same phenomenon. Thinking soon translated into a series of lectures, following which appeared a typical Schwingerian, oracle-like paper! For decades, Einstein struggled to unify the electromagnetic and the gravitational forces but he did not succeed, and now we know why. In the thirties, the electromagnetic and the gravitational forces seemed like twins, whereas the weak force had hardly any form or structure. In the mid-fifties the situation regarding the weak force was not much better, but with rare insight, Schwinger saw that if the e.m. force is to seek a companion, it must be the weak force — the two seemed so different and yet, Schwinger was able to glimpse the underlying unity; that is what genius is all about!

Schwinger's theory has not stood the test of time but two important ideas which he introduced have survived, with some modification no doubt. Firstly, Schwinger recognised the muon as a lepton of a quite different variety, *distinct* from the electron. Schwinger then added a corollary — *Just as the electron has a neutrino to go along with it, so does the muon*. Thus, according to Schwinger, there were two neutrinos, the electron neutrino and the muon neutrino.

When the muon was discovered, it was a great surprise — no one expected it and in fact it was an embarassment which Rabi echoed by asking, "Who ordered that (particle)?", a statement that has since become famous. Schwinger now (effectively) said, "Don't think that the electron is the only kind of charged lepton in Nature. There is another one, and it is the muon; further, the muon has its own special neutrino." Are there any more leptons? Today we know that there are three lepton families; and it looks like there are only three families.

To get back to the Schwinger story, the second thing he said was that the W.I. is mediated by messenger particles just as the e.m. force is mediated by the photon. However, there were important differences. At first, said Schwinger, there were two messengers in the case of the W.I. Today, these are called W^+ and W^-. Moreover, the W^+ and the W^- have mass. Schwinger thus kicked off the electro-weak unification process, and asked his student Sheldon Glashow to look deeper into the matter. Glashow did so, wrote a thesis, got a Ph.D. and there the matter rested, at least then. Nothing great had been achieved, but the first step had been taken. The year was 1960, and nobody knew then that the electro-weak unification had gotten underway.

Meanwhile, people began to wonder if there was such a thing as a muon neutrino. The only way of making sure was via an experiment but experiments involving neutrinos were known to be difficult since the neutrino is notoriously shy — it refuses to interact with matter. It was then realised that at high energies, neutrinos become less shy, meaning that they interact with other elementary particles with greater ease. The question now shifted to, "How to produce a copious supply of high energy neutrinos?" The answer emerged when high energy accelerators were developed. Finally, there was the problem of detection, but that problem was also solved.

In the early sixties, two teams of experimenters, one at CERN in Geneva, and another at the Brookhaven National Laboratory in America raced against each other to find out if the muon neutrino really existed. After high drama — a common feature in high energy experiments — the American team led by Leon Lederman was able to convincingly prove that the muon neutrino did exist. Lederman won the Nobel Prize for this discovery.

9.10 Enter charm

The year is 1964. Glashow arrives as a visiting scientist at the Niels Bohr Institute in Copenhagen, where Bjorken is already a visitor. The two work together and make a table as below:

Lepton family	Quark family
Electron	Up
Electron neutrino	Down
Muon	Strange
Muon neutrino	?

When they stared at the table, Glashow and Bjorken saw that there was an asymmetry — four members in the lepton family and only three members in the quark family. God, they said, could not have been partial to the leptons; there must surely be one more quark, they asserted, and went on to name it *charm*. With the charm quark added, there was balance — four leptons and four quarks. Glashow and Bjorken were mighty pleased but the world was not impressed. Remember the year was 1964 and no one took quarks seriously at that time.

Although charm attracted very little attention in the beginning, it served one immediate and useful purpose — that of explaining the absence of what is called *neutral currents*. We shall go into neutral currents later, but here it is enough to know that many experimenters were furiously trying to produce evidence for neutral currents but could not find any; Bjorken and Glashow were able to explain why, using the charm-quark concept. Inspite of this minor achievement, charm attracted little notice and Glashow went about describing it as an invention that awaited discovery! Speaking at a conference of experimenters in 1974, Glashow said, "There are three possibilities. One, charm is not found and I eat my hat. Two, charm is found by you (meson specialists) and we celebrate. Three, charm is found by outlanders (meaning experimenters elsewhere, especially at CERN) and you eat your hats." In less than six months, there was dramatic evidence for charm; it was found by 'outlanders' and all the experimenters, addressed by Glashow, ate hats (made of chocolate!).

9.11 The J/ψ drama

Some time prior to the event just mentioned, Glashow talked to Samios, the hero of Ω^-, about conducting an experiment to produce evidence for charm. Samios thought it was a good idea and did launch an experiment. After some effort, his team came upon a solitary bubble chamber picture that suggested that the charm did exist. But the evidence was not tight or clearcut. While with a single picture he could and did earlier announce to the world that the Ω^- did exist, this time he could not make a similar announcement; so Samios and his collaborator Palmer prayed for more events and better pictures to confirm their finding, but the pictures were slow in coming. If they had come as prayed for, Samios and Palmer might have picked the Nobel Prize; instead, the Prize went to two others.

One of these was Samuel Ting of MIT who, like Samios, was using the Brookhaven accelerator. Ting was not specifically hunting for the charm. His philosophy was simple: "There are two kinds of experimenters. The first kind do what theorists want them to do. The second follow their own ideas. I am of the second kind", he once declared. Ting was carefully looking for mesons produced when high energy protons are slammed into nucleons. He had a hunch that he would be able to find mesons that are long-lived. Theorists told him that he was foolish and wasting his time, but Ting trusted his own intuition and eventually, the gamble paid off.

In late September 1974, Ting's group found evidence for an extraordinarily sharp resonance at 3.1 GeV. The resonance was so sharp that it was unbelievable — remember, a sharp resonance means a long-lived particle. While the others were excited and elated, Ting was suspicious. He wanted to be doubly sure that the finding was true and not a result of the way the experiment was carried out; many people had reported sensational discoveries only to find later that their experiment was wrong; Ting did not want his carefully built-up reputation going down the drain due to careless measurements. He clamped censorship on his team and banned them from publicising their data, till it was confirmed and a signal was given. Ting prepared to reconfirm the findings but just then the accelerator was shut down for maintenance. When it was working again, someone else was scheduled to have access to the machine, and his turn would come only three weeks later. For Ting, this was a frustratingly long delay; meanwhile, someone else somewhere could beat him to it — this is the constant fear of scientists at the frontier, not being the first to make a discovery.

Ting's worry was not unfounded. Stanford had a new facility called the SPEAR built by the Stanford scientist Burt Richter, in which electrons and positrons went around a circle in opposite directions and collided (see Chapter 7). Mesons which Ting was looking for could be easily spotted using the SPEAR. If Burt Richter and his team got wind of the 3.1 GeV resonance, they would be able to beat Ting in no time at all. After the SPEAR was first commissioned, Richter's team did a rough scan going all the way from 2.4 GeV to 5 GeV. They took data at 2.4, 2.6, 2.8, 3.0, 3.2, ... The sharp resonance spotted by Ting at 3.1 GeV had been skipped over! After this scan, Richter's team took a break while the Stanford accelerator went through a maintenance shutdown. When it was restarted, Ting heard that Richter's team had moved over to 5 GeV and were scanning in that energy region. Ting heaved a sigh of relief and concentrated on confirming his sensational finding.

Around this time, Mel Schwartz (who, along with Lederman, discovered that the muon has its own type of neutrino) an old friend of Ting now at Stanford, met Ting at Brookhaven. Having heard rumours that Ting had found something unusual at 3.1 GeV, he asked his friend about it. Ting was cagey and denied having found any such resonance but Mel became suspicious and also felt hurt that Ting did not trust him to respect confidences. Mel went back to Stanford, convinced that Ting definitely had something up his sleeve. Meanwhile, back at Stanford, they had independently got some inkling of 'funny business' at 3.1 GeV based on some data they had earlier collected. Mel's news gave added impetus for a careful search around 3.1 GeV. At that time, the experiments were set for 5 GeV and going down to 3.1 GeV meant some problems but since the stakes were high, no difficulty was too much. And sure enough, with their superior set-up, they clearly saw the sharp resonance at 3.1 GeV. The resonance particle was named SP (3105), SP for SPEAR, and 3105 for 3.105 GeV. Later, SP (3105) was renamed ψ (psi), since elementary particles were usually named using Greek letters. It was Saturday, November 9, 1974, and it was decided to announce the discovery of ψ to the world on the following Monday.

On Sunday, November 10, Ting flew all the way from the East Coast of America where his lab was located to Stanford on the Pacific coast, to attend a routine meeting. As he went to his hotel room, he got a frantic call from his colleagues in Brookhaven that Stanford had discovered the 3.1 GeV resonance and was ready to announce it. Ting acted immediately, and the next day in the course of the meeting, he told the Stanford scientists of his discovery. Of course, it was no surprise to the Stanford group, since they had already discovered the resonance themselves. Ting and Richter had picked up the 3.1 GeV particle by entirely different methods, and it was quite clear that it was a genuine discovery. Both immediately rushed their findings to the premier journal, *Physical Review Letters*. Meanwhile, Ting conveyed the news to a particle physics lab at Frascati (near Rome) in Italy. In Frascati, they had an $e^+ e^-$ collider, designed to go upto 3 GeV. Frascati quickly pushed the energy limit to a little beyond 3 GeV and spotted the 3.1 resonance.

The *Physical Review Letters* published the papers of the three groups with lightning speed in its issue of December 2, 1974. Ting's paper submitted on November 12 said:

> We report the observation of a heavy particle J, with mass m = 3.1 GeV and width approximately zero. The observation was made from the reaction $p + Be \rightarrow e^+ + e^- + \pi$, by measuring the $e^+ e^-$ mass spectrum precisely. .....

The paper that Richter and his co-workers submitted on November 13 said:

> We have observed a very sharp peak in the cross-section for $e^+ e^- \rightarrow$ hadrons $+ e^+ e^-$, at a centre of mass energy of 3.105 ± 0.003 GeV. The upper limit to the width at full maximum is 1.3 MeV. ....

The Frascati paper submitted on November 18 began as follows:

> Soon after the news that a particle of 3.1 GeV with a width consistent with zero had been observed at the Brookhaven National Laboratory, ... it was immediately decided to push ADONE (the accelerator at Frascati) beyond its normal energy limit (2 × 1.5 GeV) to look for this particle.

All this should give you an idea of how keen the race can be in high energy physics, and how sometimes, people produce results in two days that otherwise would have taken weeks or even months. See also Fig. 9.11.

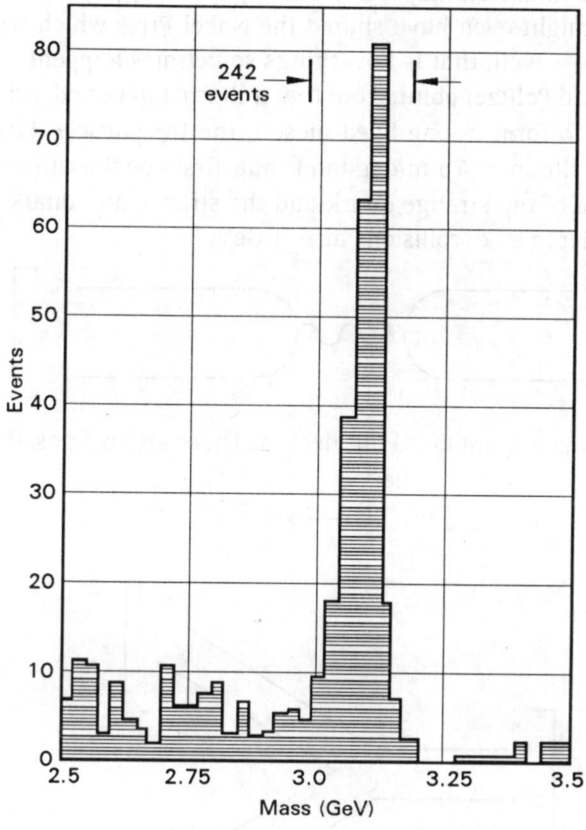

Fig. 9.11 This figure which is the outcome of the experiment of Ting and his collaborators, shows the narrow resonance spotted at 3.1 GeV and labelled by them as the J particle. Notice the paucity of events, despite long hours of logging data. This is because the Brookhaven machine was not quite suited for the observation of the J particle. By contrast, using the e^+e^- collider at their disposal, Stanford scientists were able to pick up this particle with great ease in just about two days or so. So too was the case with the Frascati scientists. Further, the width was more accurately measured by the Stanford scientists. All this shows how great Ting's achievement was — he had to struggle much harder and yet he got there first.

Ting called the new particle J, whereas Richter, as already noted, called it ψ. Today, the name J/ψ is often used; at other times, it is referred to simply as J or as ψ. The Ting–Richter discovery sent waves of excitement, and this event is often referred to as the 'November Revolution'. Why, and what is the significance of the J/ψ? The feature of J that attracted attention was its unusually long lifetime, reflected by its very narrow width — recall that Ting said that the width was nearly zero. A clue to the nature of J/ψ soon came via a paper by Applequist and Politzer, prepared a short while before the discovery was announced. For some reason, the authors felt hesitant about publishing the paper and delayed it; but once the J was announced they rushed their paper into print. Had they published it before, they might even have shared the Nobel Prize which went to Ting and Richter in 1976 — well, that is how things sometimes happen!

Applequist and Politzer pointed out that a charm quark and a charm anti-quark could combine to form a long-lived meson, the J/ψ particle. Fine, but why the unusually long lifetime? To understand that, first consider the φ meson, which is a bound state of the strange quark and the strange anti-quark (ss). This φ had been found earlier in $e^+ e^-$ collisions at ~ 1 GeV.

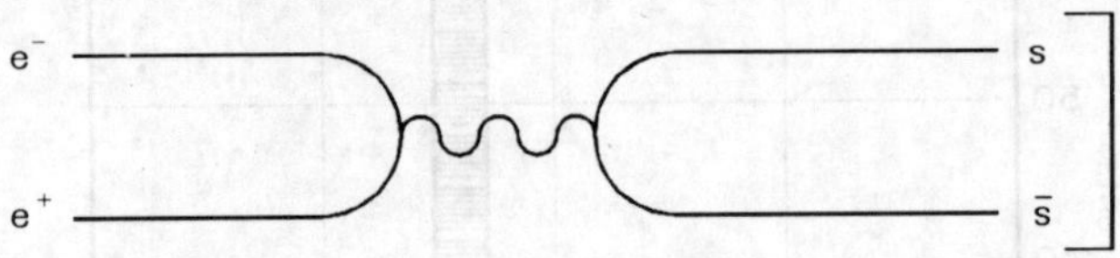

The φ meson is not immortal but decays. There are two possibilities for this:

$$\text{i.e.,}\quad \varphi \longrightarrow K^+ K^-$$
$$\text{or,}\quad \varphi \longrightarrow \pi^+ \pi^0 \pi^-.$$

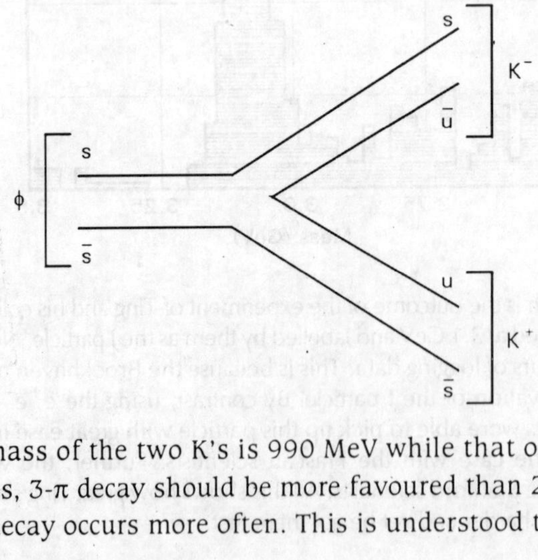

Now, the mass of the two K's is 990 MeV while that of the three π's is only 415 MeV. Thus, 3-π decay should be more favoured than 2-K decay. But it turns out that 2-K decay occurs more often. This is understood through what is called the OZI rule (after Okubo, Zweig and Iizuka). In the J/ψ there are, analogously, two possibilities:

$$\psi \longrightarrow D^+ D^-,$$

$$\text{or,} \quad \psi \longrightarrow \pi^+ \pi^0 \pi^-.$$

The former decay is kinematically forbidden because the mass of 2D is more than that of the J. The 3-π decay is not forbidden but supressed because of the OZI rule. Thus, the J/ψ decay rate is even smaller than that of the φ. A simple way of understanding the long lifetime is by means of the concept of *asymptotic freedom*, which we shall discuss later. Basically what this says is that at short distances, the quarks and the anti-quarks behave as if they are free; in other words, they do not interact much, which in turn implies that the chances for the quark and the anti-quark to annihilate into other particles is small.

The introduction of the charm quark results in a 'population explosion' among the hadrons, and supermultiplets which were earlier organised into polygons now have to be organised into polyhedra (see Fig. 9.12; you can try constructing other such polyhedra, based on Figs. 8.11 and 8.12).

9.12 The third generation

In the beginning there was the electron and its neutral cousin the neutrino. Associated with these were the up (u) and the down (d) quarks. These lepton and quark pairs form what is often called the 'first generation'. The 'second generation' is made up of the lepton pair consisting of the muon and the muon neutrino on the one hand, and the strange (s) and the charm (c) quarks on the other. There is also a 'third generation', consisting of the tau (or tauon) and the tau neutrino on the lepton side, and the bottom (b) and top (t) quarks on the quark side (earlier, the bottom and the top quark used to be referred to as beauty and truth; somehow, these romantic names have been displaced by more prosaic ones).

The third generation story consists of three parts, first the discovery of the *upsilon* (Υ) meson, which is like the J. When it was discovered, people wondered whether it was due to a new type of quark. They were not quite sure. Meanwhile, the tau lepton was discovered, and people now believed that there was indeed a third generation of quarks; promptly, the Υ was identified as the bound state of the bottom quark and bottom anti-quark. If u had a companion, namely d, and s had a companion namely c, what about b? Yes, declared the pundits, it too must have a companion, the top quark (t). People searched hard for evidence for the top quark but it proved elusive for quite a while on account of its high mass; finally, it revealed itself a couple of years ago. What follows is a brief account of the above discoveries.

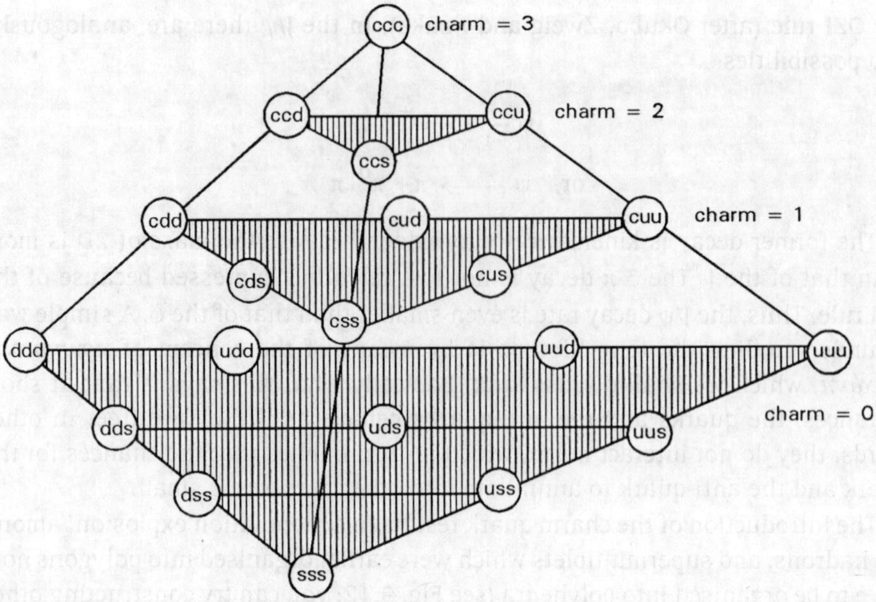

Fig. 9.12 Recall Fig. 9.1 (b). This is based on Gell-Mann's original quark scheme involving just the up, down and the strange quarks. With the discovery of the charm quark, one more dimension was needed in making such representations — polygons now became polyhedra. This is illustrated here by considering the enlargement of the decimet of Fig. 9.1 (b). The Gell-Mann deciplet forms the base of the polyhedron where charm is zero; at the higher levels, the charm is non-zero. Particles corresponding to these higher levels, which were not known earlier, have now been found.

The tau lepton (τ) was discovered by Martin Perl and William Kirk in 1974, using the SPEAR, which was used by Richter to discover the ψ. The mass of the tau is very large, ~ 1.784 GeV, which is to be compared with ~ 0.5 MeV for the electron. Perl called this lepton tau because the Greek letter tau is the first letter of the Greek word for third. Remember that lepton means a light particle. Since the tau is heavy, it is sometimes called a heavy lepton! You may notice that the term 'heavy lepton' is an oxymoron!

In 1968, Leon Lederman and others were using the Brookhaven accelerators to study the reaction

$$p + N \longrightarrow \ell^- + \ell^+ + \text{anything},$$

where N stands for the nucleon and ℓ for the lepton (see also sketch on top of page 177).

The yield of muon pairs was expected to decrease with energy as schematically indicated in Fig. 9.13(a), but the data vaguely suggested a shoulder of sorts. Nothing firm could be concluded but there was a hint of something interesting hidden. Then came the November Revolution. Lederman immediately realised

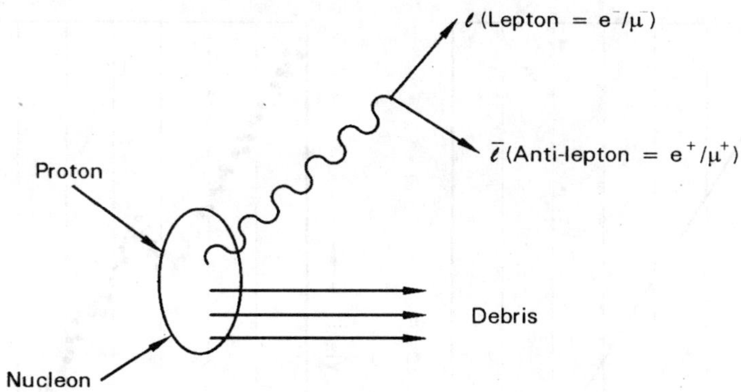

that the shoulder seen earlier could mean something very significant and promptly moved to the Fermilab where a more powerful accelerator than the Brookhaven machine was available. Better data could now be obtained, and the shoulder revealed itself as a clear resonance (see Fig. 9.13(b)); Lederman had discovered the upsilon.

Till 1974, people did not even believe in the charm quark. Now, all of a sudden, evidence had been uncovered for a third-generation lepton, and Lederman had discovered a particle that was similar to the J/ψ. Did that not clearly mean that there was a third generation quark, like bottom? What about looking for it? Many (including Martin Perl) in fact did but the evidence was not clearcut. Firm proof of the existence of the bottom quark came finally from the experiment of Nariman Mistry and others at Cornell University in America. At Cornell they had an e^+ e^- collider whose energy was just ideal for the exploration of the bottom quark. The idea behind the Cornell experiment can be understood with the help of the sketch below:

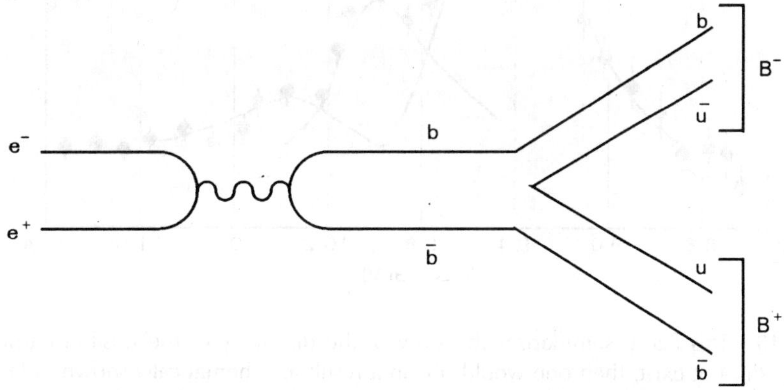

The electron–positron collision produces a virtual photon which then materialises as a (bb) bound state, i.e., the upsilon meson. The upsilon then decays into a B meson and an anti-B meson, i.e., into a BB pair. Subsequently, the B and

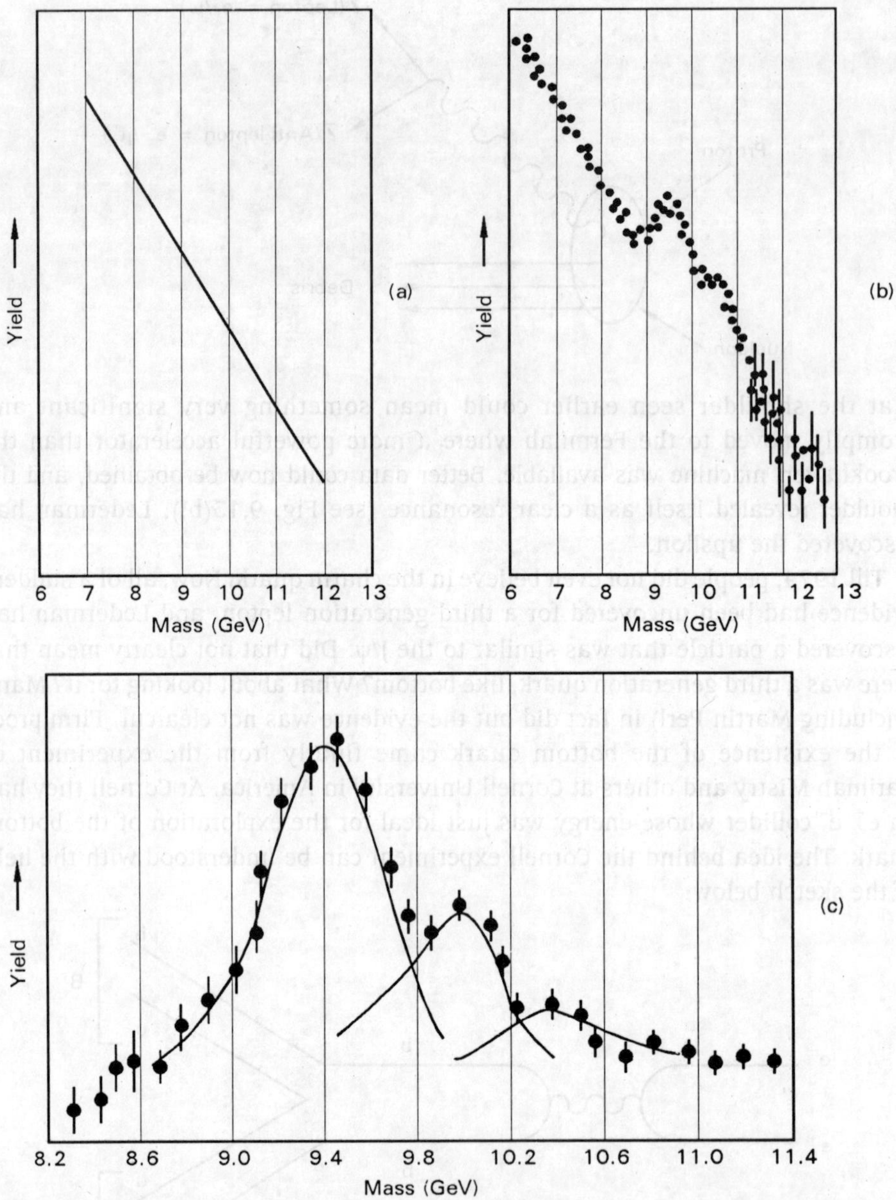

Fig. 9.13 This figure summarises the story of the discovery of the upsilon particle. If the particle did not exist, then one would obtain a result as schematically shown in (a). Instead, Lederman found something like what is shown in (b). There was a hint of a clear shoulder and perhaps more; Lederman then repeated the experiment using the more powerful Fermilab machine and the result obtained is shown in (c). The upsilon now revealed itself very clearly, along with its satellite shoulders.

the $\overline{B}$ decay via their own preferred routes. A careful study of these decays produced clear evidence for the bottom quark.

The bottom quark was discovered in 1980; fifteen years had to pass before evidence for the existence of the top quark could be obtained. This long delay was mainly due to the heavy mass of the top quark (see Table A.2.2). The question has been asked, "Where do you go when you have reached the top?" Well, one thing that could be examined is whether there are any more quark and lepton families. This has been done, and it is now clear that Nature has settled for exactly three and no more of these families. How do we know? For that, you have to wait for another chapter.

9.14 Quarkonium

The positronium, as you know, is the bound state of the electron and the positron. Likewise, one can have *charmonium*, which is a bound state of the charm quark and its anti-particle. In general, one can have such bound states of each quark and its corresponding anti-quark. Such bound states are referred to as *quarkonium*.

The physics of the positronium has been extensively studied using QED, and impressive agreement between theory and experiment has been achieved (see *QR-II*). In the case of the positronium, one can do very precise calculations because basically one knows the exact nature of the e.m. force. Such calculations have led to the positronium spectrum (see Fig. 9.14). The calculation of quarkonium spectra was more difficult because much less was known initially about the inter-quark force.

Today, the situation is better no doubt, but even so the agreement between theory and experiment is nowhere as impressive as in the case of QED, which is why Feynman referred to QED as the 'Jewel of Physics'.

9.15 Chapter overview

This chapter essentially takes the quark story a little further. If you recall, in the beginning, Gell-Mann had only SU(3). He then invented quarks to give a physical basis for the SU(3) and also for explaining baryons and mesons in terms of just a few building blocks. The guess was later made that there was some kind of a symmetry between the leptons and the quarks. At that time, three leptons were known — the electron, the neutrino and the muon; correspondingly there were three quarks — the up, the down and the strange. There was a perfect match and everyone was satisfied. Then the muon neutrino was discovered, and the lepton count went up to four; so now one more quark had to be invented, and that was the charm quark proposed by Bjorken and Glashow. The charm hypothesis received confirmation via the dramatic discovery of the J/ψ meson. Once the gate had been opened, people began to wonder whether Nature had stopped with just four quarks or were there more of them waiting to be found. The clue came from the leptons. Two new leptons were discovered, namely the tau and the tau neutrino. Therefore two more quarks were required, and these two new additions

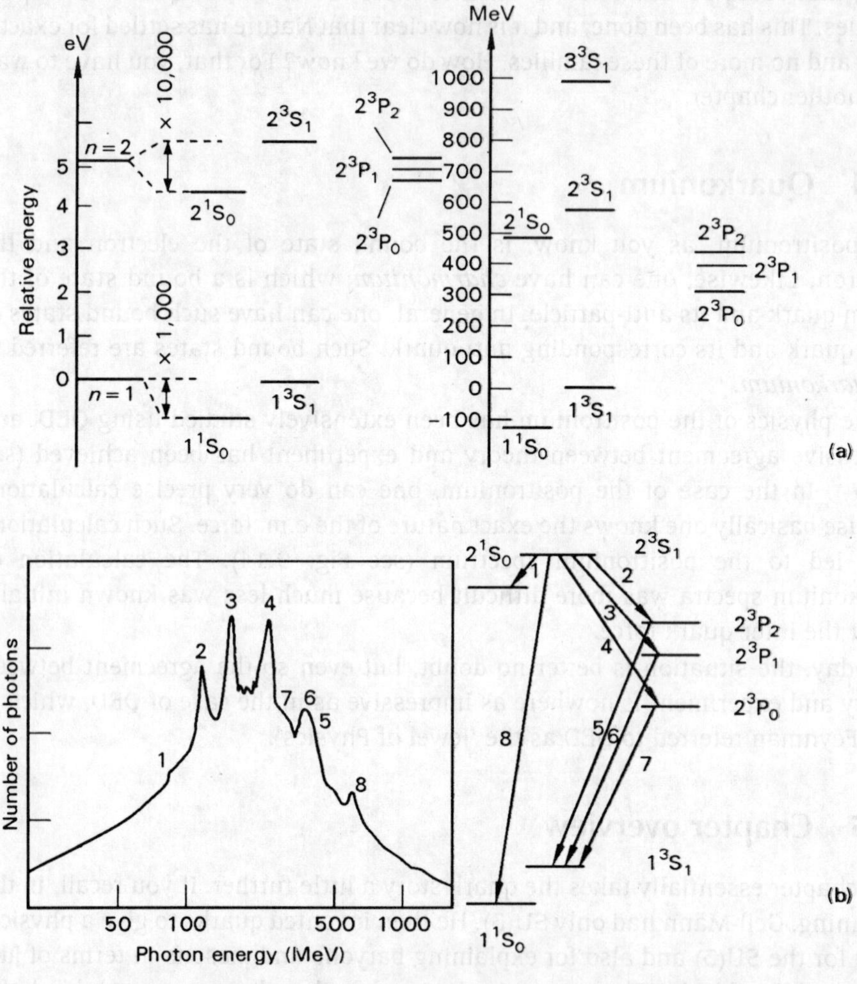

Fig. 9.14 On the left in (a) is shown the positronium spectrum as computed via QED (and confirmed by experiment of course). On the right is shown the corresponding computed charmonium spectrum. Observe the enormous difference in energy scales. (b) shows the experimental confirmation of the computed charmonium spectrum. This is done by observing the allowed electromagnetic transitions between the various charmonium states. The observed gamma ray spectrum is shown on the left in (b), and the numbers shown against the peaks identify the various transitions, as should be clear from the figure on the right.

to the quark family have been named the bottom and the top quark. Of these, the bottom was discovered first and after a long wait the top too has been netted. Any more additions? Not likely.

So one now has three generations of leptons and three corresponding generations of quarks. How exactly do they play their roles in the great drama of Nature? That story follows in the next two chapters.

Appendix to Chapter 9

A Brief Primer on Group Theory

Group theory is an integral part of physics; in high energy physics, it is a must. It was group theory that enabled Gell-Mann to bring order into the world of hadrons, and it was group theory again which played a key role in the unification of the e.m. and weak forces (which we shall discuss in the next chapter). In Chapter 4 of *Bhabha*, I gave you a glimpse of group theory; here I pick up the threads and take you a little further, setting the stage for the next chapter. You might find this appendix a bit tough and terse as well. But don't lose heart even if you don't quite follow it. Just remember the following bullet points. These will be quite adequate for a first reading. Later, you can come back to the appendix and give it a try again; the appendix will improve your understanding of the achievements which would be discussed in the chapters to follow.

- Group theory is the mathematical discipline which allows one to explore the consequences of symmetry, in detail.
- At the elementary level, one considers just the geometric symmetry of the system and studies the consequences of such symmetry.
- At more advanced levels, particularly in particle physics, one considers the symmetry of the system under what are called transformations. Elementary examples of transformations are physical translations and physical rotations.
- When the Lagrangian of a system remains unchanged after such a transformation, we say it is invariant with respect to such a transformation.
- In elementary particle physics, one studies the invariances resulting from 'rotations' in certain abstract spaces. This is done using the group theory machinery.
- The additive and multiplicative conservation laws mentioned earlier have been revealed by such studies.
- Group theory can lead to more than mere conservation laws. As was discovered while unifying the electromagnetic and the weak interactions, symmetry in fact determines the nature of the interactions themselves. This lesson has been learnt in the case of the strong interactions also.
- Thus, what you must basically appreciate is that symmetry is not merely a convenient tool for classification (the way Gell-Mann used it) but more importantly for unravelling the nature of the interaction itself. What follows does not deal with all these aspects but just tells you something about symmetry and rotations, especially rotations in abstract space. These rotations have a special name — they are called *gauge transformations*. From these are born the *gauge groups* which play a vital role in particle physics. These things will be tackled later. Here I just explain what SU(3) means and how Gell-Mann

was led by SU(3) symmetry to further speculate that there must be three basic building blocks for the hadrons, blocks which he named quarks.

In ***Bhabha***, I considered a finite group with six elements — the group of the triangle. Here, we shall be concerned with continuous groups. In Chapter 3, we considered infinitesimal transformations, and how finite transformations can be built up by the repeated application of the infinitesimal transformation; rotations in a plane provide an example.

Consider the complex plane (see Fig. 9.A.1). Imagine rotations in the plane; an infinite number of such rotations are possible; further, the rotation angle can be continuously varied. The set of rotations in this case form the Unitary Group U(1). It is the simplest example of a continuous group. Let U_1 and U_2 be any two rotations. As is easily seen, it does not matter in which order the two rotations are performed; the result remains the same. It is expressed in writing as

$$U_1U_2 = U_2U_1,$$

i.e., the rotations commute. A group in which all elements commute with each other is said to be an *Abelian* group, after the mathematician Abel who first studied their properties. Electromagnetic force respects the Abelian group U(1); this has remarkable consequences — see the observations of Feynman, quoted in ***Bhabha***.

We now move on to three dimensions. Consider first ordinary three-dimensional space. We choose three axes OX, OY and OZ. In particle physics, they are designated as the 1-axis, 2-axis and 3-axis respectively. Consider infinitesimal rotations J_1, J_2 and J_3 about these three axes respectively. We note first of all that

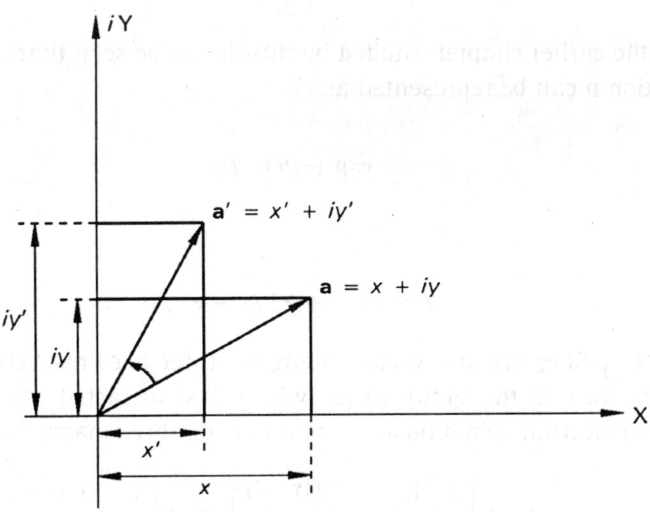

Fig. 9.A.1

in the three dimensions, rotations do not commute (see Fig. 9.A.2). Incidentally, on account of this, the group of rotations in the three dimensions is a *non-Abelian* group, unlike the group of rotations in two dimensions.

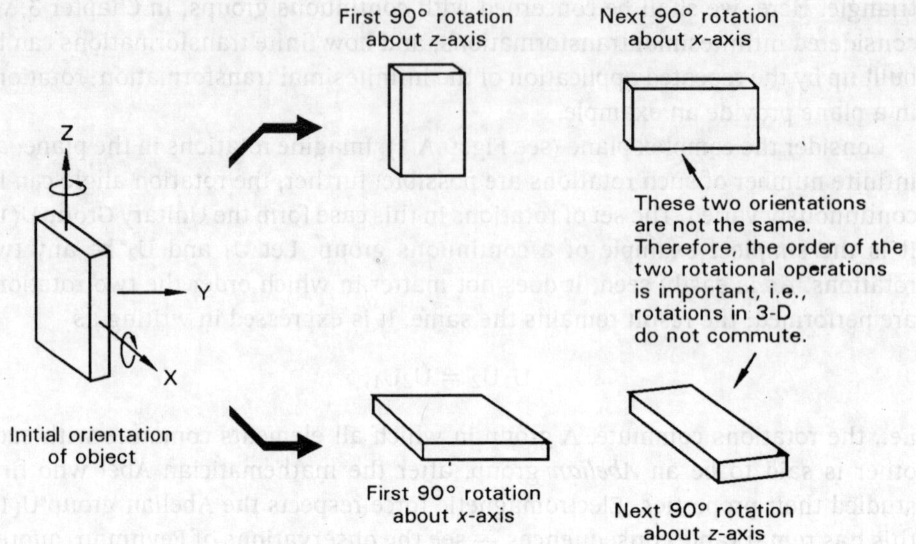

Fig. 9.A.2

Now, a finite rotation $U(\theta)$ by an angle θ about the 3-axis can be written as:

$$U(\theta) = \underset{n\to\infty}{\mathrm{Lt}} \left(1 - i\frac{\theta}{n!} J_3\right)^n = \exp\,(-i\theta J_3)\,.$$

Recall the earlier chapter. Guided by this, it can be seen that a rotation about any direction **n** can be represented as

$$\exp\,(-i\theta\,\hat{\mathbf{n}}\cdot J)\,,$$

where

$$J = \mathbf{e}_1 J_1 + \mathbf{e}_2 J_2 + \mathbf{e}_3 J_3$$

where $\mathbf{e}_1$, $\mathbf{e}_2$ and $\mathbf{e}_3$ are unit vectors along the three axes respectively.

We now turn to the group SU(2) which first attracted attention after the discovery of electron spin. Consider the following three matrices:

$$\sigma_1 = \begin{pmatrix} 0 & 1 \\ 1 & 0 \end{pmatrix};\ \sigma_2 = \begin{pmatrix} 0 & -i \\ i & 0 \end{pmatrix};\ \sigma_3 = \begin{pmatrix} 1 & 0 \\ 0 & -1 \end{pmatrix}.$$

These are called the Pauli matrices. Starting from these, one can build up a group through matrix multiplication; this is SU(2). The group is unitary because the matrices making it up are so; the S before U signifies it is a special unitary group — special in that the determinants of all the matrices equal unity.

As presently defined, SU(2) is built up from the Pauli matrices. Does it have anything to do with rotations? In the case of electron spin, yes. To see that, first recall from ***QR-I*** that only the projection of the spin vector can be measured, and this projection is usually referred to as the 3-axis. In general, one can define the spin state of the electron in terms of $\psi(\uparrow)$ and $\psi(\downarrow)$. When rotations are performed in space, the spin functions change or transform. In general,

$$\begin{pmatrix} \psi\uparrow \\ \psi\downarrow \end{pmatrix} \xrightarrow{\text{Rotation}} \begin{pmatrix} \psi'\uparrow \\ \psi'\downarrow \end{pmatrix} = \exp\left(-i\frac{\theta}{2}\hat{\mathbf{n}}\cdot\sigma\right)\begin{pmatrix} \psi\uparrow \\ \psi\downarrow \end{pmatrix}$$

where σ is treated as a vector with components σ_1, σ_2, and σ_3, each of these being represented by the appropriate Pauli matrix. The matrices $(\sigma_i/2)$, $i = 1, 2$, and 3, are called the *generators* of SU(2).

There is one important property of SU(2) worth noting here. In ordinary space, we know that rotation by 2π brings one back to the original condition. For example, on earth, if one travels along the equator, one arrives back to the starting point after going around by 360 degrees. The spin functions, whose behaviour is controlled by the SU(2) group, behave in a peculiar manner; a rotation by 2π does not restore the function to its original condition; only a rotation through 4π does so. This seems puzzling but the difficulty can be sorted out with the help of Fig. 9.A.3. Consider two ants, one of which moves along the ring 1 and the other along ring 2. Clearly, in the latter case, the ant will come back to the starting point only after a rotation of 4π. Thus, the 'space' of SU(2) is like the looped ring 2.

From the electron, we now move to the proton and the neutron. Recall that Heisenberg regarded the proton and the neutron as two states of the same particle, i.e., the nucleon (N), with different 'projections' along the '3-axis' of an appropriate space, now called the isospin space. In other words, p is like $\psi(\uparrow)$ and n is like $\psi(\downarrow)$. It does not then take much imagination to see that the SU(2) machinery can be applied to the isospin functions of the nucleon; only, one must remember that in this case the rotations are imagined to be performed in an abstract isospin space. You may be a little uncomfortable with rotation in an abstract space but you will get used to it! There will be more of this later on, and a more complicated version too. But it is not really all that difficult once you grasp the analogy.

With this background, we are now in a position to appreciate how Gell-Mann went beyond SU(3) and mere taxonomy to quarks. The first point we must note in this connection is that $\psi(\uparrow)$ and $\psi(\downarrow)$ provide a two-dimensional representation of SU(2) (for a brief background on representations, consult

Bhabha). Likewise, the p and the n also provide a two-dimensional representation of SU(2). This representation is called the *fundamental representation* of the group. The fundamental representation of SU(3) is three-dimensional. Gell-Mann now asked (effectively):

1. How many particles are necessary to construct the fundamental representation of SU(3)?
2. What are the generator matrices in this case, analogous to the ($\sigma_i / 2$) matrices in the case of SU(2)?
3. What physical attributes must one assign to these particles so that when the groups are assembled together, they would represent hadrons?

The answer to question 1 was three. The answer to question 2 came from group theory. Question 3 needed some detective work which was no problem for a genius like Gell-Mann.

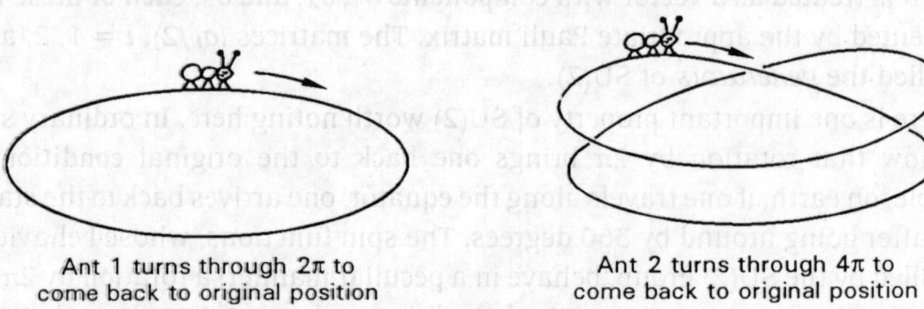

Fig. 9.A.3

The quark model described above certainly did one better than the taxonomy model proposed earlier, but there was one thing still lacking. The model did not describe the interactions between the quarks; in other words, there was still no model for the strong force. How that was built up, on the above foundation, will be described later.

10

Electrifying the Weak

This chapter describes the exciting story of how, building upon the foundations laid by Fermi, Lee and Yang, Marshak and Sudarshan, and also the pointers given by Schwinger, the trio of Glashow, Weinberg and Salam (GWS) brought the weak and the electromagnetic forces under one umbrella.

10.1 The 'technology' of physics

In the beginning, there was classical mechanics (CM). Here dynamics holds the centre stage, and Newton started it all by teaching us how to deal with the dynamics of (point) particles. Later Lagrange, Euler, Hamilton and others generalised Newton's approach so that one could deal with the dynamics of systems other than those involving just particles.

Charged particles present a very special case. Here the forces are electromagnetic, and the first lessons on how to deal comprehensively with them came essentially from Maxwell. Subsequently, Lorentz and Einstein improved our understanding further by describing how to deal with the problem when the particles move at the speed of light.

CM failed when applied to the atom. A new theory of mechanics was required which was shaped by Heisenberg, Schrodinger and Dirac — this is what is commonly referred to as non-relativistic quantum mechanics (QM). Many concepts of CM survived in QM but others had to give way to newer ones. After this came the incorporation of special relativity through the Dirac equation for the electron, and from this was born QED. With these tools, the technology of modern theoretical physics may be said to have arrived.

Dirac gave us the relativistic equation for the electron, which has a spin of 1/2. Two other similar equations are (i) the Klein–Gordon equation which applies to a particle with a spin of 0, and (ii) the Proca equation which is for a particle with a spin of 1 (the Proca equation has been mentioned in ***Bhabha***).

In classical mechanics, Lagrange showed that the primordial quantity is the Lagrangian $\mathscr{L} = T - V$, where T is the kinetic energy and V is the potential energy. The equations of motion with which one works in practical problems, follow from the Lagrangian (this has been discussed in ***QR-I***). Consider, for example, a particle of mass m moving in the potential sketched in Fig. 10.1. In this case,

$$\mathscr{L} = T - V = \frac{1}{2}m\dot{x}^2 - \frac{1}{2}kx^2 .$$

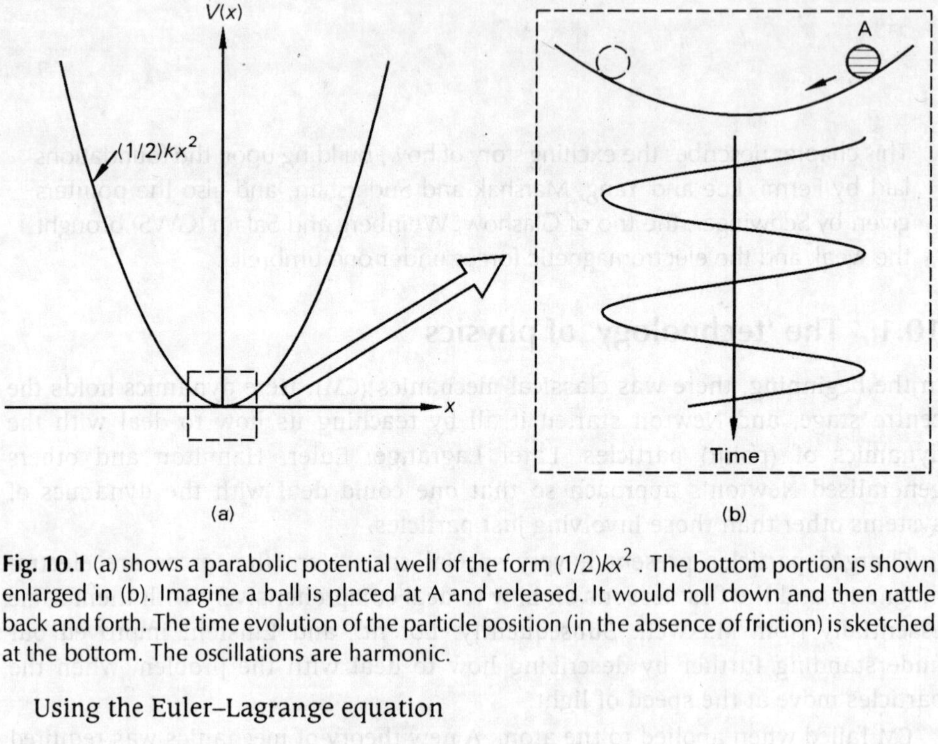

Fig. 10.1 (a) shows a parabolic potential well of the form $(1/2)kx^2$. The bottom portion is shown enlarged in (b). Imagine a ball is placed at A and released. It would roll down and then rattle back and forth. The time evolution of the particle position (in the absence of friction) is sketched at the bottom. The oscillations are harmonic.

Using the Euler–Lagrange equation

$$\frac{d}{dt}\left(\frac{\partial \mathscr{L}}{\partial \dot{x}}\right) = \frac{\partial \mathscr{L}}{\partial x},$$

one finds the equation of motion to be

$$m\ddot{x} + kx = 0 .$$

The Lagrangian formulation is applicable not only to a system of discrete particles but also to fields like say $\varphi(x)$, including those of the relativistic variety. Here $\varphi(x)$ denotes the field intensity at the point x. When one deals with a field, one talks of a Lagrangian density $\mathscr{L}(x)$, from which one may obtain, through appropriate rules, the equations of motion for the field variables. We shall not go into all those complicated details but I brought this up mainly to draw your attention to two points. Firstly, when one writes the expression for the relativistic Lagrangian density, the coefficient of the term φ^2 or $\varphi\varphi^*$ is identified as the mass of the particle associated with the field. Secondly, books describe how to write $\mathscr{L}(x)$ for quantum fields associated with particles with a spin of 0, a spin of 1/2, and a spin of 1; from these may be derived by simple rules, the Klein–Gordon equation, the Dirac equation and the Proca equation, respectively. Maxwell's equations also can be derived in a similar fashion.

When dealing with particles, normally one first considers the relativistic equation of motion for a free particle; after this one introduces perturbations and finally works out the consequences of the perturbation to any desired order. This is the discipline of relativistic QM, and I have described in some detail in *QR-II* how this operates where QED is concerned. Let me mention a few of the highlights of Feynman calculus.

Feynman invented a special calculus for use in QED but it is in fact more general, being applicable also to W.I. and S.I., with suitable modifications of course. The basic process in QED is shown in Fig. 10.2(a), and relates to an electron emitting a photon and moving on. More complicated diagrams can be generated out of this, an example being the scattering process shown in Fig. 10.2(b); this is known as Møller scattering. I remind you that in these diagrams, we allow time to flow up, i.e., vertically. Particle physicists often tend to make time flow horizontally but we shall stick to the traditional convention.

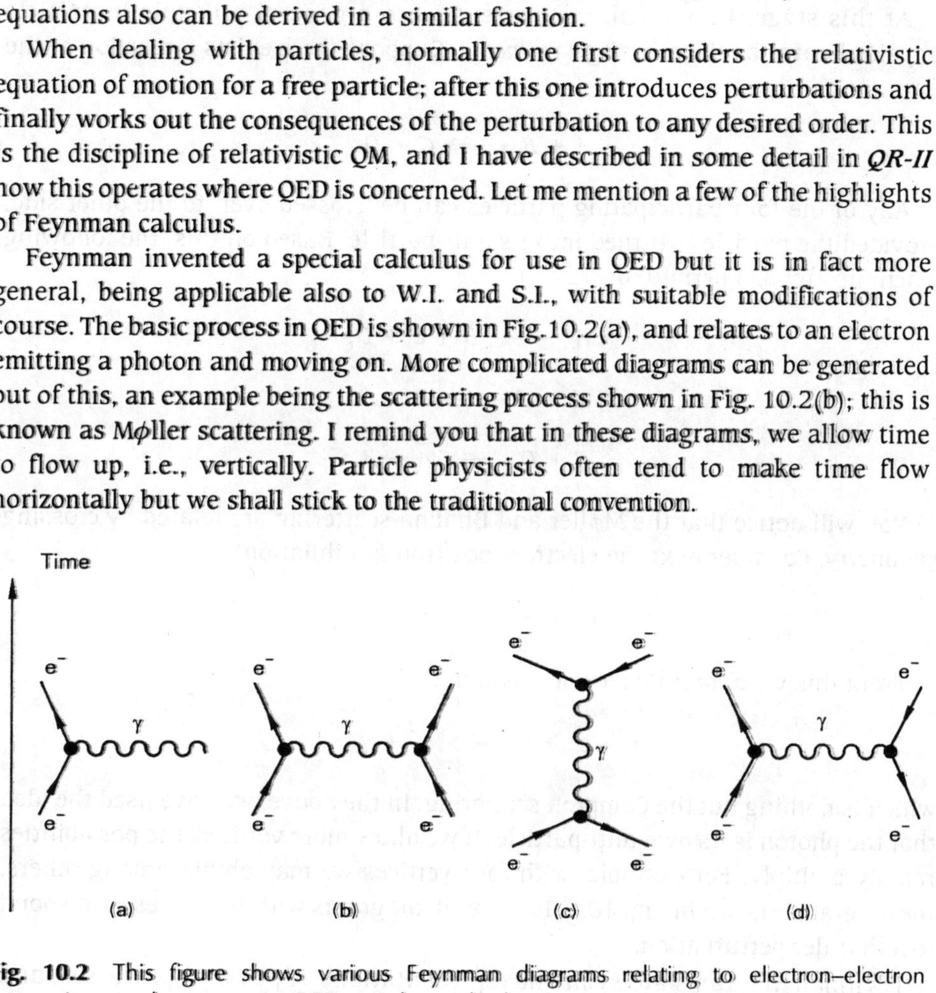

Fig. 10.2 This figure shows various Feynman diagrams relating to electron–electron interactions as they occur in QED. (a) shows the basic process, while the others represent processes generated from the basic process by going to the second order. (b) is the Møller process, while (c) and (d) correspond, as explained in the text, to Bhabha scattering. Note that although all particle lines have been identified as e^-, an electron going back in time is to be considered as e^+.

Now the Feynman diagram can be 'twisted around'. For example, we can make the diagram of Fig. 10.2(b) stand on its side as shown in Fig. 10.2(c). In this case we have some upward arrows and some downward arrows. As you will recall from *QR-II*, an electron going back in time is the anti-electron or the positron. Figure 10.2(c) corresponds to the Bhabha scattering and you will recall from the volume ***Bhabha***, that there is also another diagram which contributes to this process; this is shown in Fig. 10.2(d) — both contributions must be considered while calculating the Bhabha scattering.

At this stage, I must call your attention to a general principle in particle physics known as the *crossing symmetry*. Suppose there exists a reaction of the form

$$A + B \longrightarrow C + D.$$

Any of the four participating particles can be 'crossed over' to the other side, provided the particle is turned into its anti-particle. Based on this, the following reactions are also permitted:

$$A \longrightarrow \overline{B} + C + D,$$

$$A + \overline{C} \longrightarrow \overline{B} + D,$$

$$\overline{C} + \overline{D} \longrightarrow \overline{A} + \overline{B}.$$

You will notice that the Mϕller and Bhabha scattering are related by crossing symmetry. Consider next the electron–positron annihilation

$$e^- + e^+ \longrightarrow \gamma + \gamma .$$

From this we obtain by crossing symmetry,

$$e^- + \gamma \longrightarrow e^- + \gamma,$$

which is nothing but the Compton scattering. In the above, we have used the idea that the photon is its own anti-particle. If we allow more vertices, the possibilities rapidly multiply. For example, with four vertices we may obtain among others, the diagrams shown in Fig. 10.3. In general, diagrams with N vertices correspond to Nth order perturbation.

Feynman diagrams are symbolic representations of physical processes, and are particularly useful while making calculations. Each diagram makes a contribution that is a number. Well-defined rules are available for calculating this number, based on the structure of the diagram; that is what the Feynman calculus is all about.

One problem in carrying out the perturbation theory in relativistic QM is the appearance of infinities; some diagrams give infinite contribution. As explained in *QR-II*, the infinities come mainly from diagrams with loops. Feynman,

Schwinger and Tomonaga showed us how the problem of the infinities may be solved in QED by a process known as renormalisation.

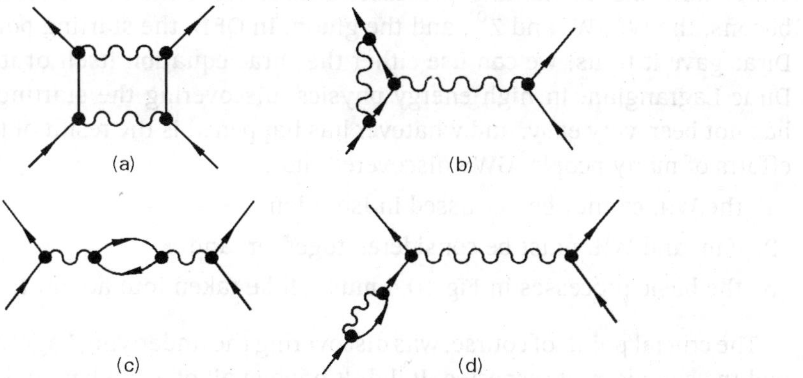

Fig. 10.3 This figure shows the Feynman diagram for electron–electron interaction with four vertices; notice that they all involve loops. From ***QR-II*** you may recall that (b) is related to the anomalous magnetic moment of the electron, (c) to vacuum polarisation and (d) to mass renormalisation.

Let us go back to the basic diagram, which is redrawn in Fig. 10.4(a). The perturbation due to the e.m. field causes the transition $|i>$ to $|j>$ in the particle states, and QM gives us rules for calculating this transition probability T_{ij} via the perturbation theory. This formula is based on the equation of motion, descended from an appropriate Lagrangian density, which may be regarded as the primordial entity for the theory. For more complicated diagrams also there is an associated transition probability, and the Feynman calculus gives us the rules for calculating these numbers. Once the transition probability has been obtained, calculating the reaction rate is the next step; the latter is not difficult, being standard QM. Then comes the comparison with experiment — typically, all experiments measure some reaction rate or the other — that, in essence, is the drill and the routine.

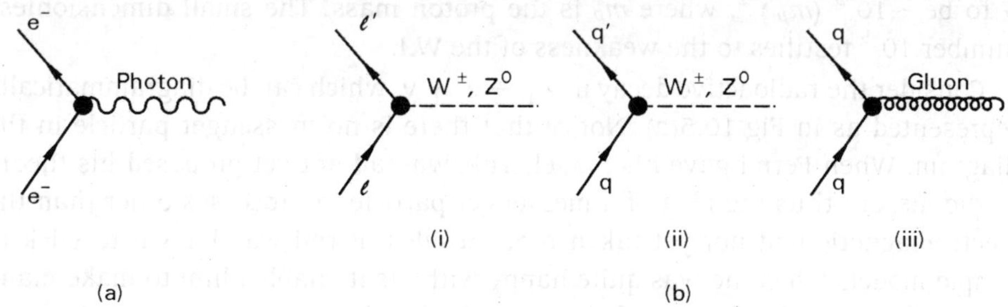

Fig. 10.4 (a) shows the basic process in QED, while the three diagrams in (b) show the basic processes needed for discussing weak and strong interactions. Here ℓ and ℓ' refer to fermions, and q and q′ to quarks.

The diagram shown in Fig. 10.4(a) is the building block for QED. Analogous diagrams needed in the electro-weak theory are shown in Fig.10.4b, (i) to (iii). They show the elementary processes associated with the intermediate vector bosons, the W^+, W^- and Z^0, and the gluon. In QED, the starting point is easy — Dirac gave it to us; we can use either the Dirac equation itself or its parent, the Dirac Lagrangian. In high energy physics, discovering the starting Lagrangian has not been very easy, and whatever has happened is the result of the combined efforts of many people. GWS discovered that:

1. the W.I. cannot be discussed in isolation,
2. e.m. and W.I. must be considered together, and
3. the basic processes in Fig.10.4 must all be taken into account.

The crucial point, of course, was discovering the underlying Lagrangian density and the symmetry it respected. It didn't happen all of a sudden, and that exciting story comes later.

10.2 Fermi's theory

We begin with Fermi, who in a sense started it all. True, Fermi did not seek to unify e.m. and the W.I., but he was the first to guess that the two forces operated in a similar manner; in other words, it was Fermi who set the ball rolling.

Fermi was preoccupied with radioactivity; in today's jargon it is a low-energy phenomenon, though it involves W.I. Thus, Fermi could get away with many things which are not possible at high energies — you will understand this better later. The essence of Fermi's theory — model would be a better word — can be represented in the following manner:

$$\mathscr{L}_{\text{Fermi}} \sim G \times (\text{nuclear current}) \times (\text{lepton current}).$$

Here G is a constant which has the dimensions of $[\text{mass}]^{-2}$. Fermi estimated G to be $\sim 10^{-5}\ (m_p)^{-2}$, where m_p is the proton mass. The small dimensionless number 10^{-5} testifies to the weakness of the W.I.

Consider the radioactive decay $n \rightarrow p + e^- + \bar{\nu}$, which can be diagrammatically represented as in Fig.10.5(a). Notice that there is no messenger particle in the diagram. When Fermi gave his model, Yukawa had not yet proposed his theory of the meson; thus the idea of a messenger particle in processes other than the electromagnetic had not yet taken root. Besides, Fermi was known to stick to simple models which he was quite happy with, as it enabled him to make many calculations and clarify many facts not understood then.

The impulse to go beyond the Fermi model did not come for nearly two decades. First there was the Second World War, and then all the excitement with QED, followed by the discovery of pion resonances (in which Fermi himself played a part), and of course also the discovery of the strange particles. In the midst of all this, who was bothered about W.I.? But then W.I. entered centre stage again via the tau-theta puzzle, parity overthrow and the Universal Fermi Interaction, all of which have been described earlier.

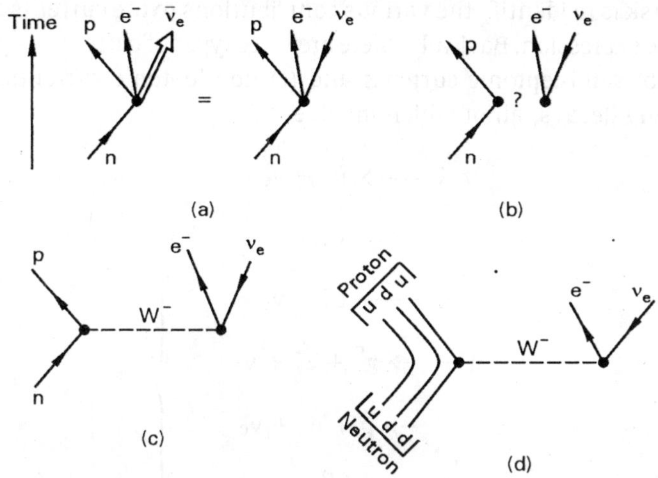

Fig. 10.5 (a) shows the beta decay of the neutron in the Fermi model. It is drawn in two fashions, one of which is 'naive'; here the anti-neutrino is shown with a double arrow pointing upwards; the preferred way is also shown, with the anti-neutrino being represented as a neutrino going back in time. (b) raises the question whether or not there is a messenger particle involved. In (c), the diagram is redrawn, with the messenger explicitly shown. (d) goes one step further, illustrating what happens in the quark picture; here the neutron and the proton are depicted in terms of their respective quark structures. Notice that the n-p conversion involves only one quark.

10.3 Improvements to the Fermi model

We now consider the various improvements that could be made to the simple model just mentioned.

1. Fermi's theory is said to be local, i.e., the current–current interaction is to be understood as meaning that the hadron current at space point x interacts with the lepton current at the *same* space point x (remember, nuclear current is nothing but hadron current). Later, this was modified to make the interaction non-local, i.e., allow for the interaction of one type of current at x with the other type of current at x'. Schematically, non-local interaction is written as

$$\int (\text{hadron current at } x) \cdot D\,(x-x') \cdot (\text{lepton current at } x') \cdot dx \cdot dx',$$

where $D(x-x')$ describes how the influence of the current at x' is transmitted to the current at x. The question mark in Fig. 10.5(b) simply means, "What about non-local interactions?" It is now known that not only must non-locality be allowed for, but should be via the W^+, the W^- and Z^0 bosons. Thus, D above would be related to the bosons just mentioned.

2. The next task is to identify the various contributions to the currents appearing in the above expression. Basically, there are three types of currents — (a) leptonic currents, (b) semi-leptonic currents and (c) non-leptonic currents. Consider the following decays, all of which involve W.I.

$$K^- \longrightarrow \mu^- + \bar{\nu}_e \qquad \text{(a)}$$

or

$$\left.\begin{aligned} K^- &\longrightarrow e^- + \bar{\nu}_e \\ \pi^- &\longrightarrow \pi^0 + e^- + \bar{\nu}_e \\ n &\longrightarrow p + e^- + \bar{\nu}_e \\ K^- &\longrightarrow \pi^0 + \pi^- \end{aligned}\right\} \qquad \text{(b)}$$

or

$$\Lambda^0 \longrightarrow p + \pi^-. \qquad \text{(c)}$$

In category (a), the decay products are leptons; the process is thus leptonic and so are the currents involved. Similarly, categories (b) and (c) illustrate semi-leptonic and non-leptonic processes, with their own characteristic contributions to the current. In general, several possibilities exist and one must carefully allow for them while analysing any particular reaction.

3. Now the whole task of updating Fermi's theory is being done by analogy with the e.m. theory. The latter has an equation of continuity. Is there any such here? Yes indeed, and it has to be suitably incorporated.

4. One can have a theory exclusively for the e.m. interaction, and in fact that is what QED is all about — it deals with the electron and its e.m. interaction in all its glory. By contrast, one cannot have a theory purely for the W.I.; the e.m. force *has* to be brought into the picture, as the following example explains.

Consider the reaction $e^+ + e^- \rightarrow \mu^+ + \mu^-$, which is possible at high energies. There are two ways in which the reaction can proceed, as illustrated in Fig. 10.6. The reaction mediated by Z^0 occurs as a weak process while the reaction mediated by γ occurs as an e.m. process. In an experiment, both processes would contribute. Thus the weak and the e.m. forces would, necessarily, have to be considered as one comprehensive package. A corollary is that the expression for the current must include not only the contribution from the ordinary charge but also the weak hypercharge. This is a consequence of Eqn. (8.2), which relates the electric charge Q, the hypercharge Y and the 3-component of isospin T_3. Due to all of this, the weak and the e.m. forces emerge as two facets of the same force.

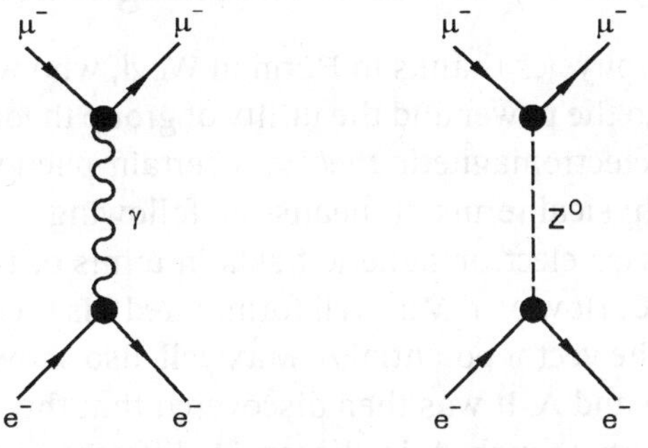

Fig. 10.6 This figure illustrates how the $e^+e^- \to \mu^+\mu^-$ process can proceed via the mediation of the photon as well as the Z^0.

5. In the absence of any external influencing agency, a photon cannot interact with another photon; in other words, there is no self-interaction for the photon, which in turn means that diagrams like those in Fig. 10.7(a) are disallowed. By contrast, the W's and the Z are permitted self-interaction and thus processes corresponding to diagrams such as those shown in Fig. 10.7(b) are permitted. This difference arises from the fact that the parent gauge symmetry of the photons is U(1), which is an Abelian group, whereas the corresponding symmetry for the W's and Z is SU(2), which is a non-Abelian group. This nuance about Abelian and non-Abelian groups has been explained in the appendix to the previous chapter.

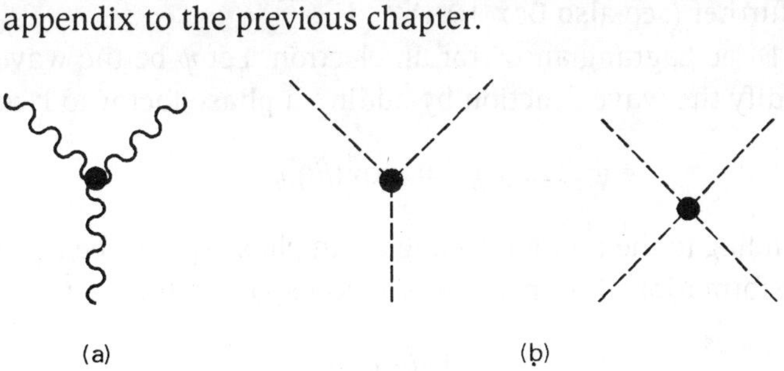

Fig. 10.7 This figure deals with the self-interactions of the messenger particles. (a) shows such a process for the photon; however, it is not allowed. (b) shows similar diagrams for the intermediate vector bosons; these processes are allowed. In the diagrams in (b), the dashed lines could represent either the W or the Z.

Points 1-5 above are what may be called a list of desirables, but they still do not quite add up to a theory. A theory must start from a basic Lagrangian based on minimal assumptions and respecting appropriate symmetry. Discovering such a theory in this case was no simple task and it happened slowly with small steps taken by many in different places. To understand all those developments, we need to digress first into gauge theories.

10.4 The gauge theory of the electromagnetic field

Gauge theories entered physics thanks to Herman Weyl, who was also one of the first to draw attention to the power and the utility of group theory in physics. Now there is, in Maxwell's electromagnetic theory, a certain phenomenon known as *gauge invariance*. In physical terms, it means the following:

Normally one discusses electromagnetic fields in terms of the electric field **E**, and the magnetic field **B**. However, Maxwell formulated his theory in terms of the scalar potential φ and the vector potential **A**. Maxwell also showed how the **E** and **B** fields are related to φ and **A**. It was then discovered that there is a flexibility in the choice of **A**, i.e., even though **A** is chosen in different ways, the physically measurable **E** and **B** turn out to be the same. This flexibility in the choice of **A** is referred to as *gauge freedom*, and making a particular choice is called *fixing the gauge*.

We have seen that in the case of colliding billiard balls, the origin of the coordinate axes could be translated at will and yet it did not make any difference to the outcome of the collision. There was a freedom and it implied that something remained invariant, which in turn meant that something was conserved. In this case there is gauge invariance. What is the corresponding conserved quantity? Till Weyl's exploration, no one knew; in fact, no one even knew that such questions could be asked and such connections discovered. All that became clear thanks to Noether's work, with which Weyl was familiar; and that was why Weyl was able to go further (see also Box 10.1).

Consider the Dirac Lagrangian $\mathscr{L}$ for an electron. Let ψ be the wave function. Suppose we modify the wave function by adding a phase factor to it as follows:

$$\psi \longrightarrow \psi' = \exp (i\theta)\, \psi. \qquad [10.1]$$

Now corresponding to the rotation by angle q in phase space (see Fig. 10.8) there is a formal transformation $U(\theta)$ in wave function space, which acts on ψ. Thus,

$$\psi' = U(\theta)\, \psi,$$

which on comparison with Eqn. [10.1] reveals that

$$U(\theta) = \exp (i\theta).$$

Box 10.1

The idea of gauge invariance was first proposed by Herman Weyl, drawing inspiration from Einstein's general theory of relativity. The fundamental concept in both the special and the general theories of relativity, is that there is no absolute coordinate frame of reference, and that descriptions of motion are relative. In general relativity (see ***Relativity***) one compares the viewpoints of a stationary observer with that of another in a freely-falling elevator. Supposing there are many freely-falling elevators, each with its own observer; how does one relate measurements made in the different accelerating frames? Einstein solved this problem by defining a mathematical relationship between nearby falling frames. This relation, which is known as *connection*, manifests as the gravitational field.

In 1919, Weyl took the next step. He said, "Einstein has linked the gravitational force to a certain type of connection. Can the e.m. force also be linked to a similar type of connection?" Weyl then constructed a theory in which there were arbitrary local dilatations and contractions, the theory itself being invariant to such local adjustments. Obviously, the theory was inspired by Einstein's work. In the theory, a separate standard of length and time had to be adopted at each point of space-time. Weyl compared the choice of scale convention to the choice of gage blocks, the polished steel blocks used by machinists as a standard of length. Writing in German, Weyl used the word *eich invarianz*, which was initially translated as 'calibration invariance'. Subsequently, the word *gauge invariance* was used which since then has become the accepted terminology.

Weyl's theory ran into problems but he did not abandon his basic quest. In 1929, after the advent of QM, Weyl succeeded in linking the electromagnetic field to local gauge invariance, this time associated with the phase of the wave function. Even then, few realised that local gauge invariance was a *dynamic symmetry, capable of generating the basic forces of Nature*. Yang and Mills were the first to emulate Weyl but the idea had few takers, at least then. In a book on electrodynamics published as late as 1965, Rohrlich writes, "Gauge invariance has no physical meaning." Barely a few years later, Weinberg and Salam changed all that! Today, local gauge invariance occupies the pride of place among internal symmetries.

At this point, I would like you to take a minute off to appreciate that the phase angle is a quantity related to the wave function, which is a complex quantity. This becomes important later on.

Now ψ is a function of spatial coordinates. What Eqn. [10.1] implies is that the phase of ψ is being changed by the *same* angle θ at all points of space; in other

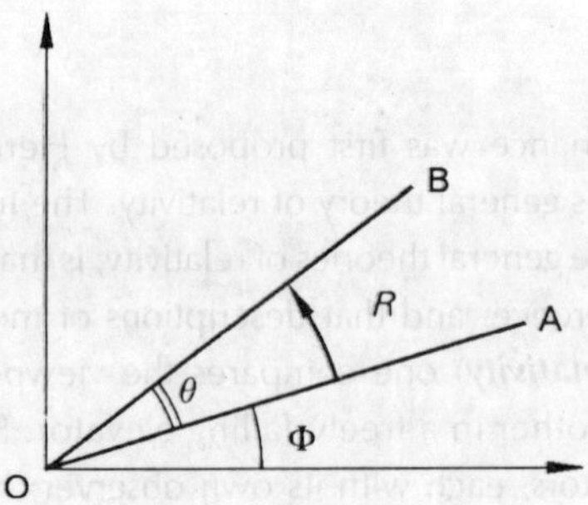

Fig. 10.8 This figure deals with the phase of the wave function. This phase can be diagrammatically represented by the orientation of a vector. Here Φ denotes the initial phase angle, and the corresponding vector is represented by OA. θ represents the change in the phase, and can be visualised as the rotation of the vector OA to the orientation OB. This phase change causes the wave function to transform from ψ to ψ', as explained in the text.

words, θ is not a function of x. Such a uniform global change of phase is referred to as a *global gauge transformation*. It turns out that the Dirac Lagrangian is invariant under a global gauge transformation. The set of such possible transformations form the Abelian group U(1). Thus $\mathscr{L}$ is invariant under the global gauge group U(1).

What is the quantity that is conserved? It turns out that electric charge is the quantity that is conserved. *Charge conservation which we all take so much for granted, follows from the global gauge invariance of the Dirac Lagrangian* — this is Noether's theorem in action in electromagnetism.

There is much more to gauge invariance than what has just been said. Suppose we decide to make the phase change different at different space points, i.e., we have

$$\psi(x) \longrightarrow \psi'(x) = \exp\{i\theta(x)\} \cdot \psi(x).$$

The transformation above is called a *local* gauge transformation (see also Fig. 10.9). Does this leave $\mathscr{L}$ invariant? No, it does not. Then why bother? There is a reason. Suppose we are able to find a suitable $\mathscr{L}_1$ such that $(\mathscr{L} + \mathscr{L}_1)$ remains invariant under local gauge transformation, then something interesting happens. There are many books describing how to find this extra Lagrangian to be added and its specific form. From our point of view, the interesting things are the following:

1. $\mathscr{L}_1$ involves a field called the *gauge field*.
2. The carrier of the gauge field is a particle with zero mass.

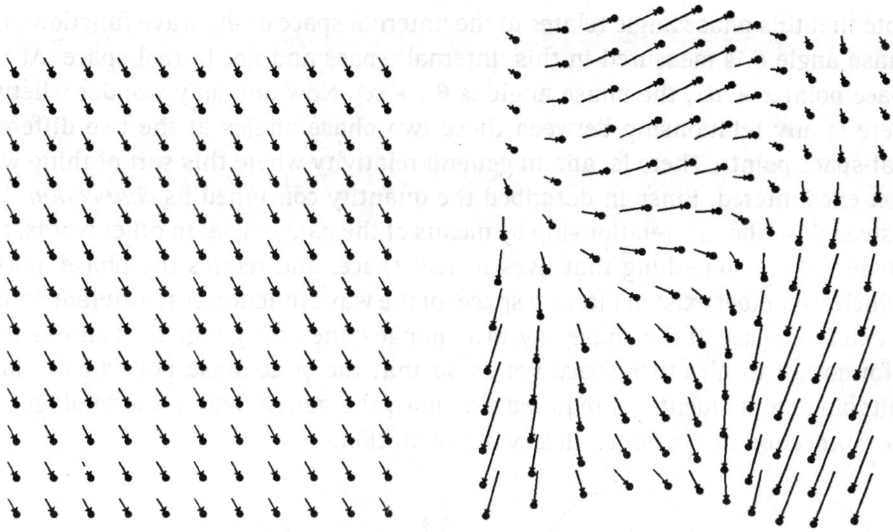

(a) (b)

Fig. 10.9 This figure is intended to highlight the difference between making the same change at all space points and different changes at different points. Imagine a rubber sheet, on which is marked a regular grid of points. Suppose all the points are displaced by the same amount and in the same direction; the pattern that would result is shown in (a). It is conceivable that such a displacement could be brought about by a suitable stretching of the sheet. Suppose now the different dots are displaced in different directions and to different extents. The resulting situation would be as in (b). Clearly, this can be done only by non-uniform stretching, which would obviously set up forces within the sheet. If something remains unchanged in the scenario shown in (a), we talk of global gauge invariance; if similarly something remains unchanged in (b), then we talk of local gauge invariance.

Translated into simple language, it means:

1. The gauge field is nothing but the (quantum version of the) **A** field of electromagnetism.
2. The carrier of the gauge field is the photon.

So you see how the imposition of local gauge invariance can lead to the appearance of a gauge field and gauge particles. Normally, symmetry or invariance leads to a conservation law. This is true of global invariances. But *local invariances on the other hand generate a force field; symmetry thus plays a vital role in dictating dynamics*. This is an important discovery, made in this century.

In short, what I am saying is that the e.m. force is due to U(1) local gauge invariance. Can one similarly visualise the weak and the strong forces as arising from some kind of local gauge invariance? Let us see what the experts did about such questions.

Before proceeding to that, I would like you to understand how *internal symmetry* and *external space* get bound together in this local gauge invariance. In Fig. 10.10, we have two space points, x and $x + dx$. At x, ψ has the phase $\theta(x)$.

Note that this phase angle relates to the 'internal space of the wave function'; the phase angle θ is measured in this 'internal' space and not in real space. At the space point $x + dx$, the phase angle is $\theta(x+dx)$. Now one may wonder whether there is any relationship between these two phase angles at the two different real-space points. There is, and in general relativity where this sort of thing was first encountered, Einstein described the quantity concerned as *connection*. We instead describe this relationship by means of the gauge field. In other words, the gauge field is something that lives in real space, and relates the phase angles (which remember exist in internal space) of the wave function ψ at different points in external space. If the phase of ψ at x changes, then the gauge field carries that information to all other space points so that the ψ at those points can make suitable readjustments as required. In short, the gauge field is the analogue of the connection in Einstein's theory of gravitation.

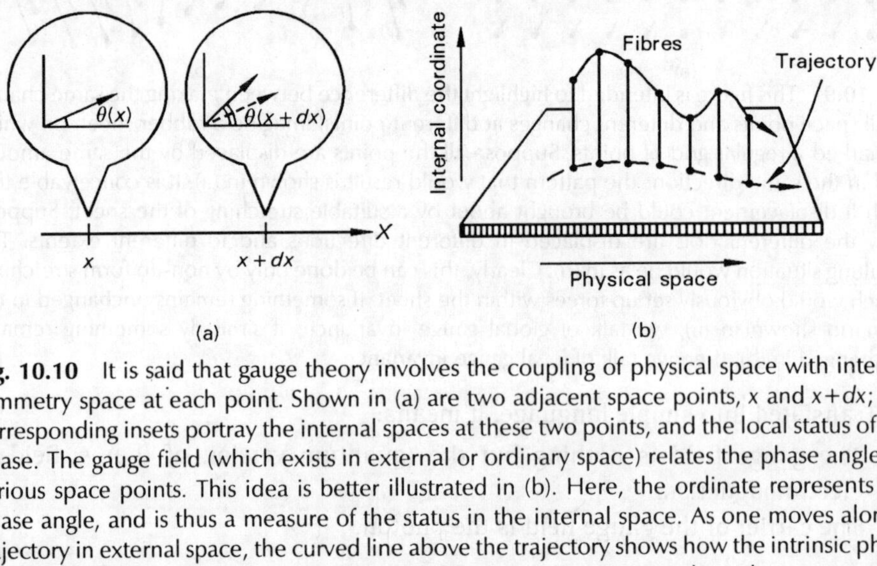

Fig. 10.10 It is said that gauge theory involves the coupling of physical space with internal symmetry space at each point. Shown in (a) are two adjacent space points, x and $x+dx$; the corresponding insets portray the internal spaces at these two points, and the local status of the phase. The gauge field (which exists in external or ordinary space) relates the phase angles at various space points. This idea is better illustrated in (b). Here, the ordinate represents the phase angle, and is thus a measure of the status in the internal space. As one moves along a trajectory in external space, the curved line above the trajectory shows how the intrinsic phase varies. Any disturbance to the intrinsic phase at one point is transmitted to other space points via the gauge field. Mathematicians would refer to the vertical line at each space point as a *fibre*, and the collection of fibres as a *fibre bundle*.

10.5 The Yang–Mills theory

In 1954 Yang and Mills asked themselves the following question: "If local gauge invariance can lead to the electromagnetic force, can it, in a similar manner, lead to the strong force and say also to the Yukawa meson?" They set out to explore this possibility, and proceeded as follows.

1. They first noted that p and n are two different isospin states of the same particle, i.e., the nucleon (N).

2. They recognised that the proton and the neutron have slightly different masses; nevertheless, they ignored this difference to start with. The mass difference, they said, was due to e.m. effects which could be patched up later.
3. They observed that isospin respects SU(2) which, in contrast to the group U(1) of the e.m. field, is non-Abelian.
4. They then wrote the Lagrangian $\mathscr{L}$ for the free nucleon.
5. Next, they considered the transformation $\psi \rightarrow \psi' = S\psi$, where S is a local gauge transformation which produces a change of phase of the isospin wave function ψ.
6. When ψ' was introduced into $\mathscr{L}$, the modified Lagrangian was not the same as the starting one, i.e., $\mathscr{L}$ did not show gauge invariance.
7. They then identified a $\mathscr{L}_1$ that must be added to $\mathscr{L}$, such that $(\mathscr{L} + \mathscr{L}_1)$ remained gauge-invariant under the local gauge group SU(2). This was a non-trivial exercise.
8. As expected, the additional term $\mathscr{L}_1$ required the introduction of gauge fields and gauge bosons or messenger particles.

When they went through all this trouble, Yang and Lee discovered that gauge particles were massless, which was very disappointing. Why? Because one expects the messenger particle of the strong force to have mass — recall Yukawa's argument about range and mass.

Yang and Mills' theory is an intellectual treat but it has little value — Nature does not work that way. Interestingly, Yang and Mills still published their work and what is more surprising, the journal did not dismiss their paper as a meaningless exercise. This was a true blessing because the so-called 'useless' work of Yang and Mills was a major motivator and a key model for many who followed. People appreciated Yang's idea of *trying to link the basic forces of Nature to local gauge invariance* — this was the important message of this paper.

The message was no doubt clear but how does one go about searching for the correct Lagrangian, the correct group and linking them all to local gauge invariance, so as to describe the weak force and the strong force? This, people discovered, was not such a simple task; a large number of people had a crack at the problem but only a few were privileged to discover the bits and pieces, which finally Glashow, Weinberg and Salam put together (as far as the W.I. was concerned). However, it took several years for that to happen.

10.6 Enter broken symmetry

Let us now forget Yang and Mills and turn instead to broken symmetry. As it often happens in physics, the solution to a long-standing problem comes from a totally different and unsuspected corner. This is precisely what happened in the present case. The solution to the problem of the massless gauge bosons that plagued the Yang–Mills theory came from broken symmetry; nobody anticipated such a development.

Symmetry breaking is a topic I have dealt with in detail in the volume ***The Many Phases of Matter***. Let us quickly brush up on some of the basics. Consider a potential energy function $V(\varphi)$ of the following form:

$$V(\varphi) = \text{Const.} + \frac{1}{2}\mu^2\varphi^2 + \frac{1}{4}\lambda\varphi^4 . \qquad [10.2]$$

Here φ is called a scalar field (see also Box 10.2). φ is thus a function of the space variable x; though we write φ, it should be understood as $\varphi(x)$. We insist that λ is always greater than 0, but allow μ^2 to be either greater than or less than 0. The form of $V(\varphi)$ in the two cases is sketched in Fig. 10.11. Incidentally, experts usually refer to the field we are considering as the φ-4 field.

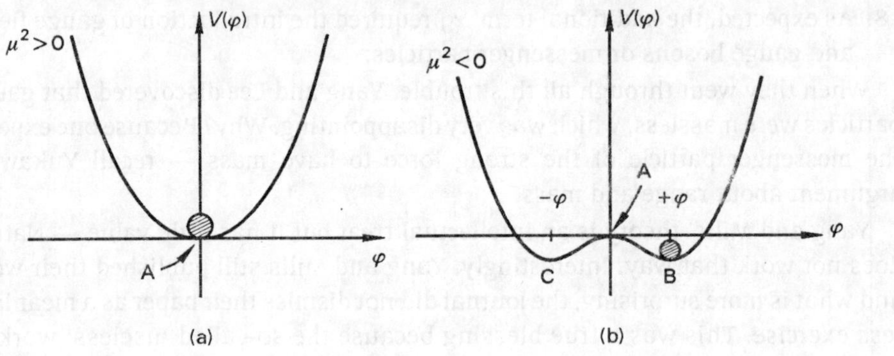

Fig. 10.11 This shows the form of $V(\varphi)$ for $\mu^2 > 0$ and $\mu^2 < 0$. In both cases, V is symmetric. However, whereas in the former case the state corresponding to the minimum, i.e., the one corresponding to A, is also symmetric, in the latter case this is not true; if, for example, the system is at B (which is a minimum), then the transformation φ to $-\varphi$, carries the system from B to C.

We now note the following:

1. For $\mu^2 > 0$, V is a minimum when φ is zero. Thus, $\varphi = 0$ is the natural or the equilibrium state of the system.
2. For $\mu^2 < 0$, $\varphi = 0$ is a local maximum and this state is unstable — imagine a marble placed at A (see Fig. 10.11(b)); the slightest disturbance, and it would roll down to one of the two valleys on either side. When it settles down in one of these valleys, it is in an asymmetric state but this state is one of equilibrium.
3. The valleys are at $\varphi = \pm\sqrt{(-\mu^2/\lambda)}$. They are the stable points for $\mu^2 < 0$.
4. If μ^2 is gradually reduced from positive to negative values through zero, then so long as $\mu^2 > 0$, the stable state is the one corresponding to A. At $\mu^2 = 0$, this

state becomes unstable, and when μ^2 becomes even slightly less than zero, the system jumps to the state at B or at C; the choice is purely random.

5. The potential $V(\varphi)$ is symmetric, meaning that changing φ to $-\varphi$ does not alter the potential. The state corresponding to A also has this property, i.e., it also does not change when φ is replaced by $-\varphi$. The potential and the ground state both possess the same symmetry when $\mu^2 > 0$.

Box 10.2

The concept of field is central to modern physics. We are normally used to electric and magnetic fields but there are actually many other types of fields, both in classical and in quantum physics.

To build up our ideas about fields, we must first consider space. We regard space to be continuous. You might ask if space can be otherwise. Good question. In the next chapter we shall see that in certain models, one considers a hypothetical space which is a lattice of points; more about that later.

So, we have this continuous space. What next? To each point **r** in this space, we associate a quantity *F*, say. This *F* is thus a function of **r**. Next we must describe the nature of this *F*. It could be a scalar, a vector, a tensor, or even a spinor, and so on. Whatever *F* may be, *F*(**r**) defines a field. A few examples would make all this easier to understand. Take the atmosphere. At each point in the atmosphere, we can talk about the temperature at that point. Thus, *F* in this case would represent temperature and since temperature is a scalar, the temperature field is an example of a *scalar* field. Suppose *F* represents the atmospheric pressure and not the temperature; then we have a pressure field which also is a scalar field. Next suppose we consider the wind velocity as a function of the spatial coordinate **r**. Then we have a velocity field which obviously is a *vector* field. Instead of the atmosphere, let us now consider a solid, say a piece of iron. Suppose *F* now represents the stress. Then *F*(**r**) would represent a stress field and since stress is a tensor, we have here an example of a *tensor* field. In short, whether the field is a scalar field, a vector field, a tensor field, depends on the quantity represented by *F*. I should mention that so far I have implicitly assumed that the value of the quantity *F* does not vary with time; in other words, here we are considering *static* fields. If the value of *F* varies with time, i.e., if the field function is of the type *F*(**r**, *t*), then one has *time-dependent fields*. Such fields are very common in electromagnetic theory.

Can there be a visualisation of these fields? Yes, in simple cases. Consider Fig. (a): here, for convenience, we regard space to be a square lattice of points. With each point is associated a scalar quantity *F*. The numerical magnitude of *F* is represented by a black dot whose diameter is proportional to the value of *F* associated to that particular lattice point. Thus, Fig. (a) is a pictorial representation of

Contd...

Box 10.2 *...contd.*

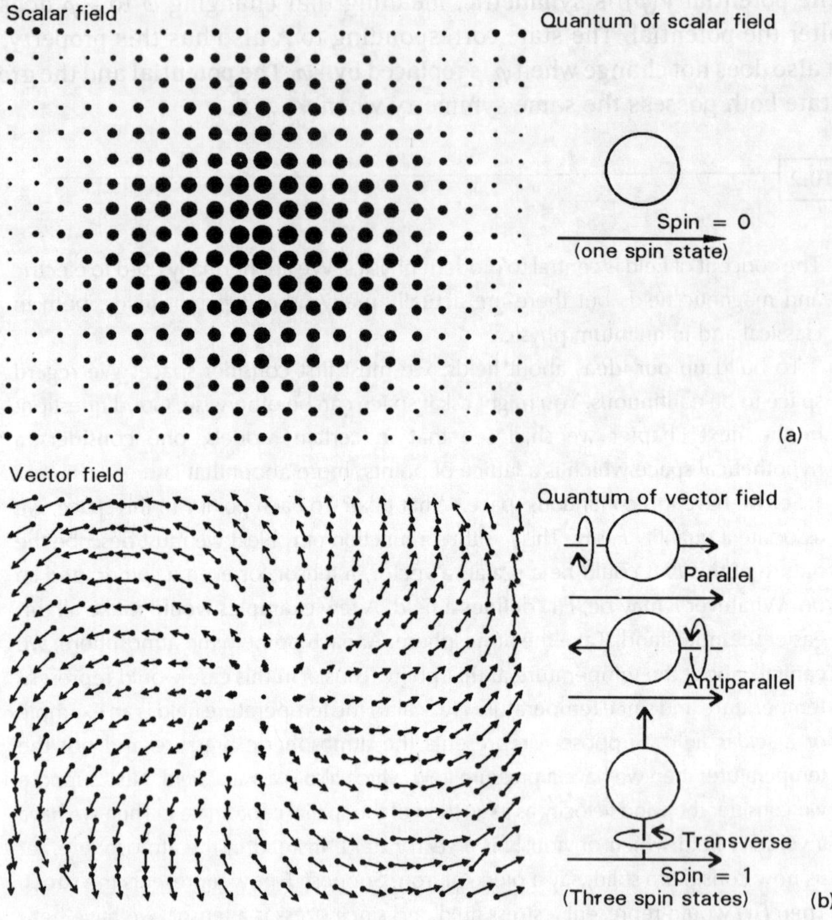

a scalar field on a square lattice. One example of a quantum scalar field would be that for a spin 0 particle. In Fig. (b), *F* represents a vector (with two components). This vector can be represented by an arrow at each point of the lattice. This vector could, for example, represent velocity — I mean the lattice could represent a 2-D array of atoms and the arrows could represent the velocity of the various atoms at any given instant of time. A spin 1 particle would be associated with a vector field.

In particle physics, the fields one considers are relativistic, and often associated with local gauge symmetry, etc. The GWS theory revolves largely around gauge fields. Recall Fig. 10.10, which explains how, while *F* (**r**) exists in real space, the value of *F* at **r** must be obtained by looking into the 'internal space'.

6. When the system is at B, the change φ to $-\varphi$ flips B to C. Thus, the state B is not invariant under inversion of φ, whereas the potential is. The same remark applies to the state C. We thus have *broken symmetry* when $\mu^2 < 0$ — i.e., *whereas the potential is invariant with respect to a particular transformation, the ground state or the equilibrium state of the system does NOT exhibit the same symmetry when* $\mu^2 < 0$.

Figure 10.12 gives a commonly used example of broken symmetry drawn from mechanics.

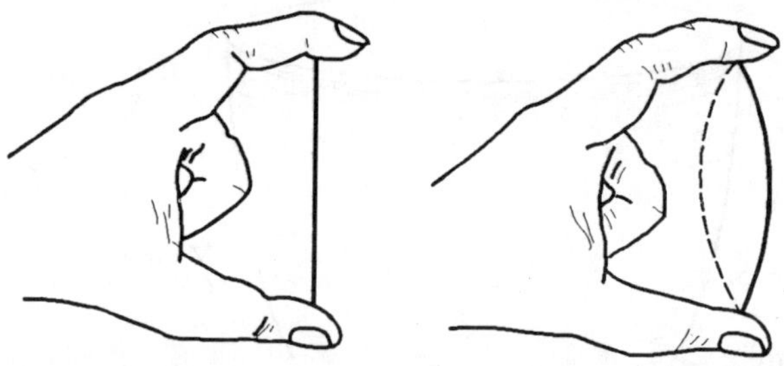

(a) (b)

Fig. 10.12 Imagine a stiff straight wire held between two fingers as shown in (a). If pressure is applied, the wire would buckle. There is an inherent circular symmetry in the problem, and in principle the buckled wire could occupy one of an infinite number of possible positions; one of these is shown in (b) with a solid line, while an alternate possibility is shown with a dotted line. Once the wire buckles and assumes a particular position, the assumed buckled position of the wire no longer has the inherent circular symmetry associated with the problem. This is an example of symmetry breaking.

You may wonder, "So what?" The answer to that is as follows. Assume first that $\mu^2 < 0$, and that the system is at B, i.e., at $\varphi = \varphi_1$ (see Fig. 10.11(b)). Now shift the origin from $\varphi = 0$ to $\varphi = \varphi_1$. A general φ can now be expressed as

$$\varphi = \varphi_1 + \eta.$$

Substitute the above equation in the expression for $V(\varphi)$. This gives

$$V(\eta) = \text{Const.} + \lambda\varphi_1^2\eta^2 + \lambda\varphi_1\eta^3 + \frac{1}{4}\lambda\eta^4. \qquad [10.3]$$

Compare with Eqn. [10.2], and remember that $\mu^2 < 0$. Equation [10.2] represents the world view from A. You can easily check that the curvature of $V(\phi)$ at A is negative. This implies a negative or unphysical mass. Equation [10.3], on the other hand, represents the view from B. Write the term quadratic in η^2 as

$$\lambda\varphi_1^2\eta^2 \equiv \frac{1}{2}M^2\eta^2.$$

Then

$$M = (2\lambda\varphi_1^2)^{1/2}.$$

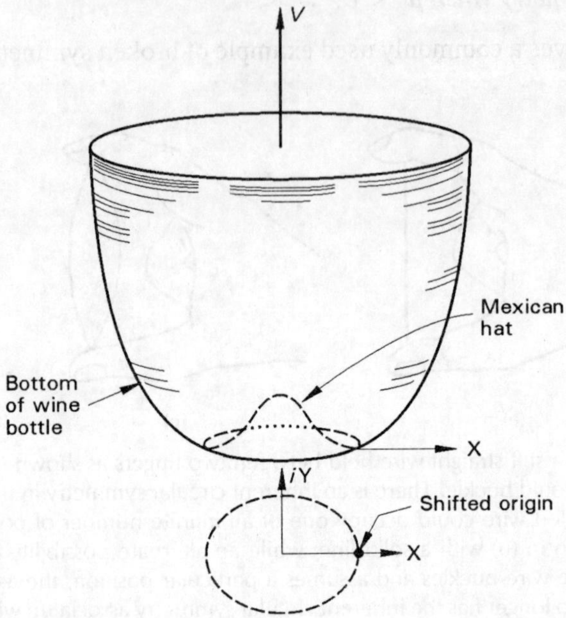

Fig. 10.13 This shows the form of $V(\varphi)$ when the field φ is a complex scalar. The form can be visualised by revolving the corresponding curve of Fig. 10.11 about the vertical axis. The whole form is often described as a wine bottle, while the bottom portion is described as the Mexican hat. Notice that the minimum is not a single point or even two points, but an infinite number of points distributed along a circle. Any position on this circle is an accepted state of minimum energy. In other words, the circle is the loci of minima of the type B in Fig. 10.11.

The field η has a mass term of the correct sign. Physically what this means is that a marble placed at B can rattle and exhibit small oscillations. From all of this, we can draw the following conclusions:

1. Both Eqn. [10.2] and Eqn. [10.3] represent the same field and therefore the same physics.
2. Equation [10.2] does not exhibit the symmetry breaking, but Eqn. [10.3] does.
3. With Eqn. [10.2], there is a mass problem (for $\mu^2 < 0$), but there is none with Eqn. [10.3].

4. Equation [10.3] is difficult to use if one wants exact results but if the fluctuations η are small, then one can use the perturbation theory to find out what happens for small deviations from the state B.

The field we considered so far is a real, scalar, one-dimensional field. We now briefly consider a complex field

$$\varphi = \varphi x + i\varphi y.$$

See also Fig. 10.13. In this case, when the symmetry is broken, there are an infinite number of points of the type B, distributed along a circle; the system will choose one point on this circle. Let us suppose that B is chosen to be that point of the circle which lies on the X-axis. Two new fields η and ξ are introduced. When the field is expressed in terms of these two quantities (as was earlier done in Eqn. [10.3]), one then finds two quadratic terms and correspondingly two masses as below:

$$m_\eta > 0, \text{ and } m_\xi = 0.$$

Thus, in the case of the complex field, symmetry breaking leads to two particles, one of which has a mass while the other is massless. The former is the *Higgs particle*, while the latter is referred to as the *Goldstone boson*. The particle with mass can be physically visualised in terms of a marble rattling in the curved portion of the 'bottom of the wine bottle', while the particle of zero mass can be visualised in terms of a marble freely rolling around the circle at the bottom (see Fig. 10.13).

10.7 Linking the Yang–Mills field to symmetry breaking

Let us now consider the Yang–Mills field theory and the symmetry breaking field theory in one breath.

1. Both are field theories.
2. The Yang–Mills theory invokes local gauge invariance; and it leads to a field mediated by massless gauge particles.
3. The scalar φ-4 field theory of Section 10.6 has a mass problem when the symmetry is broken. However, the mass problem can be solved by shifting the perspective. One says that mass is 'generated' by explicitly considering the broken symmetry aspect.
4. In the case of a complex scalar field, when the perspective is shifted there arise two particles, one with mass and one without mass.

Looking at the above, does it suggest anything? Probably not. However, some enterprising people did wonder what would happen if the Yang–Mills field and φ-4 field were considered *together*.

This brings us to what in the trade is known as the *Higgs mechanism*. It refers to an explicit method given by Peter Higgs for considering the Yang–Mills fields and symmetry breaking fields *together*, so that the problem of massless particles of the former is solved. Higgs did not produce a theory for either weak or strong interactions; he merely gave a formal technology for combining two different types of field theories in such a manner that a major undesirable problem connected with one of them disappeared.

A simplified illustration of the Higgs scheme is given in Fig. 10.14. In loose technical terms, the scalar field introduced by Higgs cancels the effects of the Yang–Mills vector field which produces masslessness in the intermediate vector bosons. In popular language (first introduced by Salam), one says that the massless particles of the Yang–Mills field 'eat' the massless particles of the Higgs field (i.e., the Goldstone bosons) to acquire mass.

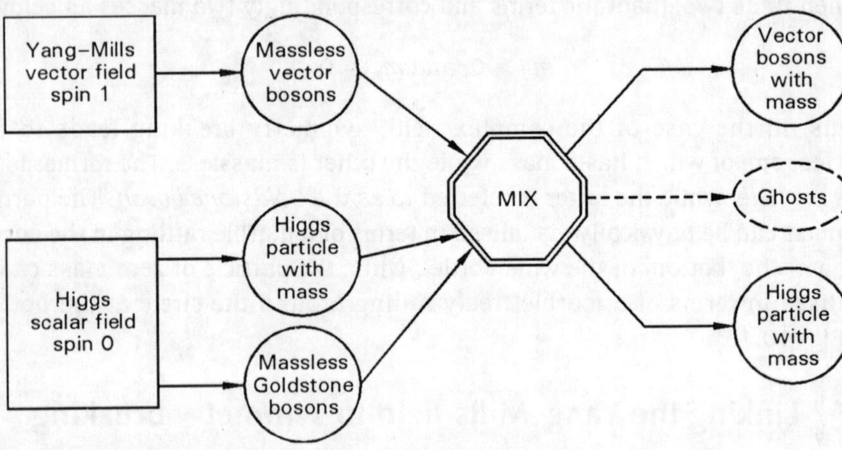

Fig. 10.14 The schema of the Higgs mechanism. The Yang–Mills fields contribute massless particles. The Higgs fields contribute massless Goldstone bosons and scalar (spin 0) particles with mass. There is a big mix-up and what finally emerge are spin 1 bosons with mass and of course the Higgs particle. The Goldstone particles disappear; it is often said that they become ghosts.

Let us consider a specific but hypothetical example of how the Higgs mechanism operates (see also Fig. 10.15). Say there is a Yang–Mills field with 2 degrees of freedom, which respects local gauge invariance and all that. By itself, this field will produce a spin 1 particle, i.e., a vector boson, which of course would be massless. Now a particle of spin 1 must, in principle, be capable of having three orientations. However, since the particle is massless, it has to travel with a speed c and therefore only two orientations are permitted. I have already discussed in ***QR-II*** such a phenomenon in the case of light, which can have only two polarisations and not three (remember, the photon has a spin of 1).

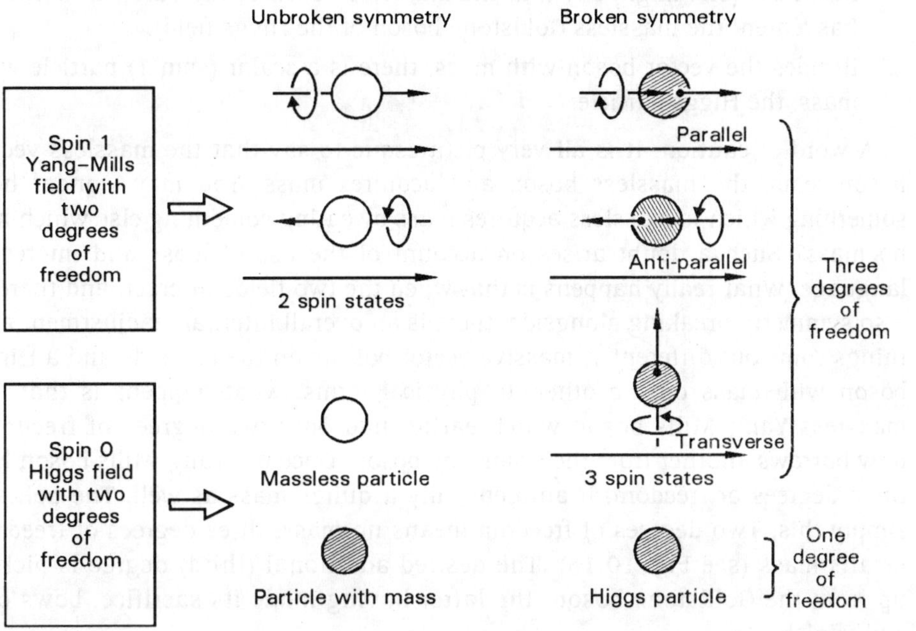

Fig. 10.15 This figure illustrates with a hypothetical example, the outcome of the interaction of the degrees of freedom of the Yang–Mills field and the Higgs field. In this example, they both contribute two degrees each. Before symmetry breaking and mixing, the Yang–Mills field leads to a massless particle of spin 1 which can have only two spin orientations of the three possible. The Higgs field yields a spin 0, massless or Goldstone particle and a spin 0 particle with mass; the latter is the Higgs particle. Both have one degree of freedom each. Consequent to symmetry breaking and mixing, the Goldstone particle surrenders its degree of freedom to the Yang–Mills boson and fades away (to become a ghost!). The Yang–Mills particle then acquires three degrees of freedom and along with it, mass. The Higgs particle is a 'bystander', watching the proceedings.

We are not interested in a massless vector boson of the above type; we want a vector boson with mass. Therefore, guided by Higgs, we introduce a scalar field with two degrees of freedom. This field is often called the *Higgs field*. If we allow symmetry breaking in this scalar (Higgs) field, it would result in two particles, one with mass which is the Higgs particle and another without mass, which is the Goldstone boson.

We now consider the two fields together, with of course a possibility for them to interact. The fields then clash and 'fight it out'. And what is the outcome? The following:

1. There is no zero mass particle now; i.e., the Goldstone boson vanishes.

2. The vector boson born of the Yang–Mills field is no longer massless; it now has mass (one might say that the massless boson of the Yang–Mills theory has 'eaten' the massless Goldstone boson of the Higgs field).
3. Besides the vector boson with mass, there is a scalar (spin 1) particle with mass, the Higgs particle.

A word of caution: It is all very picturesque to say that the massless vector boson 'eats' the massless boson and acquires mass. You may wonder how something which is massless acquires mass by eating something else which has no mass? Such a doubt arises on account of the use of loose and imprecise language. What really happens is that when the two fields interact, and there is also symmetry breaking alongside, there is an overall internal readjustment and things come out differently, massive vector bosons on the one side and a Higgs boson with mass on the other. In physical terms, what happens is that the massless Yang–Mills boson which earlier had only two degrees of freedom, now borrows another from the Goldstone boson. Once the Yang–Mills boson has three degrees of freedom, it automatically acquires mass as well. The point is simply this: Two degrees of freedom means no mass; three degrees of freedom means mass (see Fig. 10.15). The desired additional (third) degree is picked up from the Goldstone boson; the latter having made its sacrifice, bows out and vanishes!

In short, *the Higgs mechanism is a prescription for fixing the zero-mass problem of the Yang–Mills fields and producing realistic results*. Being merely a prescription, there still remains the problem of putting it to practical use and actually coming up with a proper theory for the weak or the strong interactions, as the case may be. Where W.I. was concerned, that problem was solved independently by Weinberg and Salam, based on the foundations laid earlier by Glashow, and thus was born the GWS theory.

10.8 The Glashow–Weinberg–Salam theory of electro-weak unification

As early as 1961, Glashow had conjectured that the e.m. force and the W.I. could be integrated by using a Yang–Mills field theory, based on the idea that there were three massive intermediate vector bosons, along with the massless photon, of course. Glashow said that the field must have SU(2) × U(1) symmetry. Everything was fine with Glashow's theory, except that the masses of the W^+, W^- and Z^0 came out as zero — not unexpectedly. Glashow therefore put in the masses by hand, so to say, in order to have some scheme for practical work, and Glashow's theory now slipped to the lower standard of a model. People were not impressed and at that time there were no compelling reasons to believe in the Glashow model; so it was generally forgotten. But once the Higgs mechanism became known, people furiously tried to invent all kinds of Yang–Mills fields which would do the trick. However, most people did not remember Glashow's work. Two people who picked the right combination were Weinberg and Salam who, independent of each other, started from an SU(2) × U(1) symmetry for the lepton fields, added the Higgs mechanism, and came out with what became the winner. The GWS theory had everything one could ask for — vector bosons with masses, and the photon (see also Fig. 10.16).

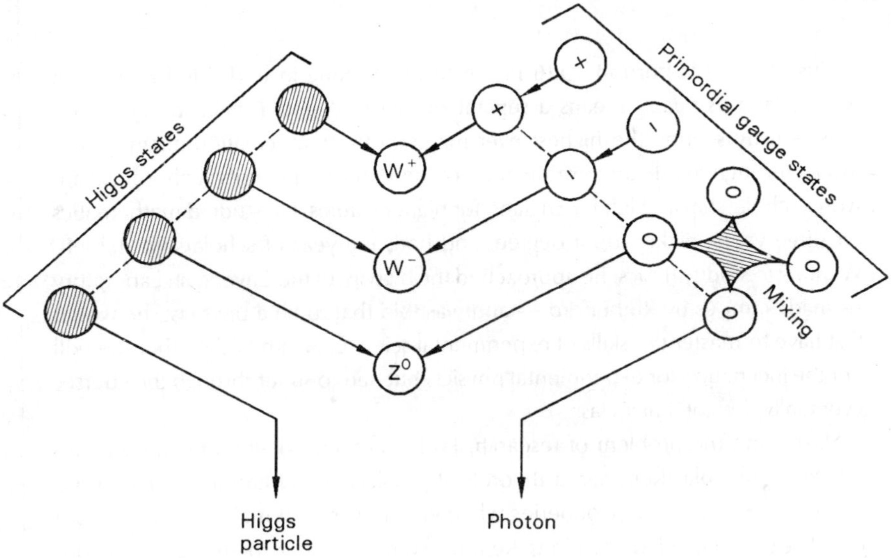

Fig. 10.16 This figure portrays the emergence of the intermediate vector bosons with mass, and the photon, in the GWS theory. To start with, there are four vector fields, associated with which are the primordial gauge states shown in the figure. Two of these mix, in readiness for creating the Z^0 and the photon. Next, the Higgs states enter the picture. When they mix with the Yang–Mills fields, the massless bosons corresponding to the W's and the Z acquire mass. The photon is left massless, while the scalar Higgs particle remains aloof as usual. Note that two of the Yang–Mills fields mix even before they tangle with the Higgs field. It is this that leads to electro-weak mixing and also to the so-called Weinberg angle.

In practical terms, what Weinberg and Salam did was the following:

1. They started with Yang–Mills fields for the two W's, the Z^0 and the photon. The fields had, in all, 4 degrees of freedom. They wrote the Lagrangian $\mathscr{L}$ which had all the desired self-interactions.
2. Next they invoked the SU(2) × U(1) symmetry and imposed local gauge invariance. This led to the corrected Lagrangian $\mathscr{L}_1$.

3. They then wrote down the Lagrangian $\mathscr{L}_2$ for the scalar Higgs fields with four degrees of freedom.
4. After this they wrote the Lagrangian $\mathscr{L}_3$ for the interaction between the Yang–Mills and the Higgs fields.
5. The composite Lagrangian ($\mathscr{L} + \mathscr{L}_1 + \mathscr{L}_2 + \mathscr{L}_3$) was then formed. Finally, Weinberg and Salam invoked broken symmetry.

That was it. And for Weinberg and Salam, it meant the Nobel Prize!

Box 10.3

Abdus Salam was born in 1926 in the town of Jhang in undivided Punjab. In Arabic, *'abd-us-salam'* means a servant of peace. A gifted child, he created a sensation by scoring the highest ever marks in the matriculation examination. Sailing through the degree examination of the Punjab University (Lahore), Salam won a scholarship and left for England for higher studies. He studied mathematics for three years, picked up a degree, and had one year of scholarship still left. Wanting to study physics, he approached the bigwigs of the Cavendish Laboratory — made famous by Rutherford — and was told that to be a physicist, he would first have to master the skills of experimental science. Salam had neither the skill nor the inclination for experimental physics, but had to suffer through the course; eventually, he got a first class.

Next came the problem of research, Ph.D., etc. He sought and was granted refuge by Nicholas Kemmer, a theoretical physicist. QED had just burst on the scene, and Kemmer was wondering whether a similar renormalisable field theory could be constructed for the pions. Kemmer wanted a theory of strong interaction, based on the Yukawa picture of mediation by pions. Kemmer asked Salam to talk to others to find out more about the problem; Salam did so, and found that there did exist such a research problem, and it seemed a good one to tackle. In two years Salam produced the required theory, and promptly became a minor celebrity. Later he recalled:

> We believed at that time in Yukawa's theory which said that pions were the final secret of the nuclear force, and went on believing it till 1951 or 1952. The expectation was that the meson theory, once renormalised, would completely solve the strong force problem. And that's why I became famous, because the theory was renormalisable. In 1950, when my proof was completed, we believed that this was the end. We expected it to be *the* theory. The end! And we lived in a fool's paradise.

Suddenly, the bottom fell through. The delta resonance was discovered by Fermi, followed by a flood of strange particles. Overnight, the strong force became more mystifying than ever, and the pion lost its central role.

Contd...

Box 10.3 *...contd.*

Meanwhile, Salam was invited to the famous Institute for Advanced Studies in Princeton. He was supposed to be an expert theorist but according to Salam's own admission, he knew nothing of physics except what he himself had done! In 1951, Salam returned to the subcontinent which meanwhile had been partitioned, with Jhang going to Pakistan. At the young age of 25, he was appointed Professor in the Punjab University (Lahore) but Salam was restless at being cut off from mainstream physics research. So in 1954, he went back to Cambridge. It was during this period that he made the trip to Seattle, and while returning, he sketched out the two-component theory in the plane — that story has been recounted in Chapter 5. Shortly after this, he was appointed full Professor of Theoretical Physics at the prestigious Imperial College, London, at the remarkably young age of 31. In 1964, he founded, with the help of a UNESCO grant, the International Centre for Theoretical Physics (ICTP) in Trieste, Italy. Soon, ICTP became a haven for third world physicists, particularly from India. For years, Salam shuttled between Trieste and London. And it was during this period that he did the work on electro-weak unification, for which he received the Nobel Prize in 1970. We now turn to the magazine *CERN Courier*, for the rest of his profile.

> Salam's deceptively soft, husky voice masked an iron will and ruthless ambition. He could be arrogant, but also used argument as an intellectual weapon to prise open a difficult problem or seek new ideas. A consummate cosmopolitan, he was equally at home with P.G. Wodehouse or the *Koran*.
>
> Despite being the only Pakistani to have won the Nobel Prize, and despite his impressive international achievements, Salam's position in his home country was ambivalent. Under the powerful rule of Ayub Khan, Salam once wielded considerable influence. However, as a member of the minority Ahmedia Islamic sect, he resigned his influential position as the Chief Scientific Adviser in 1974, when Pakistan's National Assembly under Zulfikar Ali Bhutto excommunicated the Ahmedias from Islam. The Ahmedias claim that Mirza Ahmed born in Qadian in northern India in the late 19th century, was the Mahdi, or Messiah, which is a sacrilege to Islam orthodoxy. Mainly confined to Pakistan, India and East Africa, the small Ahmedia sect is a frequent target of intolerance and discrimination. In 1979, after the announcement of the Nobel award, Salam was initially invited to Pakistan by General Zia-ul-Haq, but other pressures soon blew an icy wind of excommunication.
>
> With Salam stricken by illness (he suffered from a crippling disease for three years before he died), at Trieste it became clear he could no longer function as the Institute's Director. As a tribute to its founder, which he

Contd...

Box 10.3 ...*contd.*

would be still able to appreciate before his powers waned completely, in 1993, the Centre organised a three-day physics meeting which was attended by colleagues, admirers, and former students from all over the world. One was Frank Yang, whose talk on mirror symmetry in 1956 had deeply impressed the young Salam.

The culmination of the meeting was the award of an honorary degree of the University of St. Petersburg, Russia. The Rector of the University made the trip specially for this. After the formal ceremony, participants stood patiently in line to offer their own congratulations. There was little response from Salam but they all hoped he'd understood their message.

After famous names, it was the turn of the students. One of the last to wish him was a nervous young man from Pakistan, a young researcher who had been awarded one of the highly prized scholarships to Salam's Centre. As he bent towards Salam hunched in the wheelchair, he said, "Sir, I am a student from Pakistan. We are very proud of you." Salam's shoulders shook, and tears ran down his face.

Salam died on 20 November, 1996, in Oxford. He had many Indian friends, and actively persuaded the TIFR to search for proton decay (see Chapter 12). He toured India shortly after receiving the Nobel Prize, and was warmly received wherever he went. Another characteristic trait was that Salam firmly believed in God and was not shy of proclaiming it.

What does one do with this composite Lagrangian? That is not difficult to answer. After all, the Lagrangian is always the starting point of a theory. If one has the Lagrangian, one can select the reaction for study, and then apply the perturbation theory using the Feynman calculus, exactly as in the case of QED.

An important question arises in this context: QED had problems with infinities, and one did not know what to do till Feynman, Schwinger and Tomonaga showed the way. What about this case? Are there similar infinities? If so, how does one get over them? Well, it turns out that there are indeed infinities and they are nastier than those encountered in QED. But it has also been shown that they can be overcome by renormalisation.

To briefly recap about renormalisation in the context of QED, the infinity problem is associated with the loop-type Feynman diagrams. In QED, the problem of the loop diagrams can be eliminated. For a Yukawa-type theory, Salam had much earlier shown that renormalisation can be achieved, though the problem is a lot harder than in the case of QED. There was also a hint that a pure Yang–Mills theory was renormalisable. Demonstration of renormalisability in the GWS theory was, however, a different cup of tea altogether. One thing about photons is that they do not interact with each other; by contrast, the W's and the Z^0's can, and that makes the infinity problem harder in electro-weak unification theories. But the problem was solved first to some extent by Veltman and later completely by his student Gerard 't Hooft. The latter showed that the bare Yang–Mills theory suitably combined with the Higgs mechanism does become renormalisable.

Veltman has explained this as follows (see also Fig. 10.17). Consider a non-renormalisable theory (like the Fermi or the early Glashow model). It may give reasonable and even good answers in low orders but it blows up in high orders. In the present case, the blow up is associated with processes involving the self-interactions of the vector bosons — recall my earlier remark that unlike the photon, the W's can interact with each other. Thanks to this possibility, one can, analogous to the diagram in Fig. 10.17(a), have a diagram as in (b). However, such a diagram leads to a scattering probability greater than unity, which of course is absurd since probability can never exceed unity. The solution lies in taking account of diagrams like those in Fig. 10.17(c) which, incidentally, involve the Higgs boson.

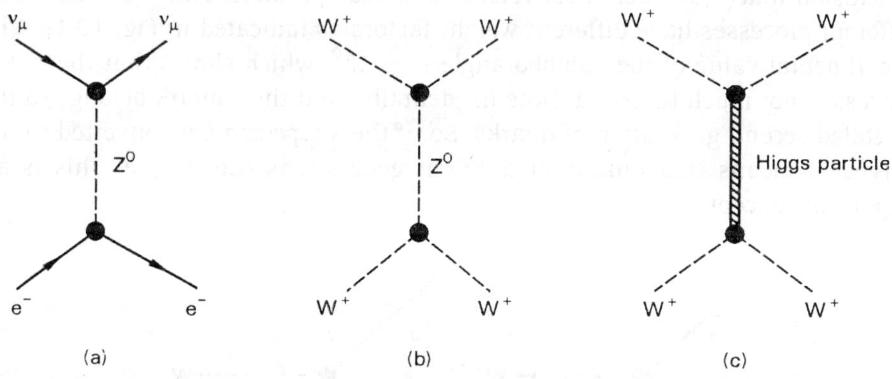

Fig. 10.17 (a) shows the Feynman diagram for the $e^+ e^- \to \nu_\mu \nu_\mu$ process. (b) shows a similar diagram involving the W's, representative of the self-interaction possible among the vector bosons. The scattering probability for this event comes out as greater than one, which is clearly absurd; the solution is to take also into account diagrams involving the Higgs particle, like the one in (c).

With 't Hooft's demonstration, the GWS theory was 'fully debugged' and ready for use. It was an interesting theory, and renormalisable, but did it really describe Nature? To be sure of that, one needs experiments. Before we turn to experiments, there are one or two points that need to be disposed of.

10.9 The quarks sneak in

Many weak interactions take place in the world of hadrons, beta decay being a simple example. Now hadrons are made up of quarks, and so the question naturally arises: do quarks have any say in the picture? After all, there are also non-leptonic contributions to the current. Indeed, the question is a good one. The first to make room for quarks were Cabibbo and Maiani, whose work led to the introduction of an important parameter known as the *Cabibbo angle* θ_C. At the time this work was originally done, only three types of quarks were known, i.e., u, d and s. Now the d quark has a charge of $-(1/3)e$ while the u quark has a charge $+(1/3)e$. The d can become u with the help of a W^- as below:

$$d \longrightarrow u + W^-.$$

Notice that charge is conserved, as it must be. Of course when I talk of d to u conversion, what I mean is that the d quark inside a hadron gets converted to the u type which too resides in a hadron. You may ask, "What about a similar s to u conversion?" That precisely was Cabibbo's question too, and he came to the conclusion that (a) such a conversion is indeed possible, and (b) these two different processes have different weight factors as indicated in Fig. 10.18. The experimental value of the Cabibbo angle is $\sim 13^0$, which shows that the s → u process is not much favoured. Note incidentally, that the s quark belongs to the so-called second generation of quarks. So, if the s type can get converted to the u type, it means that quarks of different generations can couple. This is an important concept.

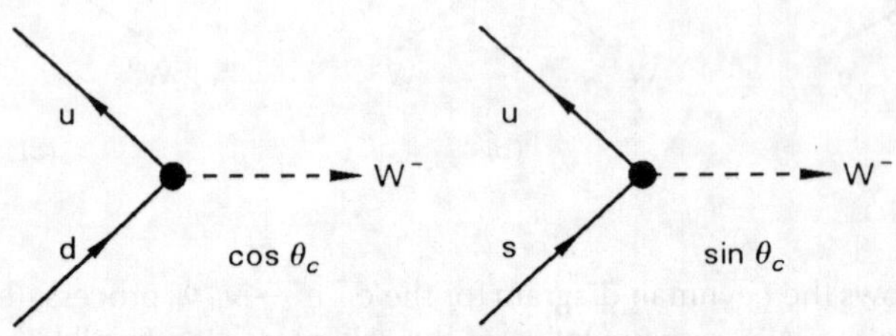

Fig. 10.18 The Cabibbo angle θ_C and the weight factors for the d → u and the s → u processes.

The Cabibbo angle can be better understood in modern perspective as follows, thanks to the work of Glashow, Iliopoulos and Maiani (GIM). Today, we recognise three generations of leptons and quarks as below:

$$\text{Lepton generations} \quad \begin{pmatrix} \nu_e \\ e^- \end{pmatrix}, \begin{pmatrix} \nu_\mu \\ \mu^- \end{pmatrix}, \begin{pmatrix} \nu_\tau \\ \tau^- \end{pmatrix}.$$

$$\text{Quark generations} \quad \begin{pmatrix} u \\ d \end{pmatrix}, \begin{pmatrix} c \\ s \end{pmatrix}, \begin{pmatrix} t \\ b \end{pmatrix}.$$

Where leptons are concerned, W.I. respects the separate identities of the three generations, i.e., W^- can participate in generation-preserving transitions as below:

$$e^- \longrightarrow \nu_e + W^-,$$

$$\mu^- \longrightarrow \nu_\mu + W^-,$$

$$\tau^- \longrightarrow \nu_\tau + W^-.$$

There is, however, no lepton cross-generation coupling like, for example,

$$e^- \longrightarrow \nu_\mu + W^-.$$

Cabibbo's work gave the first hint of such cross-coupling where quarks are concerned. GIM extended the work of Cabibbo and Maiani by invoking a fourth quark, i.e., the c quark. At that time, this was a bold and revolutionary move. People were not quite ready to accept the quark itself, and here were GIM increasing the number of quarks by one more (an idea already mooted by Glashow and Bjorken). But the November Revolution definitely settled the charm issue. Today one unhesitatingly accepts three generations of leptons and quarks. Since then, the tau, and the bottom and top quarks have been discovered as well. Seen in this light, the Cabibbo angle is a measure of cross-generations of coupling between quarks.

10.10 Detection of neutral currents

We now turn to the experimental verification of the GWS theory. When I say GWS theory, I mean the three generations of leptons and quarks, *a la* GIM and their successors. We can draw up the following list by way of demanding experimental verification:

(a) Detection of neutral currents.

(b) Detection of the intermediate vector bosons, i.e., the W's and the Z^0.

(c) Detection of the Higgs particle.

(d) Detection of interference effects.

(e) Correct prediction of various reaction processes, etc.

Most of this has been achieved. The Higgs particle is yet to be detected but all the same, the GWS theory is now deemed fully acceptable. The first major breakthrough really came from the detection of the so-called neutral currents.

Consider the diagram in Fig. 10.19(a), associated with the reaction

$$\bar{\nu}_\mu + \mu^- \longrightarrow \mu^- + \bar{\nu}_\mu \, .$$

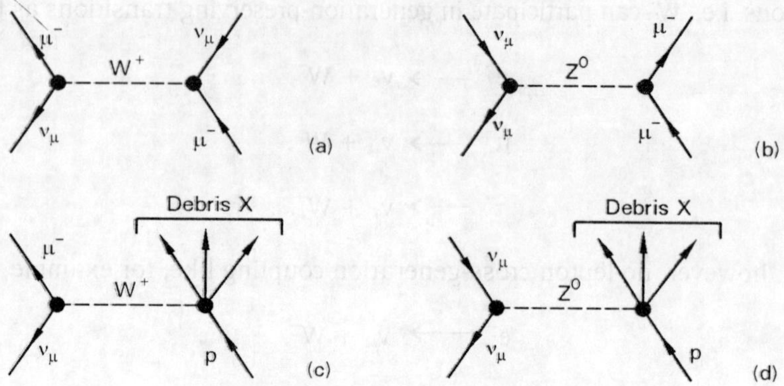

Fig. 10.19 Shown in (a) and (b) are two ways in which a muon anti-neutrino can interact with μ^-. In one case, the mediation is via the W^+; thus the current involved is the charged current. In the other, it is the Z^0 which is involved; this corresponds to a neutral current process. Similarly, in the interaction of $\bar{\nu}_\mu$ with a nucleon (e.g., the p), two processes are possible, as shown in (c) and (d).

As can be seen in the figure, the mediating particle is the vector boson W^+. Another possibility is the reaction

$$\bar{\nu}_\mu + \mu^- \longrightarrow \bar{\nu}_\mu + \mu^- .$$

The diagram corresponding to this reaction is shown in Fig. 10.19(b). Here the mediator is the neutral boson Z^0. One says that while the process in Fig. 10.19(a) involves a charged weak current, the process in (b) involves a *neutral weak current*. There are similarly, two options where the reaction of the anti-neutrino with the nucleon is involved; these are shown in (c) and (d). The GWS theory implies the existence of the neutral current, and therefore the experimental discovery of the neutral current marks the first major landmark in the acceptance of the GWS theory. The dramatic story of the actual detection of the neutral current comes later. At present, clear experimental evidence is available not only for the process in Fig. 10.19(b), but also for the reactions

$$\bar{\nu}_\mu + N \longrightarrow \mu^- + x,$$

$$\nu_\mu + N \longrightarrow \bar{\nu}_\mu + x,$$

the second of which occurs via Z^0 (Fig. 10.19(d)). The detection of neutral currents occurred in 1973 (see also Fig. 10.20: histogram of neutral events).

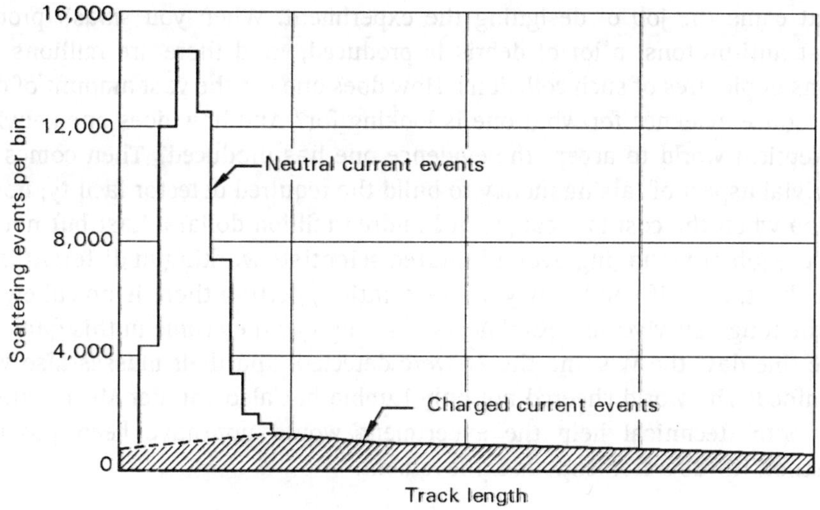

Fig. 10.20 Demonstration of the existence of neutral currents. This data does not refer to the original experiment, but to one done several years later when conditions had improved to the point that instead of just a handful of neutral current events as in 1973, one had them in thousands; this is typical of the way things improve and progress in high energy physics. What is shown above is a histogram of events corresponding to different track lengths. There are, of course, distinctive signatures that help one to distinguish between the charged and neutral events. In general, neutral current events have short tracks and the charged current events have long ones.

10.11 Detection of the intermediate vector bosons

The detection of the W^+, W^- and the Z^0 is a *tour de force*, masterminded by Carlo Rubbia, for which he eventually got the Nobel Prize. In many senses, the story is a repetition of the one enacted in the mid-fifties, when Emilo Segre went after the anti-proton and finally collected the Nobel Prize. In both cases, the discovery was no surprise (unlike in the case of the J/ψ). Rubbia knew exactly what he was looking for, and went about it in the most systematic manner possible.

The fact that he knew what he was looking for did not make the job any easier. First of all, at the time he started this hunt, even the right accelerator was not available. Now, to produce the desired bosons, Rubbia needed to arrange for proton–anti-proton collisions, with centre-of-mass energies ~ 500 GeV. There was no accelerator in the world then capable of this. But there was in CERN a proton synchrotron which could be modified. It was not enough to get the protons and the anti-protons. The anti-protons had to be bunched together suitably so that a sufficient number of them would be available for collision with the protons. This was a major technical problem which was solved by van der Meer of CERN, of course on Rubbia's initiative. van der Meer shared the Nobel Award along with Rubbia.

Next came the job of designing the experiment. When you smash protons against anti-protons, a lot of debris is produced, and there are millions and millions of pictures of such collisions. How does one sift the vast amount of data and produce evidence for what one is looking for? And how does one convince the skeptical world to accept the evidence one has produced? Then comes the non-trivial aspect of raising money to build the required detector facility, not an easy job when the cost is a couple of hundred million dollars! Last but not the least, the job of managing over a hundred scientists working in different labs! So, in short, even if you know your destination, getting there is no cakewalk. Only the tough survive, and Rubbia was as tough as they come in this game.

One fine day, the W's and the Z^0 were detected, and their masses also were determined. The world cheered not only Rubbia but also van der Meer, without whose vital technical help the experiment would not have been possible. Altogether, it was a tremendous achievement.

10.12 Electro-weak interference

The work of GWS showed that the e.m. force and W.I. are two facets of the same force, which today we refer to as the electro-weak force. Although in low energy phenomena it is the e.m. force that dominates, the two forces do in fact always operate together; however, only at high energies is their joint presence visible. The relative contribution of the two is determined by an angle θ_W called the *Weinberg* or the *weak mixing angle* (see also Fig. 10.21).

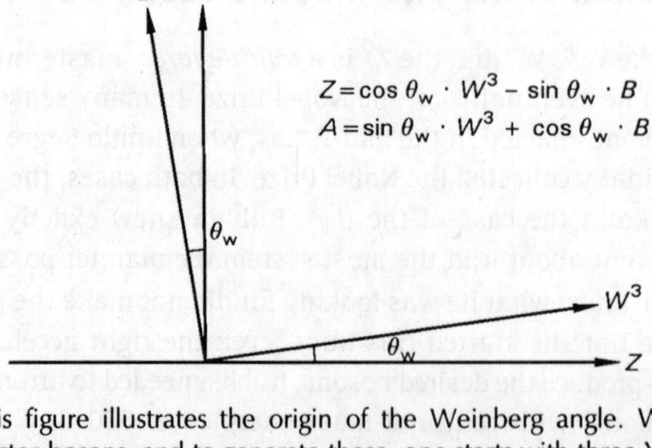

Fig. 10.21 This figure illustrates the origin of the Weinberg angle. We need the three intermediate vector bosons, and to generate these, one starts with three Yang–Mills fields. In addition, one must have a vector field for generating the photon. What happens in the theory is that one of the Yang–Mills fields (known as W^3 and which is neutral) mixes with another neutral field (known as *B*) to generate two new fields corresponding respectively to the *Z* and the photon (*A* field). This mixing is characterised by the Weinberg angle. The figure above illustrates schematically the mixing of W^3 and *B* to produce the *Z* and the photon (*A*), as well as the Weinberg angle.

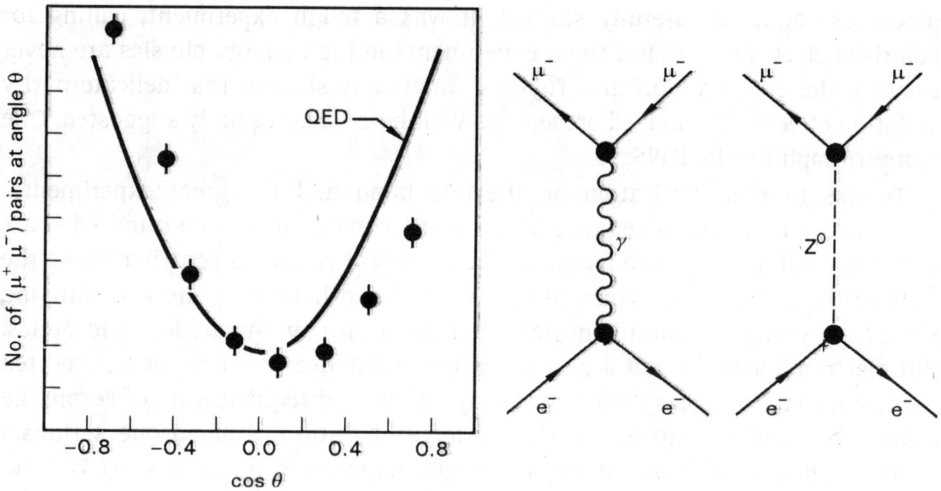

Fig. 10.22 This figure shows the angular distribution of the reaction products in the reaction $e^+e^- \rightarrow \mu^+\mu^-$. The Feynman diagrams for the process are also shown. As can be seen, the process can be mediated via the photon as well as the Z^0. The former involves e.m. interaction whereas the latter involves W.I. The solid line in the figure corresponds to a pure $(1 + \cos^2\theta)$ distribution which results from QED. The data points clearly do not fall on the QED curve, indicating that there are electro-weak interference effects.

When the two forces operate together, there are interference effects. The GWS theory motivated experimenters to look for such subtle interference effects, and deduce therefrom, the Weinberg angle. A typical example of such a result is shown in Fig. 10.22, which depicts the angular distributions of the muons emitted in the high energy reaction $e^+ + e^- \rightarrow \mu^+ + \mu^-$. The two channels through which this reaction can proceed are also indicated; one of these is e.m. in origin and the other involves the weak force. The energy of the electron and the positron is 15 GeV, and the two beams are made to collide to achieve the desired reaction. Naturally in such a situation the two forces interfere, and as a result the angular distribution deviates from what pure QED predicts. The GWS theory explains the observations well.

A question of related interest is parity violation effects in neutral currents. Detecting the existence of the neutral currents was by itself a great achievement, but soon Stanford scientists went much further by devising a delicate experiment to determine the parity violation effects in neutral currents. To understand what this means, go back to the Feynman diagrams in Fig. 10.22. One of them involves the photon and clearly there is no parity violation here because the e.m. force respects parity. But the reaction can also occur via the mediation of the Z^0 which is an offspring of the weak force; and here parity violation effects can enter. So, what the Stanford scientists Prescott and Taylor did was to come up with an experiment in which the different behaviour of the photon-driven and the Z-driven processes could be carefully studied. It was a tough experiment, calling for polarised electrons, etc. But then, experiments in high energy physics are never easy! In the end, Prescott and Taylor dramatically showed that delicate parity violating effects were indeed present, as Weinberg had originally suggested. One more triumph for the GWS.

Thanks to renormalisation on the one hand and the great experimental discoveries just summarised, the GWS theory, coming about one hundred years after Maxwell, may well be hailed as the twentieth century's contribution to the saga of the basic forces. When Maxwell developed his theory, he was unifying what then seemed separate entities, namely, electricity, magnetism and optics. But electromagnetism had a partner, whose existence was not even suspected then. Due to the discovery of radioactivity and the subsequent work of Fermi, the weak force was recognised as a basic force of Nature. Still no one seriously suspected it to be the other side of the electromagnetic force; thanks to GWS and the many others who preceded them, we now do. Professor Rajasekaran of the Institute of Mathematical Sciences, Chennai (Madras), sums it up as follows:

> Earlier discoveries of Oersted, Ampere and Faraday had unified electricity and magnetism into a single science of electromagnetism and Maxwell then unified optics with electrodynamics.
>
> Laws of Electrodynamics
>
> $$\nabla \cdot \mathbf{E} = 4\pi\rho$$
>
> $$\nabla \times \mathbf{E} + \frac{1}{c}\frac{\partial \mathbf{B}}{\partial t} = 0$$
>
> $$\nabla \cdot \mathbf{B} = 0$$
>
> $$\nabla \times \mathbf{B} - \frac{1}{c}\frac{\partial \mathbf{E}}{\partial t} = \frac{4\pi}{c}\mathbf{J}.$$
>
> All this is old stuff. Maxwell's equations have stood the test of time for these 100 years. Will they remain in isolated glory forever? The answer has to be in the negative. Progress in physics generally consists in the discovery of more general laws of wider applicability which reduce the older laws to some restricted domain.
>
> This is precisely what we are witnessing now. The successful development of gauge theory (of weak interactions) shows that there exist more general laws which subsume Maxwell's laws of electrodynamics. There exist more general fields $\mathbf{E}_i$ and $\mathbf{B}_i$ (the index i running over the set of gauge bosons) and these fields are the generalisations of the usual electric and magnetic fields and describe not only electrodynamics but weak interactions also .... These fields $\mathbf{E}_i$ and $\mathbf{B}_i$ satisfy laws of motion which are similar to Maxwell's laws of electrodynamics.

$$\nabla \cdot \mathbf{E}_i + ... = 4\pi\rho_i$$

$$\nabla \times \mathbf{E}_i + \frac{1}{c}\frac{\partial \mathbf{B}_i}{\partial t} + ... = 0$$

$$\nabla \cdot \mathbf{B}_i + ... = 0$$

$$\nabla \times \mathbf{B}_i - \frac{1}{c}\frac{\partial \mathbf{E}_i}{\partial t} + ... = \frac{4\pi}{c}\mathbf{J}_i.$$

That about wraps up all the technical stuff; and now, on to the juicier part!

10.13 How it all happened

From around 1915 onwards, Einstein strenuously tried for nearly forty years to unify the electromagnetic and the gravitational forces. To him these two forces seemed natural twins — we can guess why — yet he did not succeed in linking them; we now know why. However, as Salam later recalled, Einstein's valiant struggle was a source of great inspiration to others who came later and pursued unification.

It was Oscar Klein who in 1938 first suggested that the W.I. could be mediated by messenger particles. Klein's theory also established a connection between electromagnetism and the W.I.. Looking back, Klein's vision is almost premonitory. Klein wrote:

> The role of these particles (i.e., the messengers), and their properties being similar to those of the photons, we may perhaps call them 'electro-photons', namely, electrically charged photons.

The Second World War followed and Klein was soon forgotten; today he is hardly remembered.

As I have remarked earlier, during the first decade after the War, W.I. attracted hardly any attention — it was all QED and the strange particles. But once parity fell and the Universal Fermi Interaction emerged, it was a different story, with the weak interactions staging a strong comeback. Soon after the parity crisis, Schwinger suggested that the e.m. and the weak forces were two facets of the same force. In a paper entitled *A Theory of Fundamental Interactions* published in 1957, Schwinger proposed that the weak force is transmitted by the photon, which is massless, and two other charged bosons which today we refer to as W^+ and W^-. It was in this paper that Schwinger also boldly declared the muon to belong to a different family of leptons.

Schwinger's basic idea was good but his theory was not alright. Guided by wrong experiments (which Sudarshan had criticised), Schwinger took the W.I. to be of the *V* and *T* form instead of the correct (*V–A*) form. So Schwinger's theory fell by the wayside; even then, he asked his student Sheldon Glashow to "think about it". Glashow did as asked but nothing much came out of it, except a Ph.D. degree. However, the seeds had been sown. Glashow had a theory of sorts for the weak interaction and electromagnetism but it was full of holes; however, Glashow was not aware of that at that point of time.

In early 1959, Glashow went to England and stopped over at the Imperial College where Salam was a professor. Attracted by Schwinger's ideas, Salam and Ward (also of the same college) had been trying not only to unify the e.m. and the weak forces, but also construct a renormalisable theory. Both were skilled in tackling infinities but the infinities of this problem were tougher nuts to crack. Meanwhile Glashow arrived, glibly announcing that he had a theory for the W^+ and W^- and that his theory was renormalisable. Salam later recalled:

> My God! This young boy was claiming that his theory was renormalisable! It cut me to the quick! Both of us (Salam and Ward) were *the* experts on renormalisability, wrestling for months with the problem — and here was this slip of a boy claiming that he had renormalised the whole thing!

Glashow was of course wrong and Salam showed how. Glashow was now at the crossroads. His idea was bold but his theory was wrong. Should he stay with it or should he move to safer pastures? Glashow decided to hang on, and it paid off.

In 1960, Glashow went to Paris, where he met Gell-Mann who gave him a patient hearing. Gell-Mann supported the unification idea but warned Glashow that the reception would be cold. "What you are doing is good but people will be stupid about it", he said. Encouraged, Glashow wrote another paper which later established his claim to the Nobel Prize. Unlike Schwinger, Glashow now realised that the desired unification cannot be brought about with just the photon and two charged massive bosons; instead, *three* massive bosons would be needed, apart from the photon, of course. Glashow thus made room for Z^0, and this was a major step forward. In his paper, Glashow cautiously says:

> At first sight, there may be little or no similarity between the e.m. effects and the phenomena associated with weak interactions. Yet certain remarkable parallels emerge with the supposition that the weak interactions are mediated by unstable bosons.

Having introduced the Z^0, Glashow explains why he did so; he wanted the weak and the e.m. forces to work independently but hand in hand — they weren't quite united but organised into a team.

Glashow's theory was a union of two gauge theories — one the regular e.m. gauge theory, and the other a Yang–Mills type theory for W.I. Yang and Mills used the idea for the p and the n; Glashow used a similar idea for leptons and that was another great step forward — the idea of SU(2) × U(1) symmetry was also born.

No one paid much attention to Glashow's work because the Z^0 idea ran into problems; if the Z^0 did exist then there must be neutral currents. Glashow himself pointed this out. However, experimentalists, at least at that time, were quite sure that neutral currents simply did not exist. They had seen millions and millions of bubble chamber pictures and not even once had they come up with a picture characteristic of an event involving neutral currents.

Today we know why they did not see any trace. Now Glashow did not have a theory but rather a model. With a Yang–Mills field for the vector bosons, he naturally ended up with massless particles; so, to proceed further, he 'put in the masses by hand', as they say in the trade. Another major flaw was that the model was not renormalisable.

Having published this paper, Glashow moved to Caltech which was Gell-Mann's base. There they worked together, intending to construct a Yang–Mills theory for weak and strong interactions as well. This work resulted in a paper, but it proved to be a washout, except for the opening part which set the grammar for quantum field theory.

The next part of our story relates to the symmetry-breaking idea, and this time many personalities are involved. In 1958, the long-standing mystery of superconductivity was solved by Bardeen, Cooper and Schrieffer (BCS), for which they received the Nobel Prize. It was Bardeen's second Nobel Prize; he is the only one so far to have received two Nobel awards in Physics. Meanwhile, some Pundits were not too happy with the BCS theory for it appeared to violate the gauge invariance of electromagnetism. Among the first to realise this was Yoshiro Nambu of Chicago. He was listening to a lecture by Schrieffer who was visiting Chicago, and suddenly the idea hit him: "My God, the BCS theory is not gauge invariant!" Later he had long discussions with Schrieffer and concluded that since the BCS theory was successful in explaining all experimental data, Nature, for some reason, must have suspended the operation of gauge symmetry in superconductors. Thus was born the idea of symmetry breaking in superconductors; much earlier, Landau had proposed a similar idea in the case of superfluids — see ***Many Phases***. People then began to wonder, "Could broken symmetry be operating in high energy physics too?" Many delved deeper, Jeffrey Goldstone being one of them. Goldstone, an Englishman, was then in Wisconsin, USA. Studying a field theory in which symmetry was broken, he came to the conclusion that there must be produced as a result, a particle with zero mass — this is what we now call the Goldstone particle. A massless particle as a messenger means of course a force of infinitely long range, which is not exactly good news. In other words, a gauge theory with symmetry breaking would not work for weak interactions.

Goldstone's paper attracted some attention but many were suspicious of it. One such person was Philip Anderson, a solid state physicist of the famous Bell Labs (now, alas, extinct in its original form) who later won the Nobel Prize for his work on electron localisation in disordered solids (this phenomenon is now known as Anderson localisation). Though a solid state physicist, Anderson used to follow particle physics literature. Earlier, at Harvard, he had taken courses from Schwinger. He had always kept in touch with Schwinger's work and was thus familiar with Schwinger's attempt to unify the weak and the electromagnetic forces. Schwinger was suspicious of Goldstone's work and Anderson got wind of it. So he examined Goldstone's work in depth, which he could do because of his great expertise in superconductivity. Anderson then came to the following conclusions:

1. Gauge symmetry *is* broken in superconductors.
2. However, it does not lead to a massless particle — see also the appendix to this chapter.

Anderson was convinced that not only was Goldstone wrong, but also, broken symmetry was the key to endowing mass to the massless particles of the Yang–Mills field. Making a bold conjecture, he wrote:

> It is likely, then, considering the superconductivity analog, that the way is now open (for a theory) without any difficulties involving either zero mass Yang–Mills gauge bosons or zero mass Goldstone bosons. These two types of bosons seem capable of 'cancelling each other out' and leaving finite mass bosons only ... We conclude then that the Goldstone zero-mass difficulty is not a serious one because we could probably cancel it off against an equal Yang–Mills zero-mass problem.

Anderson had squarely put his finger on the solution to the long-standing problem but most particle physicists simply ignored Anderson's paper; the few who read it were doubtful about its conclusions.

Meanwhile, Salam and Weinberg came at that time to Wisconsin for a conference and they had long chats with Goldstone because they were intrigued by the latter's result. They did not want the massless particle, i.e., the Goldstone boson, but it seemed to be there like a 'snake in the grass', as Salam once described it. After the conference, Salam and Weinberg went back to England — Weinberg was visiting at the Imperial College at that time. Both tried hard to pick holes in Goldstone's arguments but they could not. A few telephonic conversations with Goldstone, and there came the paper by Goldstone, Salam and Weinberg which legitimised the Goldstone theorem. They were not aware then of Anderson's work.

Two people who read Anderson's paper and took it seriously were Abraham Klein and Benjamin Lee of Pennsylvania. They presented Anderson's results in the lingo of particle physicists, and quickly, the particle physics community woke up to the fact that perhaps, the cure to the problem plaguing the Yang–Mills theories might after all be broken symmetry. One of those who studied the problem in some depth was Peter Higgs of the UK. Higgs demonstrated what Anderson had clearly conjectured. Meanwhile, the Belgian physicist Brout, again a solid state physicist, aired similar views and he even telephoned Salam trying to get the latter excited.

One now had a general prescription for curing the disease of massless particles plaguing the Yang–Mills field theories, but that still left the problem of electro-weak unification wide open. Brout himself had a crack at it but did not succeed. Many others tried, with the same result. No one knew the correct gauge group. Way back, Glashow had guessed it to be SU(2) × U(1), but few remembered that work. As Glashow himself recalled during his Nobel lecture,

> These workers (Higgs and Kibble) never thought to apply their work on formal field theory to a phenomenologically relevant model. I had had many conversations with Goldstone and Higgs in 1960. Did I neglect to tell them about my SU(2) × U(1) model, or did they simply forget?

One wonders why Glashow himself did not have a go!

Higgs published his paper in 1964, but three more years were to pass before the desired unification actually occurred. The man who achieved it first was Steven Weinberg. When Higgs' paper appeared, Weinberg did not pay much attention to it. He was preoccupied then with strong interactions; soon he began to use the Higgs technology in a manner of sorts, but for strong interactions. Suddenly one day in 1967, Weinberg realised that he was applying the 'right idea to the wrong problem'. Later he said,

> I realised that the whole mathematical apparatus I was constructing was just what I needed to understand something I hadn't really been thinking about, namely, weak interactions and the intermediate vector bosons ... I suddenly realised, "My God! This is the answer to the Weak Interactions!"

And quickly came the paper, which established Weinberg's claim to the Nobel Prize.

Salam, too, was looking at the same problem. He was of course into W.I right from the fifties. Moreover, his student Robert Shaw had discovered the importance of gauge fields, quite independently of Yang and Mills. And then in the late fifties, Salam and Ward had developed the non-renormalisable version of the SU(2) × U(1) theory. After a few years came the Higgs paper. Thomas Kibble of the Imperial College made a deep study of Higgs' work, and drew Salam's attention to its importance. Once Salam understood the work of Higgs, he knew in a flash what had to be done; and it was done. Independent of Weinberg, Salam created the electro-weak theory. Meanwhile, Salam got a copy of Weinberg's work. Comparing it with his own, he was happy and relieved to find that the conclusions were identical. He knew that they both had it right.

In the world of particle physics, Higgs is given much credit (deservedly no doubt) for spelling out how the problem of the zero mass of the Yang–Mills field theories could be solved. Anderson, of course, called attention to this possibility much earlier but to this day his contribution is ignored by particle physicists. So, like Sudarshan, Anderson is sore. His gripe is that particle physicists generally look down upon solid state physicists, dismissing their work as lacking in fundamental importance. Unfortunately this is true, and some do tend to be supercilious in this manner; it is an old tradition, going back to Pauli!

So by 1968, Weinberg and Salam had completed what Glashow had begun much earlier. But did their work create waves? Not really. Weinberg explained his theory at an international conference in 1967 but the reaction was of the 'so what' type. Salam, too, unveiled his version at an international symposium, only to be ignored.

Why did this happen? Firstly, there was no knowing whether this theory was renormalisable. Secondly, there was no experimental evidence in its favour. Both came later, and after that, the GWS theory was duly hailed. Commenting on the absence of enthusiasm at that time, Sidney Coleman said, "Rarely has so great an accomplishment been so widely ignored."

We now turn to the story of renormalisation. Amongst the earliest to worry about renormalisation were Iliopoulos of Greece and Maiani of Italy. Iliopoulos did his Ph.D. in France and then moved to Harvard, USA, where he met Maiani. They both were to work with Glashow whose base was Harvard. Earlier, Maiani and Cabibbo had conjectured that the divergences of the Yang–Mills theory of weak interactions linked the strong and the weak interactions together in some way — after all, the current featuring in the theory of W.I. needed not only a leptonic component but a hadronic component as well. Iliopoulos was now determined to do some serious physics as he put it, and chase the infinities out. The trio, i.e., GIM, worked hard, and in the process, they found that they perforce had to make a place for the charm quark which had been conceptualised earlier by Glashow and Bjorken. Glashow went to neighbouring MIT where Weinberg was visiting at that time and told him about the charm idea but Weinberg was not charmed! Instead he was repelled by the idea of adding more particles — simplicity meant decreasing the number and not adding to it. Not able to sell the charm idea to Weinberg, Glashow then tried Ting, also at MIT. But Ting too refused to buy! It is ironical that Ting later discovered something that gave great support to the idea of the charm quark.

But the charm quark idea did have one achievement to its credit, that is, before the discovery of J/ψ and all that. Glashow was able to explain why experimenters were not able to see neutral current events, working with strange particles. If neutral currents had to be sought then one must design a different kind of experiment — that, in brief, was his message. As a sequel to the GIM paper, Glashow and Iliopoulos struggled to produce a paper which they claimed showed how divergences might be made to cancel each other in Yang–Mills theories. In June 1970, Glashow went to France to present this work at a conference. He remarks:

> There we are, laboriously cancelling the infinities, when this guy Veltman comes up to us. And he says, "My young student 't Hooft has solved the problem of renormalisability of gauge theories. All the work you have done for the past year is a waste of time!"

Glashow was stunned but Veltman was right. Let us now hear 't Hooft on the subject:

> In 1971, Veltman suggested that I investigate the renormalisation of the Yang–Mills theory. The rules for constructing the needed Feynman diagrams had already been formulated by Fadeev, Popov, Fradkin and Tytin (of Russia) and independently by Bryce S. DeWitt of the University of Texas at Austin, and Stanley Mandelstamm of the University of California at Berkeley. I could adapt to the task, the powerful methods for renormalisation studies that had been developed by Veltman.
>
> Formally the results were encouraging, but if the theory was to be a realistic one, some means had to be found to confine the Yang–Mills fields to a finite range. I had just learned at a summer school how Kurt Symanzik of the German Electron Synchrotron Laboratory and Benjamin W. Lee of the Fermi National Accelerator Laboratory had successfully handled renormalisation of a theoretical model in which global symmetry is spontaneously broken. It therefore seemed natural to try the Higgs mechanism in the Yang–Mills theory, where the broken symmetry is a local one.
>
> A few simple models gave encouraging results. In these selected instances all infinities cancelled, no matter how many gauge particles were exchanged and no matter how many loops were included in the Feynman diagrams. The decisive test would come when the theory was checked by the computer program for infinities in all possible diagrams with two loops. The results of that test were available by July 1971; the output of the program was a string of zeros. Every infinity cancelled exactly.

The computer checks were done by Veltman at CERN which had good computer facilities. Gerard 't Hooft's results were soon confirmed by others, notably Benjamin Lee and Jean Zinn-Justin of France. The GWS theory was safe; it was renormalisable, and both Salam and Weinberg could heave a sigh of relief. As Sidney Coleman put it, 't Hooft's work turned the Weinberg–Salam frog into a Prince Charming!

All that remained to be done was the experimental verification. The first major success was of course the detection of neutral currents. The story of the discovery of neutral currents is a kind of 'on again, off again' sequence. The moment Z^0 entered the picture, which was with Glashow's early paper, it was clear that neutral currents must be there; but experimenters scoffed at the idea; besides, they had other 'more important' things to do! Then came a series of theoretical developments, culminating with the papers of Weinberg on the one hand and Salam on the other, and finally came 't Hooft's proof of renormalisation. Even then, the fancy of experimenters was not captured. In 1971, Weinberg wrote a paper explicitly calling for the search for neutral currents, as that would be an important piece of evidence in favour of the GWS model which was otherwise neat and free from theoretical objections.

Merely writing a paper wasn't enough; it had to be promoted! And so Weinberg talked to many people, one of whom was Carlo Rubbia, then at Harvard. As Rubbia later recalled, "(Weinberg) took a lot of time telling me how nice his theory was!" And as a result, Rubbia was 'brainwashed', as he put it, into searching for neutral currents. He arranged a collaboration called HPW (Harvard, Pennsylvania and Wisconsin — these were the three participating Universities), and the experiment was to be done in the new Fermi National Accelerator Laboratory (FNAL), or Fermilab as it is better known.

We now cross the Atlantic and take a look at what is going on in CERN. In the mid-sixties CERN decided to go in for a giant bubble chamber which was built in France. Known as the Gargamelle, it took six years to construct and was specially designed for neutrino experiments. By now you know that neutral currents require neutrino experiments but the Gargamelle team members had other kind of experiments in mind; the search for neutral currents did not at all figure in their list of priorities. In November 1971, a CERN theoretical physicist received a copy of Weinberg's paper calling for the search of neutral currents. The paper created excitement among the theorists in CERN, and they organised a meeting with the experimentalists in charge of the Gargamelle experiments. They spoke passionately about the electro-weak theory, but the Gargamelle team refused to accord the desired priority for the search of neutral currents. However, the Gargamelle project director, Musset, had other plans. He decided that he and some others would personally devote their time to pick out neutral-current candidate bubble chamber pictures, from the millions that Gargamelle would routinely produce.

Early in 1972, Musset came across a picture that took his breath away. He was staring at what he believed was the first evidence of the existence of neutral currents. But then, it is not easy to convince the community with just one picture; so he became determined to search for additional evidence. Years ago, when Musset was searching for the eta meson he was beaten to the discovery by the

Americans. He knew that the Americans (Rubbia and others) were on the trail of the neutral currents and this time he wanted to come first.

So, there were two in the game now, Rubbia and company in America (i.e., the HPW collaboration) and Musset and his small group in CERN. There was a third contender also, and that was Robert Palmer of Brookhaven. He was on a sabbatical leave at CERN in 1971, and his project was to analyse Gargamelle data using computers instead of painstaking human effort. He too ran into an event which to him smacked of neutral currents but his CERN colleagues did not agree that the event represented evidence for neutral currents. So he could not make the momentous announcement the world was waiting for. Later Palmer lost out, and left with painful and bitter memories. He recalled:

> I have not altogether forgotten that it was I who first saw these neutral currents and it was I who one night made that plot and believed, alone in all the world, that God, if He exists, had decided to have a neutral current.

Let us get back to Musset. Lagarrigue who worked with him to build the Gargamelle, was now convinced that neutral currents must not be relegated to the back-benches, but be given priority. So he wrote to the Director General of CERN that the search for neutral currents must receive high priority; this was given, but there was no agreement amongst the members of the Gargamelle team as to the best way of carrying out the search. Finally, two groups were formed — one that would study neutrino–electron events and the other that would examine neutrino–hadron events; Musset belonged to the latter team.

On New Year's Day 1973, Franz Hasert of Germany and a member of the neutrino–electron subgroup, came across an event that was a textbook example of a neutral current event. This needle was in a haystack of over seven lakh pictures! By the end of January 1973, Musset went to America and spoke at the annual meeting of the American Physical Society, about the event he had found earlier from a study of neutrino–hadron collisions. Musset came back and, knowing that the Americans also were racing along, intensified his search for further events and by June he thought he had enough evidence to write a paper.

On July 3, Franz Hasert and his group submitted a paper to *Physics Letters* reporting the single event they had observed from a study of neutrino–electron events. Musset meanwhile, was working on the draft of his paper. Rubbia was at that time visiting Europe and he informed Musset that the Americans had about a hundred candidate neutral current events. Rubbia made an offer to Musset which in effect was: "If you mention our work in your paper, we will do likewise." Musset flatly refused and promptly dashed off his paper to *Physics Letters*. And as if that was not enough, Musset gave a lecture in CERN on the discovery. The two papers, of Hasert and his group and of Musset and his group were both published back to back in the 3rd September issue of *Physics Letters*. Robert Palmer who also reported a single event like Hasert et al. sent his paper to *Physics Letters* in May itself but the editor's office held it up for some unknown reason; it was finally published in the issue of 17th September. Though he had sent his paper at least two months earlier, Palmer was sidelined and priority was given to the CERN group which submitted their paper later. It is no wonder that Palmer is bitter!

Rubbia went back to the US after his brief European tour, fully determined to go ahead and publish, since he knew that the Europeans were almost there. But he had problems the moment he landed back. He was an Italian working in America, and the US Immigration Service asked him to leave the country on some technical grounds. Rubbia had no choice but to leave. His colleague Sulak personally took the HPW paper to the office of *Physical Review Letters* on August 3rd, to register their claim to fame. Meanwhile, Cline and Mann, two important collaborators in the HPW effort, began to get cold feet; they were not sure of the data they were getting ready to announce to the world and wanted further checks; this meant more modifications to their experimental apparatus and, therefore, also more time. This problem is always there in high energy physics experiments. One works hard and gets an important result; one wants to be the first in the world to announce it; getting there first and receiving kudos is what the excitement is all about. At the same time, one does not want to find that someone else who repeats the experiment later, reports that the first experiment was wrong. So, experimenters are always torn — should they go ahead and race to print, or should they do more checks before publishing? But what if someone else scoops them in between? We saw earlier how many experiments in β decay were wrong and finally had to be redone to establish the $(V-A)$ form. Then there was the case of Ting who was in agony — more cross-checks or race to print? It was the same story again.

A small digression at this stage, with a reminiscence by Salam of those exciting times. In his Nobel lecture, Salam recalled:

> I still remember Paul Matthews and I getting off the train at Aix-en-Provence for the 1973 European Conference and foolishly deciding to walk with our rather heavy luggage to the student hostel where we were billeted. A car drove up from behind us, stopped, and the driver leaned out. This was Musset whom I did not know well personally then. He peered out of the window and said: "Are you Salam?" I said, "Yes." He said, "Get into the car. I have news for you. We have found neutral currents."

Let us now revert to the mainstream story. As I mentioned earlier, in September 1973, the *Physics Letters* came out with the CERN papers. The Americans expressed doubts about the findings of the Europeans. "Is the effect real?" was the question that went around. Meanwhile, the paper personally submitted by Sulak to the *Physical Review Letters* was rejected as not being convincing enough.

News reached Rubbia in Europe and he, a powerful figure in European circles, called on the CERN Director General and drew the latter's attention to how CERN scientists had gone into print with questionable data; the DG was far from pleased. Rubbia had bought some time, with the Americans having sown the seeds of doubt (remember, his paper had bounced). However, the Gargamelle team refused to budge and remained convinced about its discovery; Musset in fact went about giving talks on the subject.

Back in Fermilab, modifications were made to the HPW setup and fresh data was taken; this time there was no doubt; the neutral currents *were* there. A paper was sent to the *Physical Review Letters*, and it appeared on April 8, 1974, confirming what the CERN scientists had reported months earlier. So the CERN scientists were right after all, and for once the Europeans had beaten the Americans; and for Musset, this was sweet victory!

The discovery of neutral currents is such an important event that its discoverers should have received the Nobel Prize; but they did not because sadly, Lagarrigue died suddenly in 1975 and Musset died in a climbing accident in the Alps shortly thereafter.

Once neutral currents were established, the GWS theory was 'in'; no one doubted it anymore, and it was only a question of time before the intermediate vector bosons were found. This happened in 1983 but even before that, Glashow, Weinberg and Salam had been awarded the Nobel Prize in 1979.

10.14 Chapter summary

- Classical mechanics can be based on the Lagrangian, and equations of motion can be derived therefrom, both for particles and for continuous fields (relativistic and non-relativistic).
- In quantum mechanics, many of the concepts carry over from CM, including the Lagrangian equations of motion, etc.
- In QM, one can use this machinery for dealing, for example, with particles of spin 1/2, spin 0 and spin 1.
- Thanks to Feynman, one also has a machinery via the Feynman calculus, to start from the Lagrangian for the quantum field and work one's way through various orders of the perturbation theory.
- Serious study of W.I. commenced with Fermi's theory/model. In today's language, Fermi's theory has the current–current form with contact interaction.
- Fermi wrote his Lagrangian by analogy with the electromagnetic case. However, he did not make any provision for the analog of the photon; there were no messenger particles in Fermi's model.
- Useful calculations could be made with Fermi's theory in the lowest order but they blew up in the higher order, reminiscent of Dirac's QED.

- The next major development was the downfall of parity, and the discovery of the Universal Fermi Interaction.
- The stage was set in 1958 for achieving a breakthrough, and it came in 1967–1968.
- In the intervening period, the following developments occurred:
  - (a) It was recognised that one needed W^+, W^- and Z^0.
  - (b) The underlying fields were identified to be of the Yang–Mills type.
  - (c) Yang–Mills theories were found to produce zero-mass particles.
  - (d) The role of broken symmetry was discovered.
  - (e) The Higgs mechanism was proposed as a means of getting out of the impasse.
- Based on the above, Weinberg and Salam independently discovered what is now known as the GWS theory. The basis for this is the SU(2) × U(1) symmetry, discovered earlier by Glashow.
- 't Hooft proved that the GWS theory is renormalisable; thus meaningful perturbative calculations could be done using the Feynman calculus.
- The GWS theory incorporates all desirable features, such as:
  - (a) Parity violation effects in W.I.
  - (b) Allowance for currents due to quarks.
  - (c) Generation crossing, symbolised by the Cabibbo angle.
  - (d) Mild mixing of e.m. and weak effects, via the Weinberg angle.

Experiments have fully vindicated the GWS theory, though the Higgs particle is still to be found. However, few doubt its existence; it is only a matter of time before it is detected.

The GWS theory took care of the electromagnetic and the weak forces. But what about the strong force? We pick up the threads of that story in the next chapter.

Appendix to Chapter 10

In this appendix we shall try to understand how Anderson used the example of superconductivity to anticipate the Higgs mechanism.

I have described in companion volumes the discovery as well as the phenomenon of superconductivity. Kammerlingh Onnes found in 1908 that when mercury was cooled below 4.2 K, its electrical resistance became vanishingly small — this is the practical manifestation of superconductivity, and the phenomenon is exhibited by many metals like aluminium, lead, etc. It is true that below the superconducting transition temperature, electrical resistivity beomes identically equal to zero; but that does not mean the metal is merely a perfect conductor; it is much more than that. Well, what is it then? Ah, that's where the *Meissner Effect* comes in.

Let us suppose we have a metal which becomes a perfect conductor below a temperature T_c— note I am not saying a superconductor but a perfect conductor. Now let us imagine we perform a series of experiments as in Fig. 10.A.1 (a) to study the magnetic behaviour of this hypothetical perfect conductor.

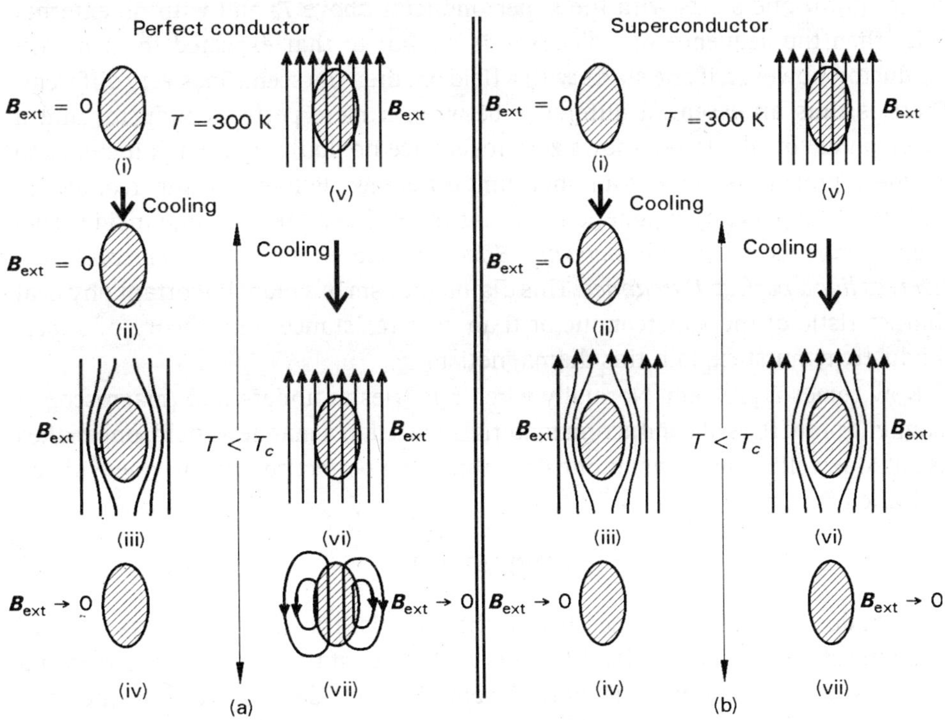

Fig. 10.A.1

We perform two series of experiments; in the first, we start with the sample at room temperature (taken, as always, as 300 K) and then cool it below T_C. Next we apply a magnetic field $\boldsymbol{B}_{ext}$. This fails to penetrate the metal, but then that is the property of the perfect conductor, as will be explained a little later. Now switch off the external magnetic field; there is no field anywhere, including within the material. The sequence of states one encounters is illustrated in Figs. a(i) to a(iv).

Let us continue our experiments, doing things somewhat differently this time. We start as before at 300 K and apply $\boldsymbol{B}_{ext}$. At room temperature, the material is not a perfect conductor (remember the transition occurs only below T_C); therefore the field penetrates the sample as shown in Fig. a(v). Now we cool and take the material below T_C with the field still on. The situation then is as shown in Fig.a(vi); notice that the field persists inside, having found a berth earlier! Now switch off $\boldsymbol{B}_{ext}$; the field inside stays put, refusing to quit; the material behaves like a magnet (see Fig. a(vii)).

Let us now imagine similar experiments to be performed on a metal that becomes a superconductor below T_C (say a real metal like lead, for example). The sequence one observes in this case is depicted in Fig. 10.A.1(b), i to vii. Compare carefully the two scenarios in Figs. (a) and (b), remembering that what one actually observes in Nature is what is shown in Fig. (b) and not what is shown in Fig. (a). If one starts with the superconductor above T_C and with no external field, then the sequence one observes is similar to that expected for a perfect conductor; however, if one starts with a field on, then the scenario is very different. There is thus an essential difference between a mere perfect conductor and a superconductor; the latter shows zero resistance no doubt, but when it comes to magnetic properties, there is an important difference: Below T_C, a superconductor mercilessly expels any magnetic field that might have taken refuge inside. This phenomenon is called the *Meissner Effect*; it shows that *the superconductor behaves like a perfect diamagnet*. This diamagnetism is a more important physical characteristic of the superconductor than zero resistance. Any theory of superconductivity must explain this diamagnetism.

Now comes a problem. Normally when one tries to understand (macroscopic) magnetic properties of conducting materials placed in a magnetic field, one makes use of Maxwell's equations. If one does that for a perfect conductor, one ends up with the result

$$(d/dt)\,(\boldsymbol{B}) = 0.$$

What does this mean? It means that the material tries to preserve its internal magnetic state, whatever it is. If there is no field to start with, then the material tries to stay the same way; if there is some field inside to start with, then the material is reluctant to expel it. If now you carefully study the scenes depicted in Fig. (a), you will notice that this is precisely what happens. When $\boldsymbol{B}_{ext}$ is switched off, how does the perfect conductor sustain the magnetic field as in Fig. a(vii)? This is a question.

Now let us turn to the superconductor. In this case not only do we have $(d/dt)(\boldsymbol{B}) = 0$, as we must since the resistance vanishes, but also the *additional* condition $\boldsymbol{B} = 0$.

The latter is a sort of overriding condition. In other words, the superconductor simply does not like any magnetic field to reside within it; it is intolerant and mercilessly boots it out!

Recall that the condition $\dot{\boldsymbol{B}} = 0$ follows from Maxwell's equations. What about the additional equation $\boldsymbol{B} = 0$? That does not follow from Maxwell's equations. In other words, *Maxwell's equations fail to describe the magnetic behaviour of a superconductor* — make a careful note of that.

Before the BCS theory (which came in 1958), this difficulty was taken care of via an equation written down specially for superconductors known as the *London equation*. The BCS theory gave the correct explanation of the Meissner effect and thus passed an acid test. Let us now try to understand magnetic field expulsion by the superconductor. Figure 10.A.2 shows a superconductor in the form of a semi-infinite slab. Next we imagine that a magnetic field is applied as shown in Fig. 10.A.1(b). Inside the superconductor the field is supposed to be zero; however, the field does not abruptly vanish as one crosses the physical boundary from the outside to the inside. Rather, the field drops off exponentially, as shown, λ representing the *penetration depth*. Typically, λ is of the order of a few Ångströms.

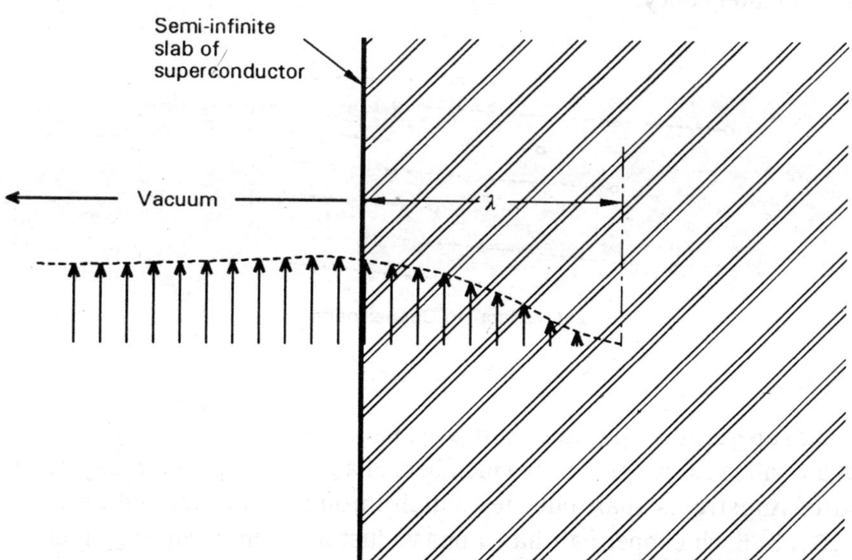

Fig. 10.A.2

For a perfect conductor, λ is infinite. Jumping a bit and using language we are used to, one could say that while in a perfect conductor the electromagnetic force has infinite range, in a superconductor the range is finite. In turn this implies that while in a perfect conductor the carrier of the field (the photon) has zero mass, in a superconductor the photon behaves *as if* it has mass. Remember the popular thumb rule: finite range means the carrier particle must have mass.

Let us explore further, explicitly bringing in local gauge invariance, BCS theory and all that. In the normal state, the material respects Maxwell's equations, which means that the local gauge invariance symmetry holds; but below T_c, Maxwell's equations do not work. Then there is the additional fact that superconductivity is due to symmetry breaking (I have discussed this elsewhere in terms of a φ-4 potential). Anderson analysed all this and showed that Maxwell's equations do not work in the superconducting state precisely because of the breaking of the local gauge invariance symmetry. And this breaking of the local gauge invariance symmetry makes the photon acquire mass (within the superconductor of course). Thus he clearly anticipated the Higgs mechanism (though with specific reference to the superconductor), but there were few takers then for his ideas; indeed, even later not many acknowledged his prescient suggestion.

In terms of the electron wave functions, the picture is as follows: Suppose a field is applied to a metal in the normal state. The magnetic field then rotates the phase of the wave functions of the individual electrons to suit its convenience. This phase rotation is different at different space points, consistent of course with local gauge invariance. In other words, the magnetic field dictates and the electrons mutely obey!

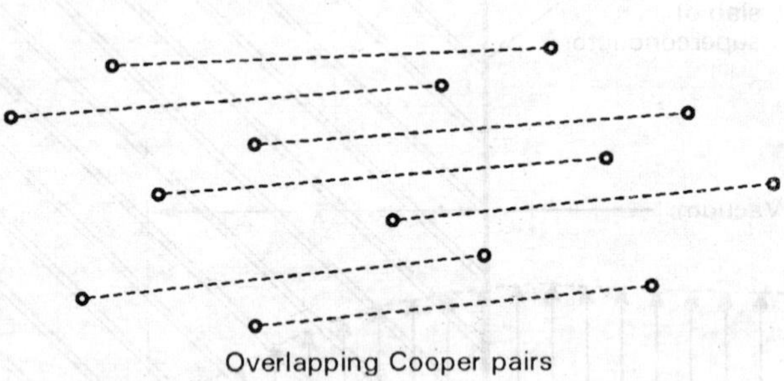

Fig. 10.A.3

In the superconducting state, things are very different. First, the electrons form what are called *Cooper pairs*. Two members of a given Cooper pair may be several hundred Ångströms apart but they march through in disciplined union — see Fig. 10.A.3. Each Cooper pair has a phase, just as a single electron had earlier. However, there are numerous Cooper pairs and they all overlap heavily. Thus their phases get locked, so to speak, and all this leads to a *macroscopic* phase for the superconductor (if you recall from other earlier volumes, both superfluidity in liquid helium and supercurrent flow in superconductors are due to spatial variations of this macroscopic phase).

When a (weak) magnetic field is applied to the superconductor, a conflict arises between the macroscopic phase and the applied magnetic field. The latter wants to have its way and do its bit of phase twisting. Earlier it could get away with it because the electrons were not united and each went its own way. But now, not only are the electrons paired but the pairs too are in substantial unison. In short, because of the macroscopic phase locking as it were, the magnetic field is unable to assert itself; this is what goes on behind the breaking of the local gauge symmetry. Suppose the strength of the magnetic field is increased, then there comes a stage when the electrons are not able to overwhelm the invader; superconductivity is destroyed and the magnetic field has its way as in the normal metal; in fact, the metal reverts to its normal state, though the temperature might be below T_C! Unity is strength, even in the microscopic world. How remarkable that Nature teaches us lessons wherever we turn!

In summary, in a normal metal, there is local gauge invariance and Maxwell's equations apply. In the superconducting state this symmetry is broken, and the electromagnetic force is no longer able to exert itself over an infinite range; its range becomes very much shorter. Consequently, the photon behaves as if it has acquired a mass. This is the point to which Anderson referred, drawing the important lesson that the solution for the massless particles of the Yang–Mills field theories is to marry those theories with symmetry breaking. Higgs rediscovered all that, and later both Weinberg and Salam followed in the footsteps of Higgs. Everyone remembered Higgs but no one remembered Anderson, and that was his grouse!

11

Splicing Flavour with Colour

Rutherford discovered the nucleus. After this came Chadwick's discovery of the neutron and it then became clear that the nucleus is made up of neutrons and protons. What holds these particles together in the nucleus? Heisenberg conjectured that there was a hitherto unknown force that does this job. This force had to be very strong to overcome the Coulomb repulsion of the protons in the nucleus, and so it was named the strong force. How did the strong force originate? Yukawa offered the first speculation but the full story emerged only recently.

11.1 Introduction

This chapter deals essentially with the strong force. It may not be as exciting as the previous one but is important nevertheless. Besides, the theory described here represents one of the great triumphs of the human mind.

When Rutherford discovered the atomic nucleus, people wondered how so many protons (in Uranium the number is as high as 92) could stay so close together inside the extremely tiny nucleus, without flying away from each other on account of the strong Coulomb repulsion. It was then realized that what kept the protons together was a new force, later named as the strong force or strong interactions (S.I.). It was hypothesized that this strong force acted not only between protons but also among neutrons, and between neutrons and protons as well. The first steps in understanding the strong force were taken by Heisenberg, Wigner, Majorana and Bartlett, but the real breakthrough came when Yukawa proposed a model based on the meson being the messenger particle. Later the pion was identified as the Yukawa meson and at that stage people thought, "This is it!" You may remember Salam recalling such a view (see Chapter 10). But that was not to be. With the discovery of the resonances, the strange particles, etc., people realised that Nature was far subtler than they had hitherto imagined. Subsequently, when Gell-Mann proposed the quark model, it became clear that a proper discussion of the strong interaction must be based not on the pion but on quarks.

Let us take a minute off to understand this last remark. Consider the potential energy of interaction between two neutral oxygen atoms. It is expected to have the form sketched in Fig. 11.1, and several phenomenological models have been proposed for the inter-oxygen atom potential which exhibit the form shown in the figure. Suppose one wants to calculate this potential *ab initio*, using QED. This is a reasonable demand. If one wants to do it from start to finish with QED alone, then one would treat each oxygen atom as a cluster of point charges (made up of electrons and the nucleus), the point charges interacting with each other according to the requirements of QED. The formal statement of the problem is easy, but to derive from it the effective inter-atomic potential would involve a horrendous exercise; nevertheless, in principle it can be done. The Yukawa potential between nucleons is an effective or phenomenological form for the inter-nucleon potential, based on the idea that the pion is the carrier of the force. However, both the nucleon as well as the pion are made up of quarks. Thus, if one wants to deduce the Yukawa potential *ab initio*, then one must really start at the quark level. So the question boils down to, "What is the dynamics that dictates the strong force, just as QED dictates the e.m. force?"

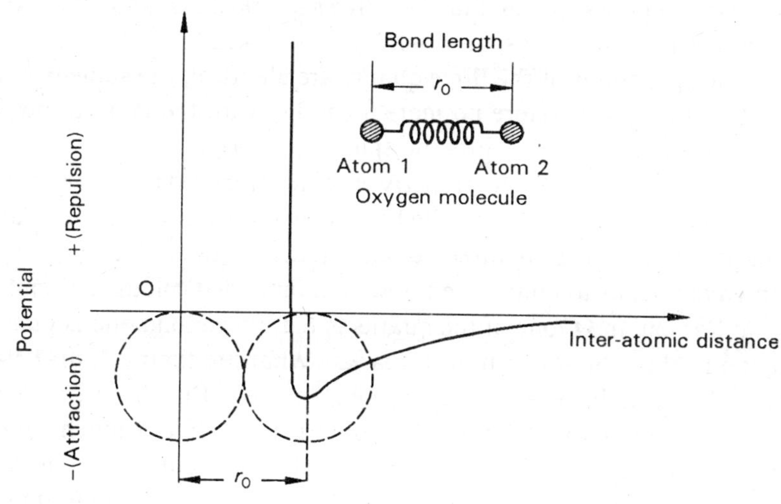

Fig. 11.1 Shown here is a sketch of the potential of interaction between two neutral oxygen atoms. At very large separations between the atoms, the potential is zero. As the two atoms approach each other, there is an attraction, as the negative value of the potential signifies. At a certain distance r_0, this energy is a minimum. If the distance between the two atoms is made smaller than r_0, then the potential rapidly turns repulsive; the two atoms now start bumping into each other, and don't like it! The separation r_0 represents the equilibrium distance of separation in an oxygen molecule.

11.2 A glitch in Gell-Mann's quark model, and the introduction of colour

Recall that Gell-Mann brought order into chaos by first invoking SU(3); at that stage there were no quarks. Next recall that when quarks were first introduced, there were only three of them; subsequently, the count increased to six. For the moment, let us consider Gell-Mann's original model in which there were only three quarks, namely the u, d and s. Even at that time, a serious difficulty was noticed in the Gell-Mann model, to understand which let us consider the Ω^-. The spin (3/2) of Ω^- is easily understood by noting that it has three quarks (sss), each of which has a spin of +(1/2). But then quarks being fermions, how can one have more than one quark in the same spin state? Would that not be a blatant violation of the sacred Pauli Exclusion Principle? This *was* a problem but a solution was quickly found, borrowing upon earlier experience in atomic and nuclear physics. The trick was to introduce another property for the quarks so that the three s quarks in Ω^-, while all being in the spin +(1/2) state, differ with respect to this second property/quantum descriptor.

Let us call this new property c, and let us suppose that c can take on three values c_1, c_2 and c_3, just as spin 1/2 can take on values +(1/2) and −(1/2). With the aid of c, we can now stipulate that the three s quarks in Ω^- should occupy the states $|\text{spin} = +1/2, c = c_1>$, $|\text{spin} = +1/2, c = c_2>$, and $|\text{spin} = +1/2, c = c_3>$. Thus while the spin states of the three quarks are identical, the same is not true of the c states; there is therefore no more a conflict with the Pauli principle; in other words, this new property c lifts the spin degeneracy.

The idea of invoking such a new property was first proposed by Oscar Greenberg and independently by Nambu who called this new property *colour*. According to Nambu, each quark comes in three colours, red (R), green (G) and blue (B). Gell-Mann gave flavour to quarks and now Nambu added colour to them! Thus according to Nambu, in Ω^- one of the quarks is red, the second one is green and the third one is blue. What about anti-quarks? What are their colours? Nambu said that anti-quarks have anti-colours, i.e., anti-red ($\bar{R}$), anti-green ($\bar{G}$) and anti-blue ($\bar{B}$). Painting the quarks with colours is itself too much and now anti-colours?! You may be turned off by the idea of anti-colours. So let me digress a little and expose you to a bit of 'colour wisdom' drawn from the illumination and the TV industries.

We learn in school that sunlight has seven colours — VIBGYOR — and all their shades. Fine. But what about colours like brown, purple and so on, which one actually sees? How does one explain them? This is where the science of colour perception comes in. These ideas are quite important for the illumination industry, the paint industry, and of course for colour television; and presently, these ideas are also important for those who study computer graphics.

One starts with the notion that the colours that one can perceive with the human eye are essentially combinations of three *primary* colours, namely red, green and blue. The primary colours can be combined in various ways to produce millions of colours — this is what is exploited in the colour TV. This model of colour perception is often abbreviated as the RGB model, and its basics are summarised in Fig. 11.2.

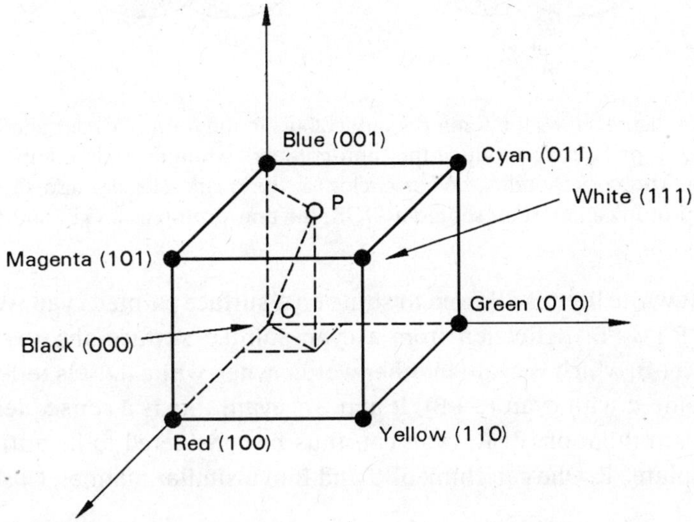

Fig. 11.2 The RGB model for mixed colours can be understood with the help of the cube shown above. Here, the three axes are marked red, green and blue respectively. Any point P in the cube represents a possible colour combination; the components of the vector OP, along the three colour axes, represent the amounts of the three primary colours present in the colour corresponding to the point P. As seen, white corresponds to equal mixtures of the three colours. The origin represents pure black, for obvious reasons, while points along the body diagonal from the origin O to the (111) corner represent various shades of grey.

Note in particular the following important colours that result from the addition of primaries (see also Fig. 11.3):

$$R + G = Y \text{ (Yellow)} \equiv \overline{B},$$

$$R + B = M \text{ (Magenta)} \equiv \overline{G},$$

$$B + G = C \text{ (Cyan)} \equiv \overline{R}.$$

Now, while the RGB model is of interest to the colour TV industry, the CMY model based on cyan, magenta and yellow is of interest to the pigment, paint and the colour copiers/ printer industries. To understand how this CMY model works,

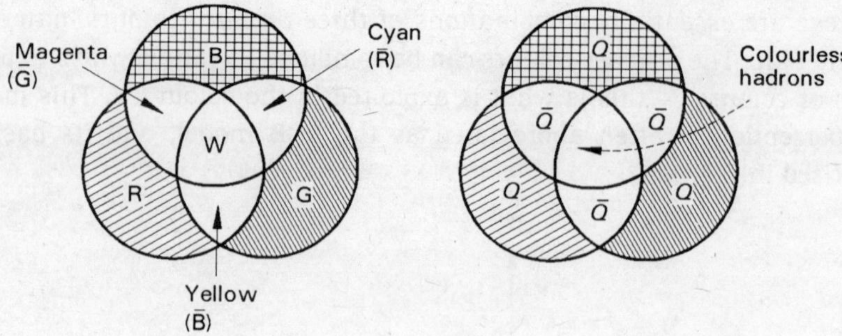

Fig. 11.3 This figure shows the three primary colours in the form of circles and how, when they combine in pairs, they produce the 'anti-colours', while in triple combination they lead to white. The corresponding colour circles for the quark case are also shown. In this scheme, the proton can be represented as RGB, the anti-proton as $\bar{R}\,\bar{G}\,\bar{B}$, and the pion as $R\bar{R} + G\bar{G} + B\bar{B}$.

imagine that white light is allowed to shine on a surface painted cyan. White light is equal to R+G+B. Reflection from a cyan-painted surface absorbs R, which leaves only G+B, which is cyan; in other words, when white light is reflected from a surface painted with cyan (G+B), it appears cyan; this is a consequence of the removal or 'annihilation' of R; cyan can thus be considered to be anti-red or $\bar{R}$, which annihilates R. One can think of $\bar{G}$ and $\bar{B}$ in a similar manner. Observe that:

$$R + G + B = W \text{ (white)},$$

$$\text{Colour} + \text{Anti-colour} = W.$$

For example, Red + Cyan = R+G+B = W. White, in our language, would be described as colourless. Now that the concept of colours and anti-colours has been duly introduced and explained with the help of analogies, let us return to quarks.

Thanks to Nambu, the number of quarks proliferated as shown in Fig. 11.4. At this stage I must stress that just because one talks of a red u quark and so forth, one must not imagine that the quark is physically red in colour! Nobody has ever detected a quark, much less with his eyes, and in fact no one can see such a tiny object. Therefore, there is no question of quarks possessing colours in the usual sense of the word. (Can you make this argument more precise?) Instead of colour, one might as well have used the word taste and declared quarks to be of the sweet, sour and bitter varieties! As Gell-Mann has admitted, words like strangeness, colour, flavour, etc., are just gags! However, the word colour has survived, and we shall use it with the above caveat.

So the first law of colour is that quarks come in three basic colours (R, G, B) and anti-quarks come in three basic anti-colours (C, M, Y). Now mesons and baryons are suitable composites of quarks and anti-quarks. Do they also have colour? None

QUARKS		ANTI-QUARKS	
Gell-Mann	Nambu	Gell-Mann	Nambu
u	u(R) u(G) u(B)	$\bar{u}$	$\bar{u}(\bar{R})$ $\bar{u}(\bar{G})$ $\bar{u}(\bar{B})$
d	d(R) d(G) d(B)	$\bar{d}$	$\bar{d}(\bar{R})$ $\bar{d}(\bar{G})$ $\bar{d}(\bar{B})$
s	s(R) s(G) s(B)	$\bar{s}$	$\bar{s}(\bar{R})$ $\bar{s}(\bar{G})$ $\bar{s}(\bar{B})$

$\bar{R}$ = Cyan, $\bar{G}$ = Magenta, $\bar{B}$ = Yellow

Fig. 11.4 This figure illustrates how, by introducing colour, Nambu increased the number of quarks and anti-quarks, compared to what Gell-Mann had originally proposed.

whatsoever; there is absolutely no evidence that mesons and baryons have colour. This leads to the second law of colour which is: All composite particles observed in Nature are colourless.

You might wonder how, for example, the proton is colourless although it is composed of quarks which have colour. The answer is indicated in Fig. 11.5. According to Gell-Mann, the quark composition of the proton is (uud). Since quarks have colour, the proton must be u(R)u(G)d(B), which would make the proton colourless, since R, G and B add up to white (colourlessness).

You might ask, "Why should the proton be u(R)u(G)d(B)? Why not u(B)u(G)d(R) or u(B)u(R)d(G)?" Indeed these combinations also are possible, which leads to the hypothesis that the colour states of the constituents fluctuate between all possibilities and in such a manner that the composite always remains colourless. Technically, one says that the hadron is in a *colour singlet state*. A colour singlet can be constructed in the following different ways:

1. By combining three quarks or three anti-quarks in such a way that all the three colours are included.
2. By combining an identically coloured quark and an anti-quark with all the three colours equally represented.

To illustrate rule 2 stated above, consider the example of $\pi^0 = (d\bar{d})$. According to the above rule, $\pi^0 = d(R\bar{R} + G\bar{G} + B\bar{B})\,\bar{d}\,(R\bar{R} + G\bar{G} + B\bar{B})$, which is a colour singlet. Thus the following rules emerge:

- Colourless baryons are obtained by equal mixtures of R, G and B.
- Colourless anti-baryons are obtained by equal mixtures of $\bar{R}$, $\bar{G}$ and $\bar{B}$.
- Colourless mesons are obtained by equal mixtures of colour–anti-colour pairs, i.e., as ($R\bar{R}$, $G\bar{G}$, $B\bar{B}$).

To recapitulate, colour is an extra property introduced to satisfy the Pauli principle. Further, colour mixtures operate in such a way as to render hadrons colourless. At this stage, I must point out that when Nambu invented colour, he also somewhat modified the quark scheme of Gell-Mann. Disliking the idea of fractional charges for the quarks, Nambu assigned them integral charges.

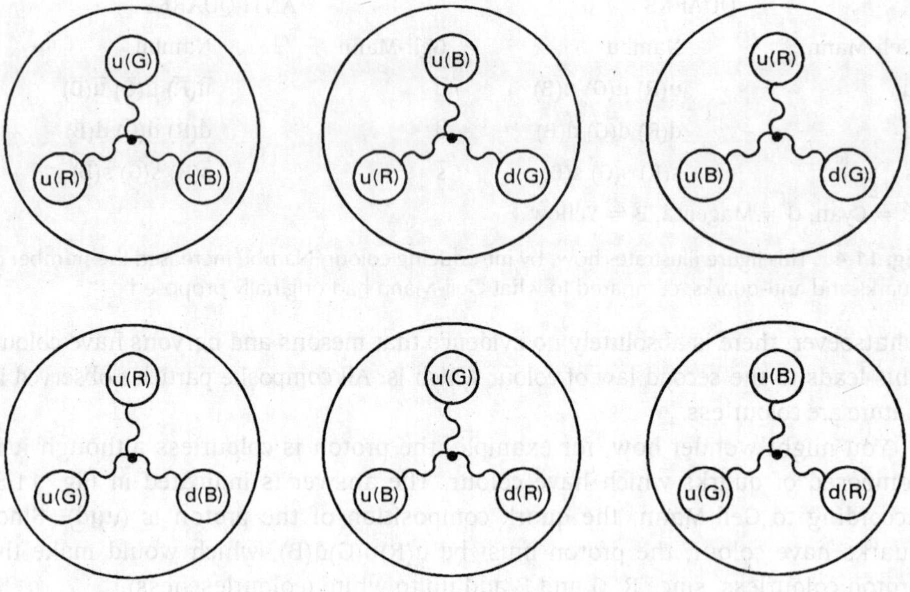

Fig. 11.5 According to Gell-Mann, the proton = u + u + d. With the introduction of colour, the question naturally arises as to what the colours of these constituent quarks are. One possibility is u(R) + u(G) + d(B). Can one have u(B) + u(G) + d(R)? Yes, of course. What about u(G) + u(G) + d(R)? No, this is not allowed. Why? Because G occurs twice, and G + G + R does not equal white. So it boils down to this — the three colours R, G and B *must* be present; subject to this, the individual quarks can be of any colour. In fact, the colours of the quarks keep fluctuating as illustrated above but at any given time, the colours always add up to 'white' which indicates a state of colourlessness for the proton.

Meanwhile, Nambu's colour idea caught the attention of Gell-Mann who promptly updated his original model to include colour. Thus there were two rival models, both featuring colour, one attributed to Nambu and the other to Gell-Mann. These two options are compared in Table 11.1.

Table 11.1 A comparison of the Nambu model and the presently accepted Gell-Mann model for quarks, with both flavour and colour. Indicated here are the electric charges of the quarks in the two models, in units of the electronic charge. Notice that in Nambu's model, the charge of a quark of a given flavour can vary with colour.

Gell-Mann				Nambu			
Colour \ Flavour	u	d	s	Colour \ Flavour	u	d	s
Red	2/3	−1/3	−1/3	Red	0	−1	−1
Green	2/3	−1/3	−1/3	Green	1	0	0
Blue	2/3	−1/3	−1/3	Blue	1	0	0

Over the years, fractional charges notwithstanding, the Gell-Mann model has managed to edge out the Nambu model because it proved to be superior in every way.

11.3 Colour field

Although colour was originally introduced to protect the Pauli principle, it turns out that the concept of colour has a deeper significance. People said that colour was like charge, and that just as particles could have electric charge, they could similarly also have 'colour charge' (in addition of course to ordinary electric charge). This automatically raises the question, "Do these colour charges produce a *colour field*, and if so, is there a quantum theory for this colour field?" The answer to both these questions is "Yes", and the quantum theory of the colour field is called *Quantum Chromodynamics* or QCD for short.

The colour force associated with the colour field not only binds the quarks within a given meson or baryon but also links quarks residing in different particles. The carriers of the colour force are the gluons, i.e., gluons are the quanta of the colour field, just as the photon is the quantum of the electromagnetic field. Like the photon, the gluons too are massless. If you think a little about this, you will realise that the masslessness of the gluons is somewhat weird. Recall that when Yukawa wanted to model the S.I., he invented a meson with a mass which was several hundred electron masses. This meson was the carrier of the strong force. Yukawa's meson *had* to have mass because the range of the strong force is very small. In QCD, the carriers of the strong force are the gluons which are massless! Yukawa described the strong force with bosons having mass whereas QCD describes the same S.I. with massless carriers — how come? This is an interesting point which we shall discuss shortly. Observe meanwhile that the W.I. also has a short range, but there one desperately strove to accommodate carriers with mass, namely the W^+, W^- and Z^0. So, for the weak force, Nature demands that the carriers have mass, but for the strong force which also is a short-range force, zero-mass mediators are preferred.

11.4 Do gluons have colour?

Consider Fig. 11. 6(a) which shows a quark–anti-quark reaction involving a gluon. Let us now examine the colour aspects. An R quark exchanges a gluon and becomes a B quark. This implies two things: firstly, the gluon must have an $\overline{\text{R}}$ character so that the R nature of the quark can be wiped out. Next, the gluon must also have a B nature which it can impart to the quark. In short, the gluon has the colour (B$\overline{\text{R}}$). This example reveals that gluons have colour charge. Contrast this with the photon which binds ordinary electric charges, but does not itself carry any electric charge.

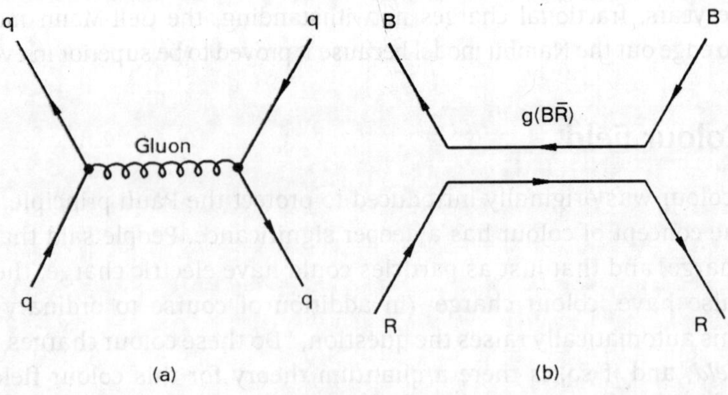

Fig. 11.6 (a) shows a quark–anti-quark interaction via the exchange of a gluon. (b) shows the flow of colour in such an interaction. For reasons explained in the text, it should be clear that the exchanged gluon must have the colour $B\bar{R}$.

We have agreed that the gluons carry colour. But how many types of colour? Group theory answers this. Consider the matrix below:

	R	G	B
$\bar{R}$	$R\bar{R}$	$G\bar{R}$	$B\bar{R}$
$\bar{G}$	$R\bar{G}$	$G\bar{G}$	$B\bar{G}$
$\bar{B}$	$R\bar{B}$	$G\bar{B}$	$B\bar{B}$

We see that nine bi-colour combinations are possible. Now pundits have declared that colour also has SU(3) symmetry which, however, has to be carefully distinguished from the SU(3) symmetry associated with flavour and which Gell-Mann had earlier invoked. To distinguish between these two, we shall label them as $SU(3)_{flavour}$ and $SU(3)_{colour}$, respectively. Since quarks have both flavour and colour, we need both the SU(3) groups; presently, we shall focus on the $SU(3)_{colour}$ group.

As a preliminary, let us hark back for a minute to $SU(3)_{flavour}$. You will recall that Gell-Mann used this group to organise particles into families. It turns out that, in a similar manner, the nine bi-colour combinations listed above can be grouped into two families:

$$9 = 1 + 8,$$

i.e., the nine combinations can be organised into a singlet and an octet, each of which transforms according to an irreducible representation (IR) of SU(3). The jargon IR has been explained in ***Bhabha***. Note that the nine bi-colours can be regrouped in accordance with group symmetry. The regroupings are listed below:

1. $\frac{1}{\sqrt{2}}(R\bar{B} + B\bar{R})$

2. $-\frac{i}{\sqrt{2}}(R\bar{B} - B\bar{R})$

3. $\frac{1}{\sqrt{2}}(R\bar{R} - B\bar{B})$

4. $\frac{1}{\sqrt{2}}(R\bar{G} + G\bar{R})$

5. $-\frac{i}{\sqrt{2}}(R\bar{G} - G\bar{R})$

6. $\frac{1}{\sqrt{2}}(B\bar{G} + G\bar{B})$

7. $-\frac{i}{\sqrt{2}}(B\bar{G} - G\bar{B})$

8. $\frac{1}{\sqrt{6}}(R\bar{R} + B\bar{B} - 2G\bar{G})$

(1–8: Octet)

9. $\frac{1}{\sqrt{3}}(R\bar{R} + B\bar{B} + G\bar{G})$ Singlet

The octet family represents the colours of the eight gluons that mediate the colour field. The singlet is colourless and ignored.

11.5 $SU(3)_{flavour}$ and quark proliferation

When Gell-Mann launched the quark model, he had only three flavours and three quarks (u, d, and s) and he invoked SU(3) to give substance to the model. Today we have six flavours and six quarks. Does SU(3) still hold good? To answer this question, let us go back to how Gell-Mann arrived at three quarks.

Based on the particles then known, Gell-Mann noted that first, SU(3) enabled a neat grouping of particles into families (based on their spin, etc.). Next, he argued that if SU(3) is a symmetry of Nature, then surely there must be three particles corresponding to the fundamental representation of the group. He thus arrived at three particles which he called the u quark, the d quark and the s quark. Some time after this, Glashow (as described earlier) managed to get the c quark admitted to the club, increasing the number to four. With four quark flavours now, the question naturally arose whether $SU(3)_{flavour}$ was adequate? The pundits said, "No problem. Remember that SU(2) was adequate for describing the isospin behaviour of the nucleon. Gell-Mann had three quarks and therefore SU(3) was OK. Four quarks now? Use SU(4) because the fundamental representation of SU(4) allows room for four particles." Then came the b and the t quarks and these were accommodated by changing the flavour group to SU(6).

All this was a long time ago. With the development of the GWS theory and the recognition that there were three lepton families and further that the lepton families and the quark families were closely related, flavour symmetry took a back seat. In other words the SU(6) idea was given up. The current philosophy is captured in Fig. 11.7. The dominant factor here is the generation. If one confines oneself to one generation, then there are two leptons and two quarks. The quarks come in three colours, linked by $SU(3)_{colour}$ symmetry. The leptons and the photon (not shown in the figure) are linked by SU(2) × U(1) symmetry.

At this stage you might wonder how one can be sure that there are only three generations. Can't the future throw up more generations? This question did bother people for a while but recent experiments at CERN have settled the issue.

Way back in the early eighties, Carlo Rubbia struggled to prove to the world that the Z^0 exists. In his long and painstaking experiment, he caught barely a handful of them. Today, the LEP in CERN is a veritable Z^0 factory churning out thousands of these particles per day. A very interesting result to come out of this factory is presented in Fig. 11.8. This shows an accurate measurement of the Z^0 resonance (like the original resonance observed by Fermi). What is interesting is that the measured data fits very well with the hypothesis that there are only three generations of leptons and not two or four. This is a very important result, firmly

Generation	Quarks Symmetry: $SU(3)_{colour}$			Leptons
1	u(R) d(R)	u(G) d(G)	u(B) d(B)	e^-, ν_e
	Electro-weak symmetry: SU(2) × U(1)			
2	c(R) s(R)	c(G) s(G)	c(B) s(B)	μ^-, ν_μ
	Electro-weak symmetry: SU(2) × U(1)			
3	t(R) b(R)	t(G) b(G)	t(B) b(B)	τ^-, ν_τ
	Electro-weak symmetry: SU(2) × U(1)			

Fig. 11.7 This figure shows three generations of quarks and leptons and the symmetry groups associated with them.

declaring that there is no need to proliferate the quark and the lepton generations, and that there are no more quark flavours. This result could not be guessed from the available theories — one had to directly probe Nature to extract this secret.

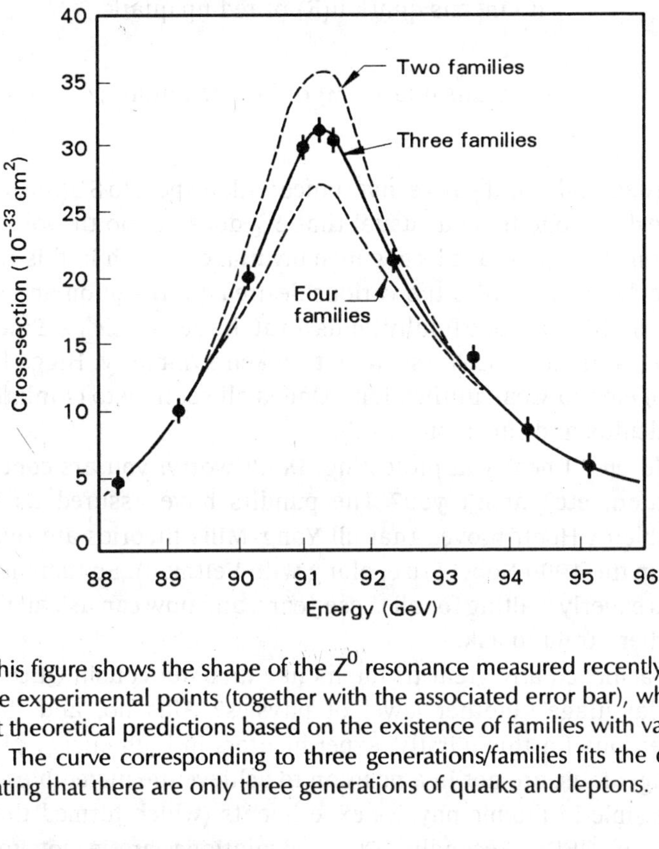

Fig. 11.8 This figure shows the shape of the Z^0 resonance measured recently at CERN. The dots show the experimental points (together with the associated error bar), while the various curves depict theoretical predictions based on the existence of families with varying numbers of members. The curve corresponding to three generations/families fits the data very well, clearly indicating that there are only three generations of quarks and leptons.

11.6 A glimpse of QCD

Now on to a brief glimpse of formal QCD. As usual, one starts with the Lagrangian for free particles, in this case, free quarks. Quarks being fermions, this Lagrangian looks like the Dirac Lagrangian; only, in this case we have to sum up all quark colours and flavours. Thus,

$$\mathscr{L}_0 = \sum_{j=1}^{6} \mathscr{L}^j ,$$

where $\mathscr{L}^j$ is the Lagrangian for quark flavour j, with j = u, d, s, c, b, t. Now quark q^j has three colours whence

$$\mathscr{L}_0 = \sum_j \sum_c \mathscr{L}_c^j ,$$

the subscript *c* denoting the three colours. Perhaps you are getting a little lost, in which case just remember that:

$$q_1^u \text{ means quark u(R) or red up quark}$$

$$q_3^t \text{ means quark t(B) or blue top quark,}$$

and so on.

Next, one demands local gauge invariance with respect to $SU(3)_{colour}$, naturally. When this is done, one finds as usual that $\mathscr{L}_0$ does not do the job. So one must add extra terms to $\mathscr{L}_0$ to save local gauge invariance, which in this case brings in the gluon fields together with the various desired quark–gluon and gluon–gluon interactions. In this way, one finally lands up at the desired $\mathscr{L}_{QCD}$. Since the gluons have zero mass, there is no hassle about broken symmetry, Higgs field, etc. (as in the case of electro-weak unification). One is all set now to crank the handle of Feynman calculus and churn out results.

"Hey, hold on," I hear you protesting. Don't worry; you are concerned about renormalisation, etc., aren't you? The pundits have assured us that QCD *is* renormalisable ('t Hooft proved that all Yang–Mills theories are renormalisable and picked up the 1999 Nobel Prize, along with Veltman). So there it is, the great QCD one was eagerly waiting for all these years, and now can ask all the questions one was earlier afraid to ask.

Has QCD achieved any great triumphs like its older cousin QED? Is it, to use Feynman's language, another jewel of physics? Alas no, and there are two important reasons for this. Firstly, experimenters in high energy physics, while extremely ingenious, are not in a position to achieve the incredible experimental accuracy possible in atomic physics experiments (which formed the basis for a precise test of QED). Secondly, QCD calculations are a lot tougher than corresponding QED calculations. Thanks to these two limitations, QCD, though a great achievement, does not in some respects enjoy the same status as QED.

11.7 Quark confinement

The development of QCD soon led to a clearer picture of quark *confinement*. When quarks were first proposed, the first question people asked was, "How can we see quarks?" People desperately wanted to catch these exotic objects with fractional charges and one experiment even claimed evidence for the discovery of fractional charges. But as the years went by, there was no confirmation and slowly the view began to develop that quarks remain confined forever in mesons and baryons. This was most puzzling indeed. With sufficient energy, one can tear an atom apart and free the electrons and the nucleus inside. With even more energy, one could tear a nucleus into its constituent neutrons and protons — at least one could knock a few of them out of the nucleus. Could not one with still higher energy manage to knock out at least one quark from a baryon or a meson? Surely when one hits a baryon very very hard, something *must* happen! Indeed it does, but wonder of wonders, the quarks manage to stay inside and *never come out singly*, even if a lot of debris is produced.

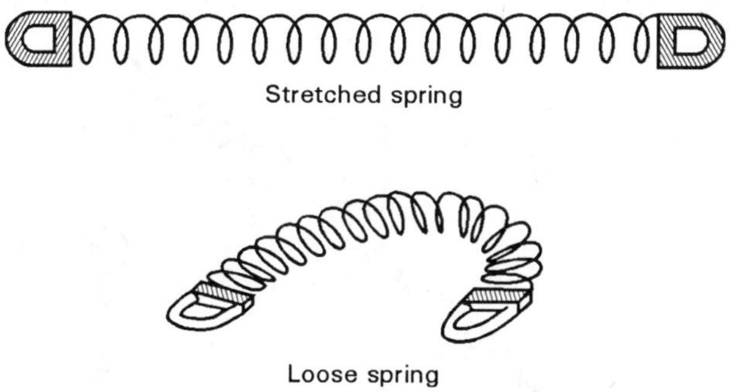

Fig.11.9 This figure seeks to illustrate with the help of the analogy of a chest expander, how quarks behave as if free when very close to each other. Imagine the two handles of the chest expander to represent two quarks, and the spring to represent the gluon which binds them. When the handles are stretched apart, there is a strong force in the spring. In fact, the more one expands, the harder it becomes to expand it further; this is the way inter-quark force also behaves. Suppose now the spring is loosely laid on the floor as shown. Then one handle can be easily moved with respect to the other. This is an analogy for asymptotic freedom.

What is the mystery behind this unwillingness of quarks to appear singly? An analogy might help. Consider a chest-expander spring. As you know, the more one pulls, the harder it is to expand the spring further. Now let us consider the other extreme. If the spring is laid loosely on the ground (see Fig. 11.9) the two ends can be easily and independently pushed around. This analogy carries over beautifully into the world of quarks. The 'spring' that binds two quarks is the colour force, and is like the chest expander. The more one increases the distance between two quarks, the harder it becomes to separate them further — the force required sort of increases exponentially. Thus it becomes impossible to tear one quark apart from another one to which it is bound; the quark pair is like a chest expander whose spring is infinitely strong.

Now consider the other extreme when the quarks are close to each other; this situation is similar to the chest expander laid loosely on the floor, and the two quarks move quite freely. To sum up, at 'moderate' energies, one explores quark behaviour at large distances; under these conditions the quarks appear to be confined. On the other hand, at extremely high energies of exploration, one deals with small inter-quark distances, and under these conditions the quarks appear to be free.

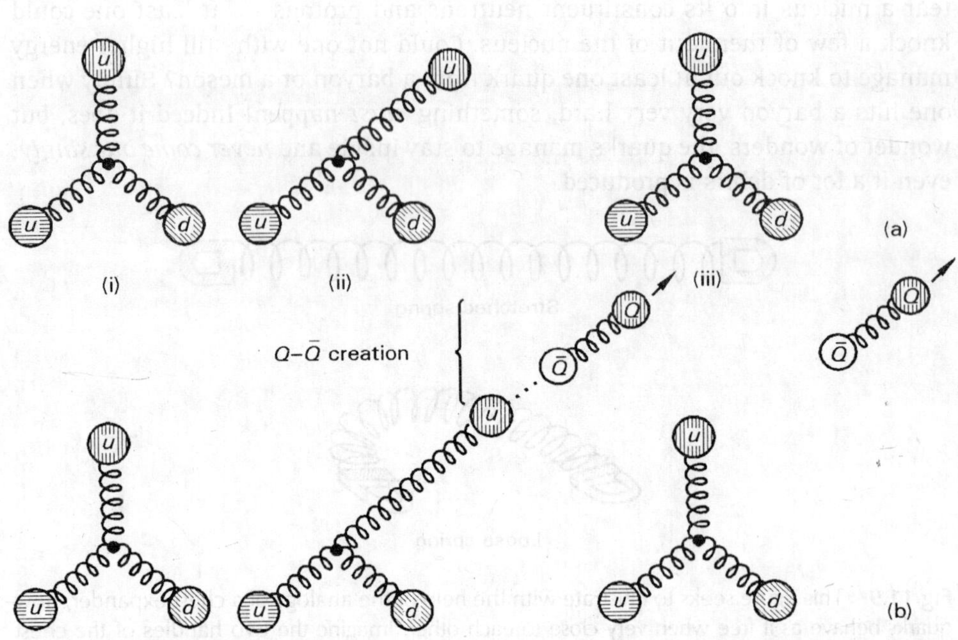

Fig. 11.10 This figure explains what happens when a proton is 'hit', say by a high energy electron. As explained in the text, at modest energies, the inter-quark 'bond' is merely 'stretched' for a while; this of course is a rather loose way of saying there is some internal excitation. Later, the system returns to its normal configuration. These processes are schematically illustrated in (a). At much higher energies, the internal excitation is so much that a quark–anti-quark pair, i.e., a meson is produced, which floats away as shown in (b). There could of course be other kinds of debris but the point is that a *free* quark is *never* produced.

This is called *asymptotic freedom*, a property whose existence was inferred by David Politzer, from a detailed study of QCD.

You might wonder, "When an atom is hit hard, electrons get knocked off. When a nucleus is hit hard, nucleons get knocked off. If quarks are bound for ever inside nucleons, then does it mean that nothing ever gets knocked off when a proton is hit hard?" Good thinking, and the answer is provided in Fig. 11.10. Here Fig. a(i) shows a proton in its normal quark configuration while Fig. a(ii) shows the state of the proton during a high energy scattering event; one of the quarks gets pushed out a bit but the overall integrity of the p remains intact. Subsequently, the stretched 'bond' shrinks and the normal configuration is restored, as depicted in Fig. a(iii). If the proton is punched even harder, the quark bond gets stretched so much that the excitation energy involved leads to the production of a $q\bar{q}$ pair. This pair, which is a meson really, then floats away and the proton recovers from the shock. This sequence is illustrated in Fig. (b). In other words, though all kinds of debris may be produced in high energy collisions, no *free* quarks ever appear. Now, isn't that remarkable?

11.8 QED and QCD compared

Let us now compare QED and QCD. Both are quantum field theories, one for describing the electron and its associated electromagnetic field, and the other for describing the quarks and the associated colour field. Both involve massless messenger particles, the photon in one case and the gluon in the other. But there are differences too; while there is only one photon, there are eight types of gluons. Further, while the photon carries no electric charge, the gluons carry colour charge. The most striking difference, however, is in respect of confinement — quarks, being confined, are never seen in isolation whereas electric charges can be seen in isolation.

The above difference may also be examined by considering the effective charge as a function of distance in the two cases. Let us first examine the effective electric charge of an electron within the framework of QED. Suppose I want to measure the charge of an electron — remember this is what Millikan did. The standard operational procedure is the following: Place a test charge $+Q$, say, at a known distance from the electron. Then measure the force of attraction between $+Q$ and e. Then knowing the value of this force and the distance between the two charges, deduce the value of the electronic charge.

Let us now try to understand what this measured value of the electronic charge really means in the light of QED. The space around the electron might be empty — it may even be a vacuum — but in QED we know that out of vacuum can pop up (virtual) $e^- e^+$ pairs. After a brief existence, the electron and the positron of the pair return to the vacuum — that is why the pair is called virtual. The point I am trying to make is that the electron whose charge we are trying to measure is not alone as we might imagine it to be, but is always surrounded by a cloud consisting of (virtual) positive and negative charges. These positive and negative charges crowd around our electron, obscuring the 'view' of the test charge $+Q$. In practical terms, the test charge does not see the 'bare' electron but a 'screened' electron. The value of the electronic charge measured by Millikan was for the "screened" electron, and it is this value which is quoted in books and tables. (Attention to all this has also been drawn in ***QR-II***.) See also Fig. 11.11(a).

Suppose we manage to make the test charge pierce the 'security' cordon and come close to our electron. In the language of high energy physics, this means using very high energies, or better, creating conditions so that the relativistic momentum transfer q^2 is large. Having penetrated the screening cloud, the effects of screening are now diluted. Consequently, the value obtained for e will now be

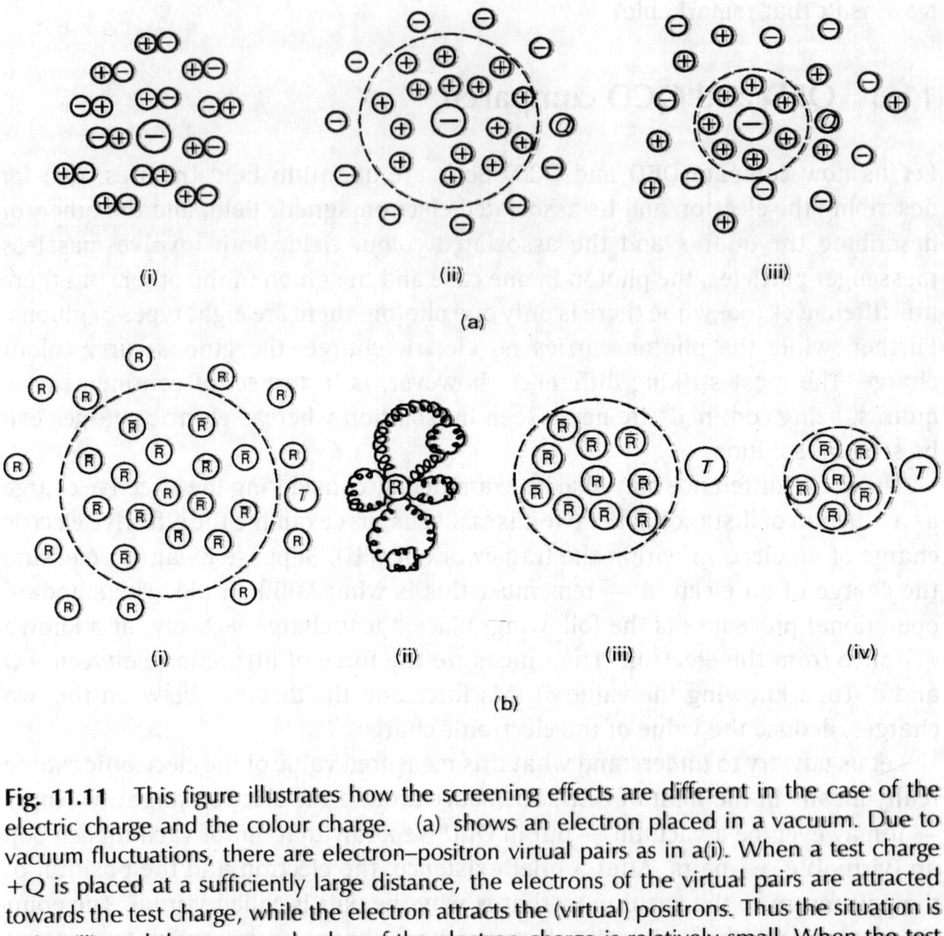

Fig. 11.11 This figure illustrates how the screening effects are different in the case of the electric charge and the colour charge. (a) shows an electron placed in a vacuum. Due to vacuum fluctuations, there are electron–positron virtual pairs as in a(i). When a test charge $+Q$ is placed at a sufficiently large distance, the electrons of the virtual pairs are attracted towards the test charge, while the electron attracts the (virtual) positrons. Thus the situation is as in a(ii), and the measured value of the electron charge is relatively small. When the test charge penetrates the screening cloud, the situation is altered to that shown in a(iii) and the measured value increases. (b) shows what happens in the case of the colour charge; here T is the test colour charge. For convenience, the quark is taken to be red (R), and the anti-quark to be anti-red ($\bar{R}$). Here there are now two competing effects. First, of course, there is the vacuum polarisation resulting in the production of virtual quarks and virtual anti-quarks (like the electrons and positrons in the QED case) as shown in b(i). This results in a screening effect which tends to increase the colour charge with decreasing distance, as in the case of the electron. But now there is a new effect: the colour charge sitting in the vacuum can continuously emit and absorb gluons — see b(ii)— and since gluons also carry charge, they change the story. The camouflage effect occurs and since it dominates, the effective colour charge reduces as one approaches closer to the quark. This is schematically illustrated in b(iii) and b(iv). Here the anti-red cloud surrounding the red quark tends to decrease the redness of the colour charge at the centre; the larger the distance, the greater is the effect. It is the dominance of the camouflage effect that leads to asymptotic freedom.

more than that measured by Millikan. In fact, the closer the test charge $+Q$ gets to the electron, the less would be the screening and therefore greater also the value measured for e as shown in Fig. 11.12(a).

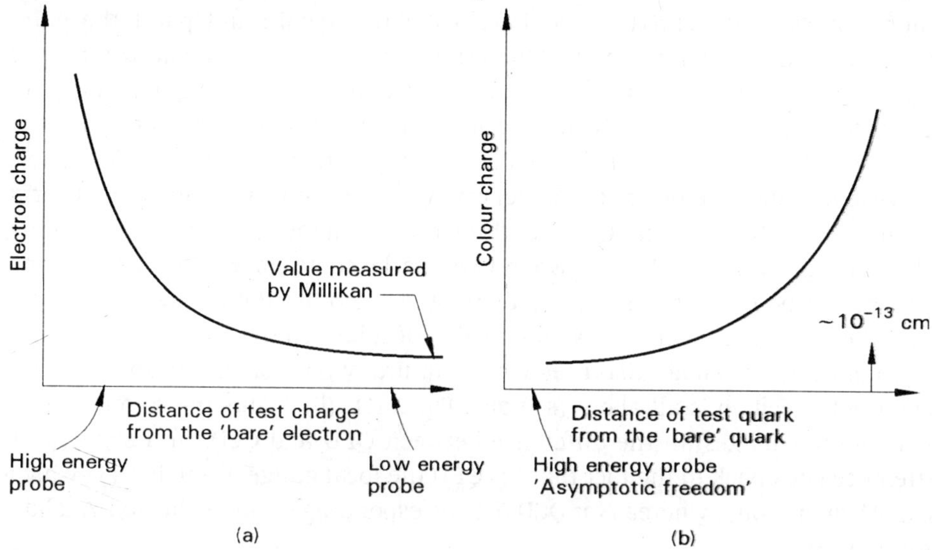

Fig. 11.12 This figure complements the previous one. (a) shows how the effective charge of the electron increases as the distance decreases — remember, in high energy physics, short distances mean exploration with high energy probes. (b) shows the corresponding scenario in the case of quarks. The striking difference basically arises from the fact that in the electrodynamic case, one deals with an Abelian group, whereas in QCD one deals with a non-Abelian group.

Let us now imagine a similar exercise involving, say, the colour charge of an R quark — it really does not matter what colour one chooses; this is just an example. Around this R quark there would be vacuum fluctuations resulting in the production of both R quarks and $\overline{R}$ anti-quarks. One would consequently expect screening effects just as in electrons. Extending the analogy further, one would also expect the measured value of the colour charge to increase as the distance between the test colour charge and the candidate colour charge decreases. Things do not work out this way at all; instead, the colour charge *decreases* as the distance decreases, in marked contrast to the electron case.

Are you wondering why? It is all due to the gluons. In the case of the quark, one not only has $q\overline{q}$ pairs popping up all the time out of the vacuum, but also gluons jumping in and out — in other words, one must consider the effects of not only the virtual quark–anti-quark pairs but also of virtual gluons; and that is what makes all the difference because gluons carry colour charge.

Armed with this information, let us now examine the screening scenario in the quark case carefully. Far away from the candidate R quark, the test colour charge sees the candidate through a $q\bar{q}$ haze, accompanied by an R gluon cloud. The former no doubt tries to diminish the measured value of the colour charge but the gluons simply overwhelm the screening cloud of the $q\bar{q}$ pairs and 'paint' the quark even redder! This phenomenon is sometimes referred to as *camouflaging* (see Fig. 11.11(b)). However, as one penetrates the double haze, the gluons cease to have a controlling effect, and conventional screening begins to operate; thus the measured value of the colour charge decreases. (Caution! Remember all this is a hand-waving description, but it is generally OK.) If you think about this for a minute, you will realise that this is nothing but asymptotic freedom, stated in the colour charge language. I mean when two quarks come close to each other, one of these can be regarded as the candidate and the other the test colour charge. Since the test charge will see a smaller value of colour charge at small distances, the colour force at small distances would naturally be weaker and that is what asymptotic freedom is all about (see also Fig. 11.12(b)). Thus the colour charge of the gluons makes all the difference between QED and QCD. In a sense, this difference goes back to the fact that for QED the local gauge group is U(1) which is an Abelian group, whereas for QCD the corresponding group is the non-Abelian group SU(3).

QCD is no mean achievement; it is a great triumph of human ingenuity. Recall in this context, Salam's observation that way back in the fifties, they thought they had the S.I. neatly wrapped up in a (renormalisable) Yukawa theory. What a far cry from those days!

11.9 Lattice gauge theory

The strikingly different behaviour of the colour charge as a function of distance introduces a corresponding difference in the practical use that can be made of QCD, as compared to QED. Recall that QED offers a scheme for using the perturbation theory without the hassles of dealing with unwanted infinities. QCD too is free from the difficulties associated with infinities, i.e., it is renormalisable. But can it be readily used in perturbative calculations like QED can? Not really, and this is entirely due to the differences displayed in Figs. 11.12(a) and (b).

Now, as you know, perturbation theory can be used when the coupling constant is small. In the case of QED, this is the case for all distances of practical interest. In the case of QCD, the story is different. The coupling constant associated with the colour charge becomes small only at extremely small distances, and to probe matter at such small distances, one must go to extraordinarily high energies or, more correctly, to very high values of q^2; this is also the situation when asymptotic freedom applies. Today's accelerators cannot reach such energies; indeed, it is doubtful if we can ever have such accelerators. What this means is that if we have to make QCD calculations to compare with present-day experimental results, then we have to come to terms with the fact that the coupling constant to be used will not be small; in turn this implies that the standard pertubation theory cannot be used; we need a *non-perturbative* theory, and that is very very difficult. So we have a theory which may be useful tomorrow but not much today!

What does one do then? How does one ensure that QCD describes today's results also? This is where the *lattice gauge theory* comes into the picture. The problem with the non-perturbative version of QCD is that the equations are almost intractable. Now there are many situations in physics where people, instead of throwing up their hands when confronted with horrendous mathematics, intelligently resort to the use of computers. Is the computer such a clever mathematician as to be able to solve equations that humans cannot? Not at all. Then how does it act as a life saver? This is something I have touched upon in the volume ***Saha*** but some of that stuff needs to be repeated here.

Because of its very high number-crunching ability as they call it, the computer can do numerical calculations with dizzying speed. It cannot deal with equations but once the problem is formulated as a numerical problem, it can, by a step-by-step procedure, churn out numbers and finally produce the required (numerical) result. This is something that people studying fluid dynamics and aerodynamics routinely do these days. What one can achieve depends basically upon the speed of the computer and its memory capacity. Thanks to impressive advances in technology, intricate problems which just a few years ago no one would even have dreamt of touching, are now routinely being cranked out on high-speed computers.

With this preamble, let us now turn to the lattice gauge theory. Space-time, as we know, is continuous (and four-dimensional). In the lattice gauge theory, however, one pretends that space-time is discrete as illustrated in Fig. 11.13. The liberty is taken of introducing such an artefact to facilitate numerical calculations in a computer. As computers become more and more powerful, one can, in principle, shrink the lattice spacing so that it approaches a space-time continuum.

So now we have the lattice. What do we do next? To fix ideas, let us imagine we are investigating the interaction between electric charges, within the framework of a lattice version of QED. In general, gauge theory calculations simulated on a lattice are referred to as lattice gauge theories. In QED we know that the presence of a charge induces quantum fluctuations in the vacuum. While considering the interaction of one electric charge with another, we must necessarily consider such quantum fluctuations because they lead to screening, etc. Now the thumb rule is: if the fluctuations are small, then a perturbation theory is adequate (where analytical calculations are concerned). If, on the other hand, the fluctuations are large, then one must resort to a non-perturbative approach. In QED, large fluctuations are rare while in QCD, they are the rule. Thus, for examining the electric field a perturbation theory is adequate, while for the corresponding 'chromic' field a non- perturbative approach is a must. The computer can simulate both kinds of situations, though on a discrete lattice. It is for this reason that people have explored lattice gauge theories, despite the objections that might be raised.

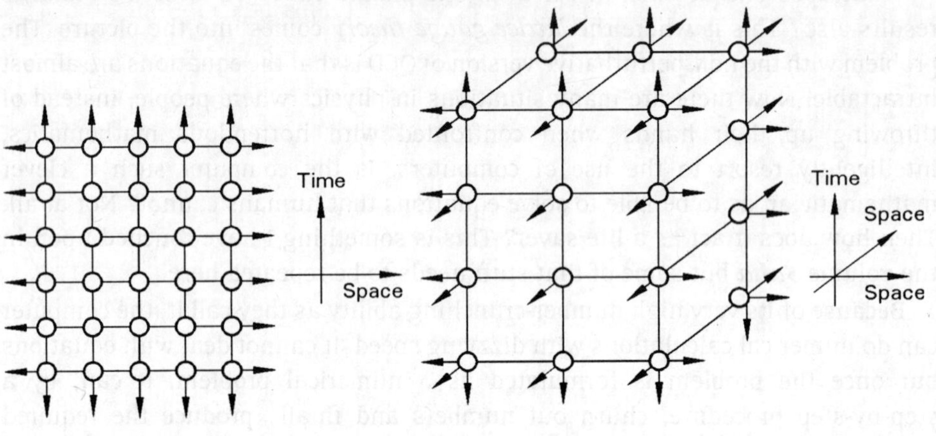

Fig. 11.13 Space has three dimensions and time has one. Both are continuous and thus in relativity, space-time is a four-dimensional continuum. In (analytical) QCD also, space-time is a continuum, naturally. But QCD calculations are difficult in the non-perturbative regime, which is the regime of interest to experimentalists, at least at the present time. How then can predictions be made which could be compared with experiments? Numerical simulations via the lattice gauge method is the approach adopted. In this approach, space-time is approximated as a discrete lattice. (a) shows a 2-D space-time lattice, while (b) shows the 3-D version. The 4-D version cannot be depicted since we live in 3-D space. In the lattice gauge models, space-time exists only at the lattice points. Clearly, this is an artefact. But its virtue is that it makes calculations possible. Abstraction is brought closer to reality by making the lattice mesh finer and finer so that for all practical purposes, the lattice behaves like a continuum.

The lattice gauge theory is quite complicated, and is a fallout of the renormalisation group theory developed by Ken Wilson for discussing phase transitions (see ***Many Phases***). It is not possible here to give even a sketchy view of the lattice gauge theory because of the complications involved; we shall therefore content ourselves with highlighting the important findings (the interested reader can consult the article entitled 'The Lattice Gauge Theory of Quark Confinement', by Claudio Rebbi, *Scientific American*, February, 1983).

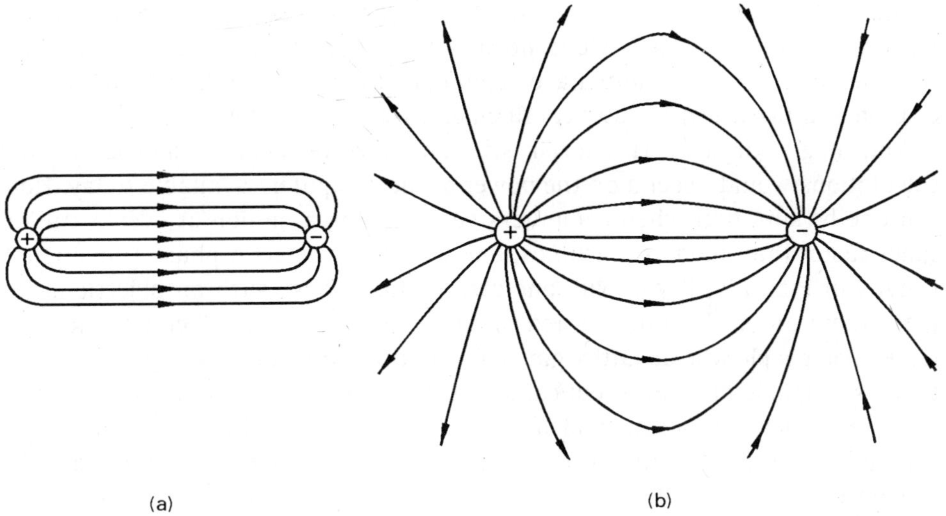

Fig. 11.14 (a) shows schematically the electric field pattern between two electric charges, as revealed by lattice gauge QED, when the mesh size is large. This is typical of a confinement situation and contrary to what is expected. Clearly the result is an artefact. The latter arises because the mesh is coarse and quite unrepresentative of actual space-time. In a coarse mesh, large quantum fluctuations dominate and they lead to confinement. Suppose the mesh is made fine; the field pattern is now like in (b), which is what we expect. Thus, lattice QED shows a 'phase transition' from confinement to non-confinement as the mesh is made finer. By contrast, lattice QCD does *not* show a transition. No matter how fine the mesh, colour charges are *always* confined. This is a powerful (though numerical) demonstration of the quark confinement concept.

Lattice gauge studies have proved to be quite illuminating, and in the case of QED, the following results have been obtained:

- When the lattice is coarse, large fluctuations dominate.
- As a result, electric charges experience confinement, like quarks do.
- The electric field pattern in such a case has the form shown in Fig. 11.14(a). Obviously, this is in contradiction to reality.
- However, when the lattice becomes finer, confinement disappears and single charges can be observed. Correspondingly, the electric field has the form shown in Fig. 11.14(b).
- This disappearance of confinement, as the lattice is made finer, is like a 'phase transition'.

In summary, provided the lattice is fine enough (and thus a good approximation to space-time continuum), all the features of standard QED are reproduced by the lattice gauge version of QED.

We now turn to the QCD calculations on a lattice. Here large fluctuations *always* dominate, be it on a coarse mesh, fine mesh, very fine mesh or even a super-fine mesh. Hence, in contrast to lattice QED which showed a 'phase transition' with respect to confinement behaviour, no such transition is expected, no matter how fine the lattice mesh. Pundits have demonstrated that this indeed is the case — QCD on a lattice *always* shows confinement, which is comforting.

You might argue: "The mesh size one can reasonably choose for the calculations would depend on the power of the computers available today. How can one be sure that when computers become, say, a million or even a trillion times faster than today's machines, one would not detect a phase transition?" Legally or logically, if you will, one can no doubt raise such an objection. But then, there are such things as reasonable expectations, gut feelings, etc. And remember, people who do lattice gauge simulations are themselves quite cautious and also capable of being skeptical about the results churned out by machines. However, when all is said and done, the high energy physics community is generally agreed that numerical simulations *do* convincingly demonstrate the existence of quark confinement. What remains to be done is to establish by analytical procedures, the conclusions arrived at via simulation studies. In many branches of physics, a definite result first achieved numerically has subsequently been confirmed by analytical reasoning. Likewise, one may hope that soon, mathematical proof would be forthcoming that no phase transition (of the type discussed above) is possible in QCD, as the lattice tends to a continuum.

Thanks to lattice gauge simulations, QCD is here to stay, even though at present analytical calculations in the non-perturbative regime are forbiddingly difficult.

11.10 Chapter summary

This chapter deals essentially with the theory of strong interactions whose existence was first surmised by Heisenberg and later given some form by Yukawa.

- First there was the SU(3) model of Gell-Mann.
- Gell-Mann then invented three quarks to give substance to the SU(3) scheme of particle taxonomy.
- The u, d, and s quarks of Gell-Mann explained many facts previously considered puzzling. However, the scheme appeared to violate the Pauli principle.
- Nambu sought to save the Pauli Principle by introducing a new quantum number, called colour. However, in Nambu's scheme, the quarks had integral electric charges and not fractional electric charges as in Gell-Mann's scheme.
- Gell-Mann then modified his quark model to accommodate colour. This, in turn, gave rise to the concept of $SU(3)_{colour}$ symmetry, in addition to the previously known $SU(3)_{flavour}$ symmetry.
- According to the colour principle, quarks come in three colours. Anti-quarks also come in three anti-colours.
- Baryons and mesons are colourless.

- As the quark family grew to its full strength of six, colour symmetry retained its importance.
- It was realised that quarks are the agents that generate the strong force. This then gave rise to the idea that like its weak and electromagnetic counterparts, the strong force must originate from an appropriate local gauge invariance.
- In turn this gave rise to a proper quantum theory for the strong interaction based on quarks, just as the electromagnetic interaction arises from electric charges. This theory is called quantum chromodynamics (QCD) since it describes the interaction between particles with colour 'charges'.
- QCD is a straightforward Yang–Mills type of gauge theory, based on colour symmetry. The gauge bosons are the gluons. There are eight of them. They are massless and carry a colour 'charge'. The underlying symmetry group is non-Abelian, and therefore there are complications, unlike in an Abelian group.
- Quarks are never found free; they are always confined in hadrons. This is the idea of quark confinement.
- Quark confinement is closely related to the inter-quark force. It is very strong when the distance is large but very weak when the distance is small, quite contrary to the behaviour of electric and gravitational forces.
- The ability of quarks to behave as if almost free at short distances is referred to as asymptotic freedom.
- QCD gives a good account of all the peculiar features of strong interactions.
- There are two regimes in QCD, the non-perturbative and the perturbative. The former applies when the inter-quark distances are large (or q^2, the square of the relativistic momentum transfer, is small). Present-day experiments fall under this regime. The perturbative regime is where q^2 is large.
- Perturbative QCD is rather like the good old QED, and is straightforward to apply. By contrast, non-perturbative QCD is quite tough as the mathematics is complicated.
- People therefore use numerical methods while exploring the non-perturbative regime. This is done via the lattice gauge approach. In this, the space-time continuum is approximated by a lattice.
- As a cross-check, people first applied the lattice method to QED and made sure that the method is reliable.
- In the case of QCD, it has been demonstrated (numerically) that quark confinement does result from the theory. It remains to prove this conclusion analytically.

So there it is, the theory of strong interactions, which was sought for nearly six decades. What is left now? To find that, you have to turn to the next chapter.

12

What Next?

This chapter wraps up the present volume, not only reviewing the progress thus far in physics and placing it in the proper perspective, but also giving a foretaste of what will come in the next volume, i.e., Part II of this book.

Looking back at the twentieth century, it is amazing how much ground has been covered (in physics). As the nineteenth century was winding to a close, the electron had just been discovered, and so also were radioactivity and X-rays. No doubt all these were tremendously exciting discoveries, but who at that time would have even dreamt of what has happened since then? And so, the question arises (as it perhaps did when the twentieth century was born): What next?

Before dealing with that question, it is perhaps useful to place the events described earlier in this book, in their proper perspective with respect to the development of physics as a whole. We will start with Galileo and Newton, particularly the latter, who gave modern physics its first major push forward. We owe a deep debt of gratitude to Newton for the many things he gave us — the laws of motion, the law of gravitation, calculus (for which he must share credit with Leibnitz), particle mechanics, and various bits and pieces concerned with heat, light and optics.

Skipping the years let us move on to the third quarter of the nineteenth century, at which time the situation was as follows: Mechanics (or classical mechanics as we now know it) was in pretty good shape, thanks to the contributions of many, in particular, Euler, Hamilton and Lagrange. One knew not only how to deal with the motion of particles, but also finite objects with complex shapes, like tops, for example. The two great laws of thermodynamics had been discovered, and the foundations had also been laid for the kinetic theory of gases. The great achievement of this theory was that it directly related heat to the motions of the constituent atoms. To top it all, there was the famous electromagnetic theory of Maxwell which in one grand sweep united charged-particle motion, electricity, magnetism and optics. The mood amongst physicists of that time was quite upbeat, and why not?

Yet, there were lurking signs that all was not well. Soon it became quite clear that major modifications to the existing mechanics was needed (a) when the speed of particles approached that of light, and (b) when the particles became very tiny, as in the case of the electron. The modifications required in the two cases were quite different, and each in its own way opened up a new world as the following sketch indicates.

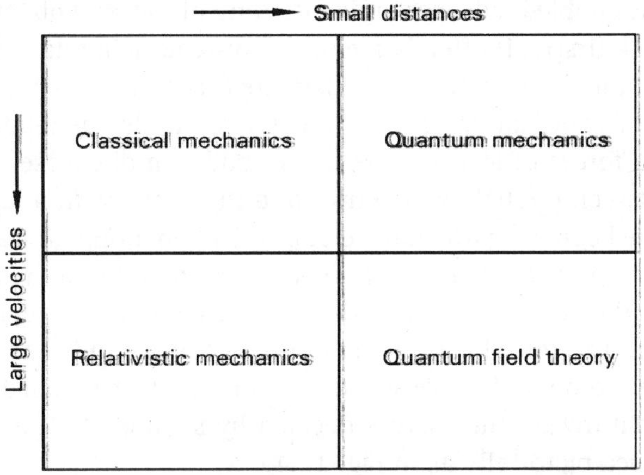

Einstein discovered the modifications needed in the high-velocity domain (see ***Relativity***), while the new dynamics required in the microscopic domain was uncovered by Heisenberg, Schrodinger and Dirac (see ***QR-I***). Dirac provided the next step, showing how non-relativistic quantum mechanics could be combined with special relativity, when dealing with the electron and purely electromagnetic forces. This, you will recall, is nothing but a reference to the famous Dirac equation. Two decades later, Tomonaga, Schwinger and Feynman moulded the Dirac equation into QED, subsequently hailed as the 'Jewel of Physics', a position it continues to hold (see ***QR-II***).

The Universe is not made up merely of electrons, and there are forces other than just the electromagnetic. So what does one do about them? What about other theories which, like QED, will enable one to tackle the weak force and the strong force, for example? These were precisely the theories produced in the second half of the twentieth century, and our preoccupation thus far has been with their description. I must mention here that the developments described in the preceding chapters have brought about a general change of style in physics, if one might say so. When I was in college, even non-relativistic quantum mechanics was considered difficult and new-fangled; besides, all the founding fathers of QM were still alive whereas now, even the founding fathers of QED are gone, not to mention people like Salam who sparked later developments.

Concerning the change of style to which I made a reference above, the most important development in this respect is the exalted place now accorded to symmetry. To appreciate this change, we go back to Newton whose great contribution was to show that diverse phenomena connected with the motion of particles could be described via the same laws, but invoking different initial conditions, as appropriate. What this means is that the falling stone and the comet moving around the Sun obey the same laws, but their motions are very different mainly because of different initial conditions.

Symmetry has enabled a deeper understanding of the way physics operates. It plays two roles — firstly, it offers a means of systematising data that otherwise seems arbitrary, and sometimes even disjointed. Secondly — and this was the great discovery of Yang and Mills — symmetry could also be made the basis for the fundamental forces of Nature. There is in addition of course the connection noted earlier between (global) symmetries and the conservation laws of physics.

In short, in the beginning, the nature of physical phenomena was understood. It was found that physical phenomena could be explained as resulting from a combination of the laws of Nature (laws which describe the regularities of phenomena), and the initial conditions (which could be quite arbitrary). Later, and this was the profound discovery of this century, it was realised that the basic regularities of Nature are themselves decided by symmetry. The entire concept can be summarised pictorially as in Fig. 12.1.

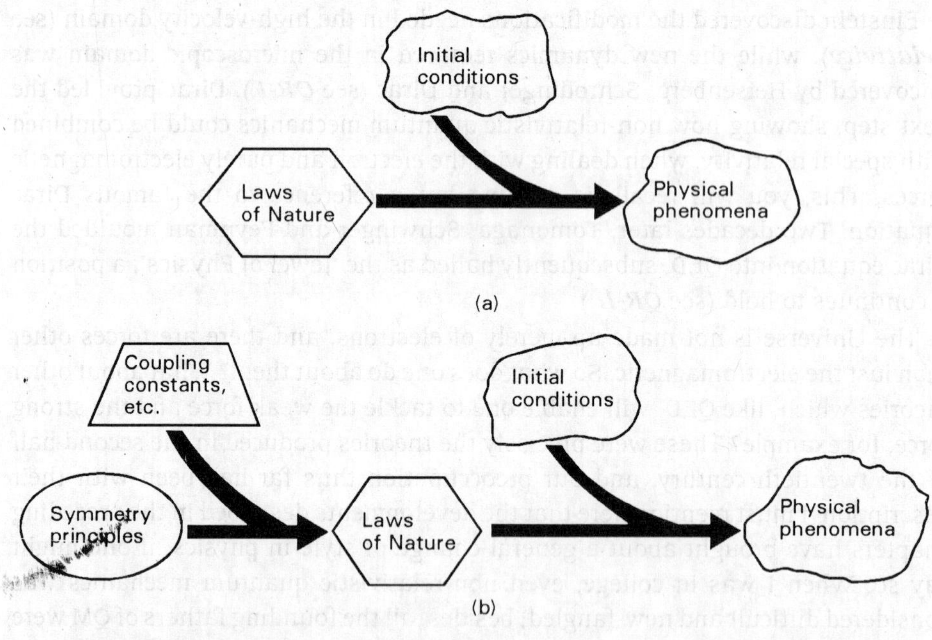

Fig. 12.1 (a) illustrates the character of physics before say 1950, while (b) illustrates how after 1950, symmetry carved for itself a special place. Notice that some of the boxes have irregular shapes, while others are regular shaped; this is deliberate, and its significance will become evident upon reference to the text.

It is of interest to note in passing that in the midst of all the furious activity centering around the elementary particles, the basic methodology of making calculations has not fundamentally changed. True, one has gravitated from the Dirac equation to formal quantum field theory but in the end, when it comes down to brass tacks, one still continues to use the rules of textbook QM. However, lately even this is beginning to change somewhat, but that story comes later (in the next volume!).

Before proceeding further, let us take a quick look at the developments in the twentieth century; this is summarised in Table 12.1 and Fig. 12.2.

Table 12.1 Some of the major milestones in 20th century Physics

(a) Pre-QED era	
1900	Discovery of gamma rays
	Planck's quantum theory
1905	Special relativity
	Einstein's theory of the light quantum
1906	Rutherford discovers α particle scattering
1911	Rutherford discovers the atomic nucleus
1913	Bohr model of the atom
1914	Discovery of the continuous β spectrum
1915	General relativity theory of gravitation
1918	Noether's theorem
1923	de Broglie's hypothesis about the dual nature of matter
1924	Bose statistics
1925–1927	Birth of non-relativistic QM
1928	Dirac equation
1931	Discovery of the positron
1932	Discovery of the neutron
1933	Fermi's theory of β decay
1934	Yukawa's theory of the meson
1936	μ meson discovered
1939	Discovery of nuclear fission
[1939–1945	Second world war]

(b) Post–QED era	
1948	QED
1952	Discovery of Δ resonance and the V particles

Contd...

Table 12.1 (Contd.) Some of the major milestones in 20th century Physics

1953	Conservation of leptons Strangeness idea
1954	Yang–Mills theory
1955	τ–θ puzzle
1956	Detection of the neutrino
1957	Parity violation discovered; $V-A$ theory; Universal Fermi interaction; Schwinger hints at electro-weak unity
1958	First suggestions about vector bosons
1960	Discovery of the first strange-particle resonance
1961	Goldstone theorem SU(3) scheme
1962	Discovery of the muon neutrino
1963	Cabibbo's theory
1964	*CP* violation Ω^- discovery Idea of charm Higgs mechanism Current algebra
1965	Idea of colour proposed
1967	Electro-weak unification: SU(2) × U(1) Idea of neutral currents
1969	Discovery of scaling Idea of partons
1970	GIM mechanism
1971	Proof of renormalisability of Yang–Mills theory
1973	Discovery of neutral currents Beginnings of QCD Salam suggests that the proton may decay
1974	Discovery of J/ψ Lattice gauge theory Idea of supersymmetry Georgi and Glashow's GUT
1975	Discovery of the τ
1976	Idea of supergravity
1977	Discovery of b quark
1983	Discovery of W^+, W^-, Z^o
1995	Discovery of t quark

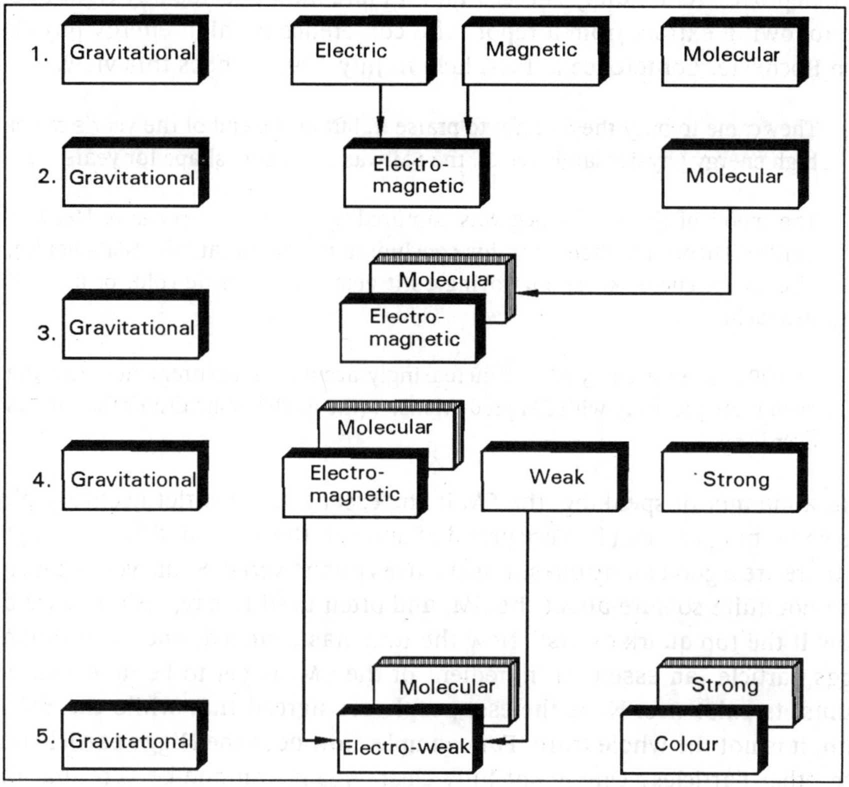

Fig. 12.2 This figure paraphrases the evolution of our ideas concerning the basic forces of Nature. To start with, one had the gravitational, the electric, the magnetic and the molecular forces. This was the situation in the middle of the nineteenth century. At that time, one did not know much about molecular forces, except that they existed and determined the binding between atoms, and, in an indirect way, such things as surface tension, etc. The next step was the unification of the electric and the magnetic forces by Maxwell. After this came the realisation that the so-called molecular forces are nothing but a manifestation of the electromagnetic force as it operates in the world of atoms. This understanding was a sequel to the development of the Bohr atom model, and the discovery of the electronic structure of atoms. Next, Fermi discovered the weak force and Heisenberg the strong force. Today's viewpoint is summarised in the last line where the gravitational force stands alone, the electro-weak is unified and the strong force is recognised to be due to colour.

As a new century dawns, the substantive achievement is the emergence of the so-called Standard Model (SM). As remarked earlier, it is an amalgam of QCD which is based on SU(3)colour, and the electro-weak theory which is based on SU(2) × U(1). Thus, the SM is often symbolically represented as SU(3) × SU(2) × U(1); it is really a mixture of two distinct but similar theories. Despite this, the

SM has proved remarkably successful, far more than one has any right to expect. The following extract from a report of a conference on high-energy physics (the 28th Rochester Conference in fact) held in July 1995, echoes this view.

> They came to bury the SM, not to praise it, but at the end of the year's major high energy physics jamboree .... the SM was in its best shape for years. ...
>
> The mood of the conference was captured by Gabriele Vereziano, Head of CERN's Theory Division, with his concluding remarks that "the SM's health, already excellent, keeps improving as it gets rid of a little cold, or a small headache." ...
>
> In 1995, after many years of increasingly accurate measurements agreeing even more precisely with SM predictions, .. (the model is) healthier than it has ever been ...

In a manner of speaking, the SM is the essence of twentieth century physics. It is something one can be very proud of and yet, the SM is not the end of physics for there are a good many questions it leaves unanswered. Some years ago, people were not quite so sure about the SM, and often used to say, "Ah, but we do not know if the top quark exists". Now the tune has changed, and even though the Higgs particle (an essential ingredient of the SM) is yet to be detected, no one doubts its existence. Nevertheless, people are agreed that while the SM is the truth, it is not the whole truth. For example, how does the Higgs particle interact with other particles? One is not fully clear. Again, can one be sure that the SM would work without any modifications when one goes from the currently accessible energy scale of about a hundred GeV to say a billion GeV, or even higher? No one knows. Besides, there are about two dozen adjustable parameters in the SM, too many for comfort. Clearly, the SM is not the end of the road.

Agreed, but have people been looking beyond, and if so, what have they found? Well, a lot of people have been active in such tasks, and we have begun to get tantalizing glimpses of what promises to make physics in the forthcoming century even more exciting!

The SM shows that physics at about 100 GeV can be derived from the postulates of gauge symmetry, using the group

$$f = SU(3) \times SU(2) \times U(1),$$

where f is a fundamental gauge group of Nature. f contains three groups, associated with each of which is a coupling constant. Let us denote these as α_S, α_W, and α_{em}. At currently accessible energies, these three coupling constants have very different values. People who want to look beyond the SM asked: "What is the origin of f? Can it be derived from simpler and more fundamental principles or did God merely say, 'Here is f, take it!' ?" If f cannot be derived from a more fundamental principle, then the observed values of the three coupling constants would have no further meaning. On the other hand, it is also plausible that f is part of a larger symmetry group G. This would mean that at very high energies, this larger symmetry group would make its presence felt, which in turn would imply that at such high energies the values of the three coupling constants would become equal. Beyond this energy, the strong, the weak and the electromagnetic forces would cease to have independent existence. Instead there would be only one comprehensive force. A theory which is based on G and subsumes all the three forces just mentioned, is called a Grand Unified Theory or GUT for short. At what sort of energies would GUT reveal itself? That question is answered in Fig.12.3.

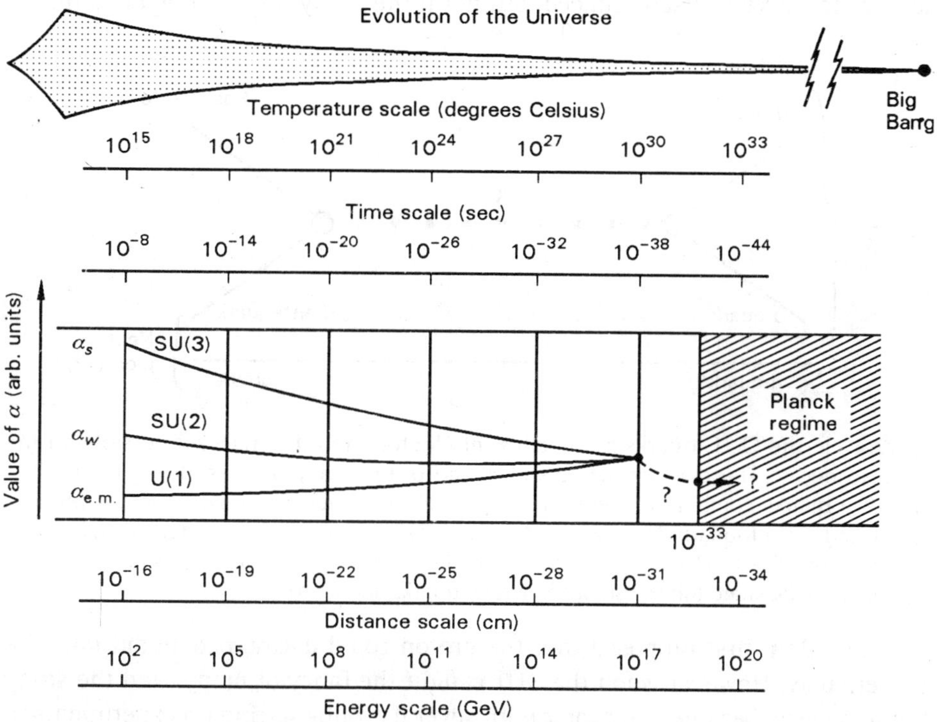

Fig. 12.3 This figure places physics on different scales of distance/energy/ time, in juxtaposition to the evolution of the Universe from the Big Bang (which occurred somewhere around 15 to 20 billion years ago). Closely related to the energy scale is the temperature scale. This figure must be studied along with similar figures in the earlier chapters. The essential point made by the present figure is the following. At about 100 GeV, the SM with three coupling constants works beautifully. As one goes up in energy, there would come a stage when the three coupling constants would become identical; such a state is described by GUT. The distance involved is $\leq 10^{-31}$ cm. When the distance becomes less than 10^{-33} cm, i.e., in the Planck regime, the GUT must be augmented by quantum gravity. Clearly, physics on such scales of length and energy cannot be explored in the laboratory. However, such physics *is* relevant, since it is these laws that have determined the early history of the Universe. Thus, there is meaning in studying such physics, at least theoretically. But what about experimental confirmation? Think about it!

Salam was the first to search for a GUT. Since then, there have been many attempts, and at one time people used to sport T-shirts which said, 'It takes Guts to develop GUT!' One of the important characteristics of GUT, no matter what version it is (i.e., what *G* group it is based on), is that it leads to interconversion between leptons and quarks, i.e., these two species can mix. In turn, this opens up the door for the proton to decay, which is a powerful prediction of GUT. Up to now it has been an article of faith that, unlike the neutron, the proton is stable, just as the electron is. GUT, on the other hand, predicts that the proton will decay as explained in Fig. 12.4. Now if the proton decays, then would not the hydrogen atom be destroyed, and would that not create all kinds of problems for us and in fact for the Universe itself? Indeed, but not if the decay time is long enough.

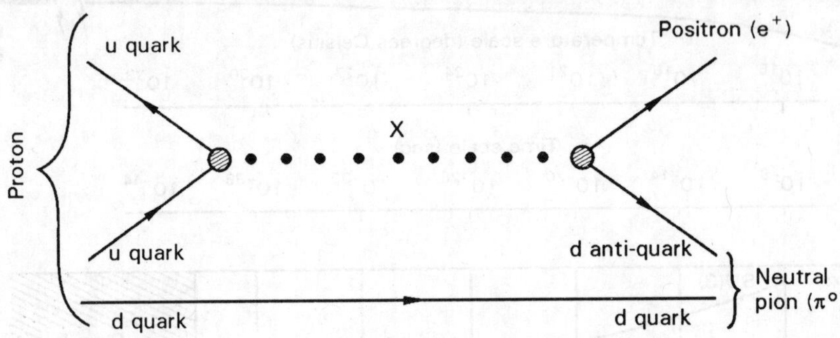

Fig. 12.4 In the GUT, quarks can change into leptons and this makes it possible for the proton to decay. The diagram above shows one of the many possible decay routes. The proton's u quarks combine to form an X particle, which then disintegrates into a positron (lepton) and a d anti-quark. The d anti-quark combines with the remaining d quark to become a neutral pion. The GUT features X and Y particles which are like the bosons of the present theories. They facilitate the transitions between quarks and leptons.

When Salam first proposed that the proton could decay, people did not take him seriously. However, soon the GUT caught the fancy of many, and the study of proton decay became a hot topic with several groups setting up experiments to detect the decay of the proton. One of the groups involved was from TIFR, which arranged its experiment deep down in the Kolar Gold Mines in Karnataka. In the beginning, there were occasional reports that an event corresponding to the decay of the proton had been observed but on careful re-examination all such claims were found to be incorrect. To date, no clear experimental observation of proton decay has been made. All that experimenters are able to say is that the lifetime must be of the order of or greater than 1.5×10^{32} years. This is to be compared with the estimate of $10^{28 \pm 1.5}$ years, of the most favoured GUT due to Georgi and Glashow, based on the choice of SU(5) for *G*. In a nutshell, people believe in the GUT and the search for a good GUT is still on; and so is the search for proton decay.

So, the GUT offers a unified theory that works at distances down to 10^{-31} cm and energies upto the 10^{20} GeV range. What happens at still smaller distances and still higher energies? What about gravity? Does it not have a place in the scheme of things? Indeed, yes, and people are already figuring out how gravity can be brought into the picture. Now in 1915, Einstein gave a theory of gravitation which, even today, is the corner-stone of cosmology. But Einstein's theory is a *classical* theory; obviously, this will not do at the kind of distances we are talking about here. What is needed is a *quantum theory of gravity* which can be neatly dovetailed with the GUT to describe physics at distances smaller than the so-called Planck length (10^{-33} cm).

Supposing one is able to develop such a theory which encompasses both GUT and quantum gravity. Of what use would such a theory be, especially as there is no way such a theory can be verified in the laboratory? Strange as it might seem, there is a use as well as a need for such a theory. Once upon a time, say 10^{-45} sec after the Big Bang (which resulted in the birth of our Universe), the Universe was very tiny and smaller than the Planck distance. To describe the physics of this baby Universe, one would need a theory of the type mentioned above. You might say: "But all that happened so long ago! Since then, the Universe has expanded to astronomical proportions. And we don't have any means of knowing whether this theory that we are talking about, really worked at that time! Suppose there were two rival theories? How does one know which of these is the correct one? It all happened so very long ago!!" Your point is well taken, but there is an aspect which the above hypothetical question ignores, and this is where the title of this book comes into the picture.

Agreed that the Universe today is huge. But doesn't the seed determine the nature of the tree? And would not the nature and the structure of the Universe today depend, at least to some extent, upon the way it was when it was very tiny? Indeed, and that is where theories that unify everything under one umbrella become relevant. Different theories which paint different portraits of the baby Universe would lead to different portraits of the adult Universe, i.e., the Universe of today. Taking advantage of this fact, one can critically examine different theories and choose the one which gives the best description of things as they are at present. That would be the ultimate theory of physics, or as they say, the Theory of Everything (TOE)! Let us accept this argument. Next question: "Is there a theory that takes one all the way from the SM to what is needed to describe the infant Universe? In other words, is there a TOE?" Well, for the answer to that, you have to wait for Part II! Meanwhile, you can indulge in your own speculations. And, good luck in your ventures!!

Index